D1653490

Sensors Applications
Volume 5
Sensors in Household Appliances

Sensors Applications

- Sensors in Manufacturing
- Sensors in Intelligent Buildings
- Sensors in Medicine and Health Care
- Sensors for Automotive Technology
- Sensors in Household Appliances
- Sensors in Aerospace Technology
- Sensors in Environmental Technology

Related Wiley-VCH titles:

W. Göpel, J. Hesse, J. N. Zemel

Sensors Vol. 1–9

ISBN 3-527-26538-4

H. Baltes, G. K. Fedder, J. Korvink

Sensors Update

ISSN 1432-2404

Sensors Applications
Volume 5

Sensors in Household Appliances

Edited by
G. R. Tschulena, A. Lahrmann

Series Editors:
J. Hesse, J. W. Gardner, W. Göpel (†)

Series Editors

Prof. Dr. J. Hesse
formerly of Carl Zeiss, Jena
Bismarckallee 32c
14193 Berlin
Germany

Prof. J. W. Gardner
University of Warwick
Division of Electrical & Electronic Engineering
Coventry CV4 7AL
United Kingdom

Prof. Dr. W. Göpel †
Institut für Physikalische
und Theoretische Chemie
Universität Tübingen
Auf der Morgenstelle 8
72076 Tübingen
Germany

Volume Editors

Dr. G. R. Tschulena
sgt Sensorberatung
Reichenberger Straße 5
61273 Wehrheim
Germany

Prof. Dr. A. Lahrmann
Fachhochschule für Technik und Wirtschaft
Ingenieurwissenschaften II
Produktentwicklung
Blankenburger Pflasterweg 102
13129 Berlin
Germany

Library of Congress Card No.: applied for

British Library Cataloguing-in-Publication Data:
A catalogue record for this book is available from the British Library.

Bibliographic information published by Die Deutsche Bibliothek
Die Deutsche Bibliothek lists this publication in the Deutsche Nationalbibliografie; detailed bibliographic data is available in the Internet at <http://dnb.ddb.de>.

printed in the Federal Republic of Germany
printed on acid-free paper

Composition K+V Fotosatz GmbH, Beerfelden
Printing Strauss Offsetdruck GmbH, Mörlenbach
Bookbinding Großbuchbinderei J. Schäffer GmbH & Co. KG, Grünstadt

ISBN 3-527-30362-6

Preface to the Series

As the use of microelectronics became increasingly indispensable in measurement and control technology, so there was an increasing need for suitable sensors. From the mid-Seventies onwards sensors technology developed by leaps and bounds and within ten years had reached the point where it seemed desirable to publish a survey of what had been achieved so far. At the request of publishers WILEY-VCH, the task of editing was taken on by Wolfgang Göpel of the University of Tübingen (Germany), Joachim Hesse of Carl Zeiss (Germany) and Jay Zemel of the University of Philadelphia (USA), and between 1989 and 1995 a series called *Sensors* was published in 8 volumes covering the field to date. The material was grouped and presented according to the underlying physical principles and reflected the degree of maturity of the respective methods and products. It was written primarily with researchers and design engineers in mind, and new developments have been published each year in one or two supplementary volumes called *Sensors Update*.

Both the publishers and the series editors, however, were agreed from the start that eventually sensor users would want to see publications only dealing with their own specific technical or scientific fields. Sure enough, during the Nineties we saw significant developments in applications for sensor technology, and it is now an indispensable part of many industrial processes and systems. It is timely, therefore, to launch a new series, *Sensors Applications*. WILEY-VCH again commissioned Wolfgang Göpel and Joachim Hesse to plan the series, but sadly Wolfgang Göpel suffered a fatal accident in June 1999 and did not live to see publication. We are fortunate that Julian Gardner of the University of Warwick has been able to take his place, but Wolfgang Göpel remains a co-editor posthumously and will not be forgotten.

The series of *Sensors Applications* will deal with the use of sensors in the key technical and economic sectors and systems: *Sensors in Manufacturing, Intelligent Buildings, Medicine and Health Care, Automotive Technology, Aerospace Technology, Environmental Technology* and *Household Appliances*. Each volume will be edited by specialists in the field. Individual volumes may differ in certain respects as dictated by the topic, but the emphasis in each case will be on the process or system in question: which sensor is used, where, how and why, and exactly what the benefits are to the user. The process or system itself will of course be outlined and the volume will close with a look ahead to likely developments and applications in the future. Actual sensor functions will only be described where it seems necessary for an understanding of how they relate to the process or system. The basic principles can always be found in the earlier series of *Sensors* and *Sensors Update.*

The series editors would like to express their warm appreciation in the colleagues who have contributed their expertise as volume editors or authors. We are deeply indebted to the publisher and would like to thank in particular Dr. Peter Gregory, Dr. Jörn Ritterbusch and Dr. Claudia Barzen for their constructive assistance both with the editorial detail and the publishing venture in general. We trust that our endeavors will meet with the reader's approval.

Oberkochen and Coventry, November 2000

Joachim Hesse
Julian W. Gardner

Foreword

NEXUS is a network of experts aimed to introduce more microsystems in the industrial applications. NEXUS operates within industrial and academic organisations, primarily in Europe, but also has significant high-level membership from other continents, including USA and Far East Asia.

In its starting period, about one decade ago, NEXUS was mainly driven by academic members, working on dissemination of information on technical developments and market possibilities. Industrial interest soon expanded the network and the current membership now exceeds 1000. Since mid-2001 the NEXUS network has been transformed into the "NEXUS Association" based in Grenoble, France. NEXUS activities continue to be strongly supported by the European Commission and MEMS/MST-related industry.

In response to the needs of the membership and the interest of the European Commission the following specific activities were carried out:

- The "NEXUS Market Analysis for Microsystems" was initiated in 1998 and the first edition published in 2000. This second edition was published in Spring 2002, covering the period to the year 2005.
- The "NEXUS Technological Roadmap for Microsystems" was published in Dec. 2000, with and updated and broadened report due in 2002/2003.
- The members are also informed regularly on new European and national research an development support activities in the areas of microsystem technology and nanotechnology.

The NEXUS User Supplier Clubs provide the means for members to collect, discuss and to verify technical and market-related information. These USC operate in the application areas of:

- Automotive
- Pharmaceutical and diagnostics
- Medical Devices
- Industrial Process Control
- Peripherals and multimedia
- Aerospace and geophysics
- Telecommunication

- Household appliances
- CAD tools
- Packaging
- Nano materials technology

The "NEXUS User-Supplier-Club Household Appliances" was founded during the first European conference on "Commercialisation of Microsystems" held in Dortmund in 1999. This Club has grown rapidly to a current membership of more than 40 from major and smaller appliance companies, sensor and MST supplier companies and from academic institutions all over Western Europe.

The Club holds 3 to 4 meetings per year, and with an exchange of information and documents by e-Mail. Specifically the USC Household Appliances meetings take place at factories of member companies, of research organisations, or in conjunction with large fairs.

The household industry in Europe employs significant numbers of people and requires high volumes, but it operates in a highly competitive area. The use of more and better adapted electronics has been a major driving force in the last decade and provides one means to reach the overall objective of reducing energy, water and detergent consumption. This coupled to the introduction of more reliable and cost effective sensors and microsystems provides a means of remaining competitive in a global economy.

The work on this book was one of the successful outcomes from the activities of the NEXUS User Supplier Club on Household Appliances. We wish the members well and hope for a successful continuation of these activities in the future.

Gaetan Menozzi
Chairman of NEXUS

Contents

List of Contributors *XVII*

1 **The Increasing Importance of Sensors in Household Appliances** *1*
A. Lahrmann and *G. Tschulena*
1.1 Introduction *1*
1.2 Sensors in Household Appliances *3*
1.3 References *8*

2 **Market Data** *9*
G. Tschulena
2.1 Introduction – The Household Appliance Industry *9*
2.2 HA Industry Data – World Market Data *10*
2.3 HA Industry Data – European Market Data *10*
2.4 Sensors and MST in Household Appliances Market Data *14*
2.5 Conclusion *17*

3 **Appliances and Sensors** *19*
3.1 Home Laundry Appliance Manufacturers – Drivers of Change: Socioeconomics and Enablers *19*
G. Wentzlaff, R. Herden and *R. Stamminger*
3.1.1 Introduction *19*
3.1.2 Technical Outline *20*
3.1.3 Basic Functions *21*
3.1.3.1 Agitation *21*
3.1.3.2 Water Intake, Water Level *24*
3.1.3.3 Temperature Control *25*
3.1.3.4 Detergent Dispensing *27*
3.1.3.5 Water and Suds Monitoring *29*
3.1.3.6 Program Sequence *32*
3.1.3.7 Operation and Display Technology *32*
3.1.4 Summary *35*
3.1.5 Acknowledgements *37*

3.2 Intelligent Combustion *37*
H. Janssen, H.-W. Etzkorn, and *S. Rusche*
3.2.1 The Situation *37*
3.2.2 Possible Measurements of Gases Characterising the State of Combustion *38*
3.2.3 Suitable Sensors for Controlling Combustion Processes *40*
3.2.3.1 Capacitive Measuring *40*
3.2.3.2 Acoustic Principles *41*
3.2.3.3 Optical Principles *41*
3.2.3.4 Measurement of Thermal Conductivity *42*
3.2.3.5 Heat of Reaction *42*
3.2.3.6 Electrochemical Cells with Liquid Electrolyte *43*
3.2.3.7 Electrochemical Cells with Solid Electrolyte *43*
3.2.3.8 Semiconductors *45*
3.2.3.9 Additional Possibilities *46*
3.2.4 Some Exemplary Solutions *47*
3.2.4.1 Combustion Control of a Boiler Open to the Room *47*
3.2.4.2 Surveillance of a Fan-assisted Sealed Boiler Using a Semiconducting Element *48*
3.2.5 References *51*
3.3 Condition Monitoring for Intelligent Household Appliances *52*
J. Goschnick and *R. Körber*
3.3.1 Introduction *52*
3.3.2 High-Level Integrated Gas Sensor Microarrays *54*
3.3.3 Gas Analytical Performance of the Gradient Microarray *57*
3.3.4 Application Examples *60*
3.3.4.1 Controlling Frying Processes *60*
3.3.4.2 Indoor Air Monitoring *61*
3.3.5 Summary and Outlook *66*
3.3.6 References *67*
3.4 Sensor Examples in Small Appliances *68*
T. Bij de Leij
3.4.1 Reinventing Appliances *69*
3.4.2 Facts and Figures *69*
3.4.3 Sensors in DAP Appliances *69*
3.4.4 Sensor Criteria *71*
3.5 Infrared Ear Thermometers *72*
B. Kraus
3.5.1 Introduction *72*
3.5.2 Sensor Unit *74*
3.5.3 Electronics and Temperature Calculation *76*
3.5.4 Calibration *78*
3.5.5 Conclusion *80*
3.5.6 References *80*

4 Sensorics for Detergency *81*
W. Buchmeier, M. Dreja, W. von Rybinski, P. Schmiedel, and *T. Weiss*
4.1 Introduction *81*
4.2 Household Laundry Products *82*
4.2.1 Conventional Powder Heavy-Duty Detergents *83*
4.2.2 Compact and Supercompact Heavy-Duty Detergents *83*
4.2.3 Extruded Heavy Duty Detergents *84*
4.2.4 Heavy-duty Detergent Tablets *84*
4.2.5 Color Heavy-Duty Detergents *85*
4.2.6 Liquid Heavy-duty Detergents *85*
4.2.7 Specialty Detergents *86*
4.3 Detergent Compositions *87*
4.3.1 Surfactants *87*
4.3.2 Builders *88*
4.3.3 Bleaches *89*
4.3.4 Auxiliary Ingredients *89*
4.3.4.1 Enzymes *90*
4.3.4.2 Soil Antiredeposition Agents, Soil Repellent/Soil Release Agents *90*
4.3.4.3 Foam Regulators *91*
4.3.4.4 Corrosion Inhibitors *92*
4.3.4.5 Fluorescent Whitening Agents *92*
4.3.4.6 Fragrances *92*
4.3.4.7 Dyes *93*
4.3.4.8 Fillers *93*
4.4 Physical Parameters in Detergency *93*
4.4.1 Introduction *93*
4.4.2 Surface Tension and Wetting *94*
4.4.3 Adsorption at the Solid/Liquid Interface *95*
4.4.4 Liquid/Liquid Interface *96*
4.5 Phase Behavior of Surfactant Systems *97*
4.6 Foaming *99*
4.7 On-line Sensorics in Detergent Development *100*
4.7.1 Product Relevant Physico-chemical Parameters and their Availability for On-line Determination *100*
4.7.2 On-line Determination of Dynamic Surface Tension by the Bubble-pressure Method *102*
4.7.3 Other Accessible On-line Parameters *104*
4.8 Sensors in Household Appliances *105*
4.8.1 Introduction *105*
4.8.2 Sensor Applications *106*
4.8.2.1 Turbidity *106*
4.8.2.2 Conductivity *107*
4.8.2.3 Water Hardness *107*
4.8.2.4 Concentration *108*
4.8.2.5 Miscellaneous *108*

4.8.3 Conclusion 109
4.9 Detergent Products with Built in Sensor Functions 109
4.9.1 Introduction: "Intelligent" Detergent Products 109
4.9.2 Automatic Dishwashing Detergent (ADD) Tablets 110
4.9.3 Laundry Detergent Tablets 111
4.10 References 113

5 Sensor Related Topics 117
5.1 Thin Film Temperature Sensors for the Intelligent Kitchen 117
M. Muziol
5.1.1 Introduction 117
5.1.2 Platinum Temperature Sensors 117
5.1.2.1 The Variety of Forms for Platinum Temperature Sensors 118
5.1.2.2 Temperature Dependent Changes in Electrical Resistance 118
5.1.2.3 Advantages of Platinum Temperature Sensors 119
5.1.3 High Tech in the Home 120
5.1.4 Future Prospects/Outlook 123
5.2 Reed Switches as Sensors for Household Applications 123
U. Meier
5.2.1 Market Aspects 123
5.2.1.1 Further Technical Development 124
5.2.1.2 Development of Reed Sensors in Household Applications 124
5.2.2 Reed Switch Characteristics 124
5.2.2.1 Introduction 124
5.2.2.2 Reed Switch Features 124
5.2.2.3 The Basic Reed Switch 125
5.2.3 Magnetic Response in a Variety of Arrangements 126
5.2.3.1 Direct Operation 126
5.2.3.2 Indirect Operation 128
5.2.4 Handling of Reed Switches in Various Sensor Applications 128
5.2.4.1 Cutting and Bending of a Reed Switch 128
5.2.4.2 Soldering and Welding 129
5.2.4.3 Printed Circuit Board (PCB) Mounting 131
5.2.4.4 Using Ultrasonics 131
5.2.4.5 Dropping Reed Switch Products 131
5.2.4.6 Encapsulating Reed Switch Products 132
5.2.4.7 Temperature Effects and Mechanical Shock 132
5.2.5 Applications 132
5.2.5.1 Measuring the Quantity of a Liquid, Gas and Electricity 132
5.2.5.2 Detecting Water Flow 133
5.2.5.3 Water Level Sensors 133
5.2.5.4 Monitoring the Level of Water Softener and Clarifier in a Dishwasher 134
5.2.5.5 Protecting Dishwasher Spray-arms 135
5.2.5.6 Preventing Condensed Water Overflow 135

5.2.5.7 Detecting the Drum Position in Washing Machines 136
5.2.5.8 Safety Control in Appliance Doors 137
5.2.5.9 Reed Sensors in Stove Applications 137
5.2.5.10 Electric Toothbrushes 138
5.2.5.11 Level Control of Jet-Dry Cleaning Liquid 139
5.2.5.12 Reed Sensors in Carpet Cleaners 139
5.2.5.13 Detecting Movement or End Positions 140
5.3 Gas Sensors in the Domestic Environment 141
U. Hoefer and *M. Meggle*
5.3.1 Introduction 141
5.3.2 Sensor Principles 142
5.3.2.1 Metal Oxide Semiconductor Gas Sensors 142
5.3.2.2 Pellistors 143
5.3.2.3 Liquid-state Electrochemical Gas Sensors 145
5.3.2.3.1 Amperometric Sensors 145
5.3.2.3.2 Potentiometric Sensors 146
5.3.2.4 Solid-state Electrochemical Sensors 147
5.3.2.4.1 Potentiometric Measurement 147
5.3.2.4.2 Amperometric Measurement 148
5.3.2.5 Optical Gas Detectors 149
5.3.3 Appliances 150
5.3.3.1 Domestic Burner Control
(Fuel Burners, Gas Condensing Boilers) 150
5.3.3.2 Sensor Requirements for Combustion Control 151
5.3.3.3 Air Quality 154
5.3.3.4 Indoor Detection of CO 156
5.3.3.4.1 Possible CO-sources 157
5.3.3.4.2 Sensors for CO-indoor Monitoring 158
5.3.3.5 Natural Gas Detection and Alarm Systems 159
5.3.3.5.1 Natural Gas Detectors 161
5.3.3.6 Other Appliances 162
5.3.3.6.1 Cooking and Frying Control 163
5.3.3.6.2 Self-cleaning of Ovens 163
5.3.4 References 164
5.4 UV Sensors – Problems and Domestic Applications 165
O. Hilt and *T. Weiss*
5.4.1 UV Radiation – A General Introduction 165
5.4.2 UV Radiation in Household Environments 165
5.4.2.1 Natural UV Radiation 165
5.4.2.2 Man-made UV Radiation 167
5.4.3 Principles of UV Detection 167
5.4.3.1 UV-Enhanced Si Photodiode 167
5.4.3.2 Crystalline Wide Band-Gap Semiconductors 168
5.4.3.3 Polycrystalline Wide Band-gap Semiconductors 168
5.4.3.4 Fluorescence Converters 169

5.4.3.5 Discharge Tubes 169
5.4.3.6 Filters 170
5.4.3.7 The Entrance Window 170
5.4.3.8 Concentrator Lens 170
5.4.3.9 Packaging 170
5.4.4 Household Applications 171
5.4.4.1 Personal UV Exposure Dosimetry 171
5.4.4.2 Surveillance of Sunbeds 171
5.4.4.3 Flame Scanning in Gas and Oil Burners 172
5.4.4.4 Fire Alarm Monitors 173
5.4.4.5 Water Sterilization 173
5.4.5 The Marked Potential of UV Sensors in Household Appliances 174
5.4.6 References 176
5.5 Displacement Sensors in Washing Machines 177
E. Huber
5.5.1 Contact and Non-contact Displacement Sensors 177
5.5.2 Measuring the Load of Washing Machines 178
5.5.3 Displacement Sensors in the Washing Machines 179
5.5.4 Sensors Design and Measuring Principles 179
5.5.4.1 Inductive Measuring Principle 180
5.5.4.2 Displacement Sensor Integrated into Damper 180
5.5.4.3 Inductive – Potentiometric Measuring Principle 181
5.5.5 Sensor Control and Signal Evaluation with Discrete Electronics 182
5.5.6 Summary 183
5.6 Low-Cost Acceleration Sensors in Automatic Washing Machines 184
R. Herden
5.6.1 Imbalance in Automatic Washing Machines 184
5.6.2 Normal Procedure for Detecting Imbalance in Automatic Washing Machines 185
5.6.3 New Procedures for Detecting Imbalance in Automatic Washing Machines 186
5.6.4 Summary 191
5.7 Fuzzy and Neurofuzzy Applications in European Washing Machines 191
H. Steinmueller
5.7.1 Introduction 191
5.7.2 Explanation of the Conventional Wash Process 193
5.7.3 Engineering the "FUZZY RULE SET" 194
5.7.4 Setting up the Actual Control Unit 196
5.7.5 Summary 197
5.8 Cutting-edge Silicon-based Micromachined Sensors for Next Generation Household Appliances 198
F. Solzbacher and *S. Bütefisch*
5.8.1 Brief Introduction to MEMS-based Sensors 198

5.8.1.1 Microsystems Technology (MST) – A Complex Game of Materials, Technologies and Design *198*
5.8.1.2 Materials *200*
5.8.1.3 3D Processing Technology *200*
5.8.1.3.1 Sensors, Actuators and Passive Components *201*
5.8.1.3.2 Bulk Micromachining Technology *201*
5.8.1.3.3 Surface Micromachining of Silicon *205*
5.8.1.3.4 Summary *205*
5.8.4 Perspectives for Future Developments *209*
5.8.5 References *210*

6 Influencing Factors – Today and Tomorrow *211*
6.1 Future Developments, Roadmapping and Energy *211*
G. Tschulena
6.1.1 General Trends *211*
6.1.2 Functional Trends for Large Household Appliances *212*
6.1.3 Sensors in Large Appliances – Current and Future Trends *213*
6.1.4 Functional Trends in Small Household Appliances *217*
6.1.5 Sensors for Small Appliances – Current and Future Developments *218*
6.1.6 On Energy Consumption in Appliances *219*
6.1.7 Heating and Climate Control *220*
6.1.8 Climate Control Current and Future Sensor Developments *221*
6.1.9 Visions for Household Appliances *224*
6.1.9.1 The Automated Kitchen of the Future *224*
6.1.9.2 The Laundry of the Future *225*
6.1.9.3 Service Robots *226*
6.1.9.4 Services for Health *227*
6.1.9.5 Home Automation *227*
6.1.10 References *228*
6.2 Smart Buildings – Combination of Sensors with Bus Systems *229*
K. Abkai
6.2.1 Introduction *229*
6.2.2 What Is Behind the "Smart Building" Concept? *229*
6.2.3 Efficiency *233*
6.2.4 Security *233*
6.2.5 Communication Systems *234*
6.2.6 Conclusion *235*
6.3 Smart Homes – A Meeting with the Future Even Now *236*
C. Kühner and *U. Koch*
6.3.1 Comfort and Security as Central Aspects *237*
6.3.2 The Service and Security Center (SSC) *238*

Appendix *241*

Index *279*

List of Contributors

Kjumar Abkai
Friatec AG
Steinzeugstr. 50
68229 Mannheim
Germany

Willi Buchmeier
Henkel KGaA
Henkelstr. 67
40191 Düsseldorf
Germany

Sebastian Bütefisch
First Sensor Technology GmbH
Carl-Scheele-Str. 16
12489 Berlin
Germany

Michael Dreja
Henkel KGaA
Henkelstr. 67
40191 Düsseldorf
Germany

Heinz-Werner Etzkorn
Gaswärme-Institut e.V. Essen
Hafenstraße 101
45356 Essen
Germany

Joachim Goschnick
Forschungszentrum Karlsruhe GmbH
IFIA Inst. für Instrumentelle Analytik
Postfach 3640
76021 Karlsruhe
Germany

Rudolf Herden
Miele & Cie. GmbH & Co.
Carl-Miele-Straße 29
33332 Gütersloh
Germany

Oliver Hilt
twlux Halbleitertechnologien
Berlin AG
Ostendstr. 25
12459 Berlin
Germany

Ulrich Hoefer
Steinel Solutions AG
Allmeindstr. 10
8840 Einsiedeln
Switzerland

Eduard Huber
Vertrieb/Projekte OEM Sensoren
Micro-Epsilon Messtechnik
GmbH & Co. KG
Königbacher Straße 15
94496 Ortenburg
Germany

Holger Janssen
Gaswärme-Institut e.V. Essen
Hafenstraße 101
45356 Essen
Germany

Ulrich Koch
Managing Director
eBuilding GmbH
Weinheimer Str. 68
68309 Mannheim
Germany

Reiner Körber
Forschungszentrum Karlsruhe GmbH
IFIA Inst. für Instrumentelle Analytik
Postfach 3640
76021 Karlsruhe
Germany

Bernhard Kraus
Braun GmbH
Frankfurter Str. 145
61476 Kronberg/Taunus
Germany

Claudia Kühner
Kachikoshi PR
Annweiler Straße 30
76829 Landau
Germany

Andreas Lahrmann
Fachhochschule für Technik
und Wirtschaft
Ingenieurwissenschaften II
Produktentwicklung
Blankenburger Pflasterweg 102
13129 Berlin
Germany

Tjerk bij de Leij
Philips Domestic Appliances
and Personal Care B.V.
Postbus 20100
9200 CA Drachten
The Netherlands

Martin Meggle
Steinel Vertrieb GmbH
Dieselstraße 80–84
33442 Herzebrock-Clarholz
Germany

Ulrich Meier
Meder electronic AG
Friedrich-List-Strasse 6
78234 Engen-Welschingen
Germany

Matthias Muziol
Heraeus Sensor-Nite GmbH
Reinhard Heraeus Ring 23
63801 Kleinostheim
Germany

Stefan Rusche
Gaswärme-Institut e.V. Essen
Hafenstraße 101
45356 Essen
Germany

Wolfgang von Rybinski
Henkel KGaA
Henkelstr. 67
40191 Düsseldorf
Germany

Peter Schmiedel
Henkel KGaA
Henkelstr. 67
40191 Düsseldorf
Germany

Florian Solzbacher
First Sensor Technology GmbH
Carl-Scheele-Str. 16
12489 Berlin
Germany

Rainer Stamminger
Electrolux
AEG Hausgeräte GmbH
Muggenhoferstr. 135
90429 Nürnberg
Germany

Harald Steinmueller
Electrolux Home Products PTY, Ltd.
19, Pope St.
Beverly SA 5009
Australia

GUIDO TSCHULENA
Sgt Sensor Consulting
Dr. Guido Tschulena
Reichenberger Str. 5
61273 Wehrheim
Germany

TILMAN WEISS
twlux Halbleitertechnologien
Berlin AG
Ostendstr. 25
12459 Berlin
Germany

TILO WEISS
Henkel KGaA
Henkelstr. 67
40191 Düsseldorf
Germany

GÜNTER WENTZLAFF
Facility Management
Hochschule Niederrhein
Reinarzstraße 49
47805 Krefeld
Germany

1
The Increasing Importance of Sensors in Household Appliances

A. LAHRMANN and G. TSCHULENA

1.1
Introduction

Household appliances make up one of the largest markets for electrotechnical and electronic products. While comparatively simple versions of sensors and microsystem products, such as temperature sensors or level sensors, have long been used in household appliances, new and improved sensors conquer the market at a breathtaking rate. The way modern sensors with intelligent control systems are used is one of the main distinguishing features between the various products and companies.

The household appliance industry produces

- large household appliances, which make up about three quarters of the annual turnover. About 60 Million units are produced in Europe and about 150 million units worldwide.
- small household appliances, (responsible for about 20 to 25% of the turnover, with another about 23 million units produced annually in Western Europe).
- electrical and gas heating and climate conditioning equipment (responsible for about 5 to 6% of the turnover, according to the ZVEI-GfK data for 1998)

These industries comprise the appliances described in more detail in Tabs 1.1 and 1.2.

More details on appliance markets are given in Chapter 2, together with some data on the sensors in question. There has been some recent research into the markets of modern micromechanical sensors in household appliances, documenting the market potential for various types of sensors in this area, including those for pressure, acceleration and tilt, thermopiles, flow and gas sensors [3]. Examples of future developments will also be given.

Some applications shown in Tab. 1.1 are described in Chapter 3 of this book. As home laundry applications are of major importance as the largest sector, Chapter 3.1 is dedicated to the major driving forces and developments in this area.

Combustion control for domestic hot water production is another important area where modern sensor technology comes in. New developments are discussed in Chapter 3.2.

Tab. 1.1 Classification scheme for household appliances.

Large household appliances	Home laundry appliances, with washing machines, dryers, combined washing machines and dryers; Dish washers; Cookers, with gas cookers, electric cookers, ovens, cooker hoods, microwave appliances; Refrigeration appliances, with refrigerators, combined fridge-freezers, upright and chest freezers Home comfort appliances, with water heaters, air conditioning, and cooling units, gas heaters, home automation systems. Also air filters, like in kitchen vapor extractor hood systems, for vacuum cleaners and for air conditioning systems.
Small household appliances	Floor care systems, with vacuum cleaners Irons Coffee machines Mixers Tooth care equipment Shavers Blood pressure equipment Electronic thermometers
Heating and climate control	Heater systems Warm water boilers Climate conditioning equipment
Security and safety	Intruder alarm Security systems Fire detection systems, with sensors for – temperature – toxic gases like CO, CO_2, exhaust gases, smoke, etc. – combustible gases like CH_4, C_2H_6 flame detection, fire detectors, caravans with gas detectors, etc.

In the small appliances sector, the picture is more diffuse. As these appliances usually involve low cost technology, sensors were only introduced fairly recently. However, some interesting developments have taken place here. For instance, thermopiles for remote temperature sensing that were introduced in the late 1990s are now found in hairdryers and in toasters made by Philips DAP (as described in Chapter 3.4). Since their introduction, the use of remote temperature sensors has been increasing, and further applications of the technology can be expected in the next decade. One interesting new appliance is the ear thermometer, e.g. from Braun, which is described in Chapter 3.5.

Applications for multifunctional chemical sensors to detect constituents in liquids as well as gases are going to expand rapidly in the near future. Chemical sensors act as "noses" and can be extremely useful in kitchens, helping in a range of tasks. from exhaust control to monitoring of cooking, frying or baking processes, as described in Chapter 3.3. Further applications of such "micronose sensors" include air conditioning and climate control as well as safety devices e.g. fire alarms or gas detectors for the prevention of health and fire hazards.

Monitoring the use of detergents is a specific household task which should be automatically controlled as far as possible in order to make use of modern microelectronic systems with sensors. Chapter 4 "Sensors for Detergency" gives a good overview of washing agent monitoring sensors.

1.2 Sensors in Household Appliances

More and more sensors and microsystem devices are being used in a wide range of household appliances, and the number is set to increase in the near future. Several technologies are used for such sensors, as described in the subchapters of Chapter 5. An evaluating synopsis of these sensors is given in Tab. 1.2 for large household appliances, for small household appliances in Tab. 1.3 and for heating and climate control in Tab. 1.4.

The sensors are also classified according to the major technologies used in their production. Technologies that allow large-scale production of reliable sensors at low cost are becoming particularly important. The compatibility of such sensors with electronic signal conditioning is an additional bonus.

We also give a rough overview of the product development status of the various sensors. Some are in use and available in large quantities, while others are still being developed. Their introduction may be imminent, depending on the combined interests of the appliance industry and the consumers.

The following sensor features are of special interest for the production of washing machines and dishwashers

- The use of pressure sensors for water level switches, or in a more sophisticated form also for foam content surveillance in washing machines and dryers.
- Introduction of chemical sensors for water quality monitoring. This includes parameters like turbidity, color, surface tension, detergent concentrations, pH-value etc. Optoelectronic systems are used to monitor the turbidity of washing water, which then determines the number of rinsing cycles (aqua-sensor system).
- Magnetic sensors for controlling the movements of the water spray arms in dish washers

Improved sensors will be used in many other household appliances, such as

- contactless monitoring systems, including e. g. micromachined thermopile infrared sensors for temperature control, which can also be used in cookers, hair care appliances and toasters.
- Air mass flow sensors can be used in fans and vacuum cleaners.
- Acceleration or tilt sensors are used in irons.
- Automatic baking control by introduction of intelligent multigas sensors (artificial noses).

Tab. 1.2 Development status of sensors used for large household appliances.

Function	*Device*	*Technology used*	*Status*
Automated baking	Multigas sensor, electronic noses	Non MST, Infrared sensing MST	UD
Colour	Spectrometers	Potential MST	UD
Current sensors	Magnetoresistive Sensors	Thin Film, MST	P
Detergent control: Cleanliness, dirt monitoring	Turbidity Conductivity	Optical	P P
Dirt content			UD
Dosing units for detergents + Rinsing aids	Macro Micro	Non MST MST	
Engine overload	Temperature		P
Foam content	Pressure Sensor	MST?	UD
Flow for liquids and gases	Flow turbines Pressure sensors...	MST? MST	P
Gas quality	Electronic noses, multigas sensors	MST? Optical?	UD
Humidity (hairdriers)	Infrared Capacitive	MST	
Humidity control	Conductivity sensor		UD
Humidity/Moisture status in dryers	Capacitive	MST	UD
Motion (Water spray arm movement)	Magnetic	MST	P
Motors	– Micromotors Ultrasonic motors	No MST products until now MST	– UD
Pressure sensors, switches	Membrane actuated relay Si-pressure sensors	Non-MST MST	P P
Proximity	Capacitive Inductive Magnetic LEDs – Lasers	MST in future?	P
Push button	Capacitive sensors Touch sensitive switches		P
Rotation, speed	Tachogenerators Magnetic	No MST MST	
Security, intruder alarm	Vibration Tilt sensors	MST? MST?	

Explanation: MST Micro System Technology
UD Under Development
P Product on the market

Tab. 1.2 (continued)

Function	*Device*	*Technology used*	*Status*
Security systems for gas heating	Gas sensors	MST?	P
	Temperature sensors	–	P
	UV sensors	Optical	DU
Temperature control	NTC, PTC resistor	No MST	P
	Ni-, Pt-film resistors	Thin film technology,	P
	Expansion capillary thermostats	MST	P
		No MST	P
	Electromechanical thermostats	No MST	P
	Thermocouples	–	P
	Infrared radiation	MST	P
Tilt	Acceleration sensors	MST	P
	Tilt sensors	MST	
Timers			
Torque			
Valves, dosing units	–	No MST until now	P
	Microvalves	MST	UD
Vision	Photodiode-arrays, cameras	Optical	UD
Vibration	Acceleration sensors	MST	
Wash quality	Tagging		UD
Water flow measurement	Turbine flow meters	Magnetic	P
Water level	Pressure switches	MST	P
	Magnetic actuated relays	MST	
Water quality (input water) Wash water	Turbidity PH-sensors Ion-FETs	MST?	First products
Weight sensors	Piezoresistors,	MST	
	strain gauges	No MST	
	Pressure sensitive membranes		

Gas sensors are a new type of sensor for which there is an increasing demand. We include a description of state-of-the-art-gas sensors in Chapter 5.3.

Long-term stability (more than 10 years lifetime) under realistic conditions is a crucial feature in all sensors, especially gas sensors.

Several sensors are in use in the heating and climate control sector, such as

- Temperature sensors and switches
- Pressure sensors and switches
- Flow sensors for liquids and for gas.

Tab. 1.3 Development status of devices for small household appliances.

Function	*Device*	*Technology used*	*Status*
Contactless temperature measurement	Infrared sensors	MST	P UD
Automated baking	Multigas sensor, electronic nose		R&D
Humidity (hairdriers)	Infrared Capacitive		
Rotational speed	Tacho generators Magnetic sensors		P UD
Distance sensors	Optical (?)		UD

Many of them are covered in the previous tables 1.2 and 1.3; some additional specific sensors are listed in Tab. 1.4.

Temperature control is one of the longest established and most important functions in household appliances. One example of modern thin film fabrication technology of platinum temperature sensors with application examples in the kitchen in hot plates and ovens is given in Chapter 5.1.

Reed relais are an example of well-established distance sensing technology. They are used in millions of appliance units and are described in Chapter 5.2.

Reed technology applications include
- measuring the flow of liquids in turbine-like meters,
- the water level determination, e.g. in coffee machines, or in monitoring the water level of softeners and clarifier in dishwashers
- controlling the movements of dishwasher spray-arms
- detecting the drum position in washing machines
- security control for appliance door detection
- electric toothbrush
- reed sensors in carpet cleaners
- detecting of movements or end positions in massage chairs, special lifts for bathtubs, hospital beds,

A survey on several gas sensing methods is given in Chapter 5.3. There is a great need for several types of gas sensors in order to detect the wide range of harmful gases in the human environment – toxic and/or explosive gases like carbon monoxide and natural gas. An uncomfortable room climate is probably a contributing factor to the so-called "sick building" syndrome. Furthermore, there are about 10 000 additional air components which can affect our health. Therefore many gas sensor working methods are in use, comprising
- semiconductor oxide gas sensors, based e.g. on tin oxide or gallium oxide,
- pellistors for the detection of flammable gases,

Tab. 1.4 Development status of devices for heating and climate control.

Function	*Device*	*Technology used*	*Status*
Temperature sensors and switches	NTC, PTC resistors Bimetals Film sensors	No MST No MST MST	
Contactless temperature measurement	Infrared sensors	MST	P UD
Pressure sensors and switches		MST	
Flow sensors for liquids and for gas.		MST?	
Valves, dosing systems		No MST today MST?	
Exhaust control systems	Lambda sensors		
Security systems	Gas sensors Temperature sensors UV sensors		
Gas quality	Multigas sensor, electronic nose		R&D

- amperometric or potentiometric liquid state electrochemical gas sensors,
- potentiometric and amperometric solid state electrochemical gas sensors,
- optical gas sensors.

Some application areas of gas sensors are described in chapter 5.3, including gas and fuel powered domestic burner control, air quality sensing, indoor detection of CO, and natural gas detection. Several further applications of gas sensors are still in the development stage, e.g. for cooking and frying control, or for controlling the self-cleaning procedure (pyrolysis) of ovens.

In Chapter 5.4, optical ultraviolet radiation sensors are described, including UV-enhanced silicon-based pn diodes, detectors made from other wide band gap materials in crystalline or polycrystalline form, the latter being a new, less costly alternative. Other domestic applications are personal UV exposure dosimetry, surveillance of sun beds, flame scanning in gas and oil burners, fire alarm monitors and water sterilization equipment surveillance.

Displacement sensors developed for application in the dampers of washing machines are described in Chapter 5.5. Measuring the displacement of washing machine drums allows direct control of any unbalance in the washing machine. Thus, vibrations will be reduced and the lifespan of the washing machine prolonged. Furthermore, the electrical power needed for the drives can be reduced, and the washing process can be adjusted to the load by automatic selection of an appropriate washing program.

It is also interesting to look into the future use of sensors in household appliances. An attempt to do this is made in Chapter 6 where the influence factors in this broad field are analyzed. These include socio-economic data of the end users (like age of the population) and their preferences (like savings of energy, water and detergents), ease of use and cost of ownership, as described in Chapter 6.1. One specific and interesting future development can be foreseen – the integration of home appliances into heretofore strictly separated areas, such as

- telecommunication and internet services
- home entertainment
- computing
- heating and ventilation, and even
- building controls

For a successful integration, the domestic appliances and all the other functions must be able to communicate with each other, which requires compatible software as well as hardware. Such digitalization may occur not only at appliance level, but also at the sensor level, as described in Chapter 6.2.

An example of such an integrated home project still under construction is given in Chapter 6.3, which describes how in South West Germany about 1200 households will be interconnected in a cost-friendly way to share comfort, service and security-related functions.

1.3 References

1 G. Tschulena, Micro System Technology in Household Appliances, mst news 2 (2001) 36–37.
2 The NEXUS Technology Roadmap for Microsystems, edited by H. Zinner, (2000), Chapter II.6 on "Household Appliances", compiled by G. Tschulena.
3 NEXUS Market Analysis for Microsystems, 2000–2005, edited by Reiner Wechsung, April 2002, with specific contribution on "Household Appliances application market".

2
Market Data

G. Tschulena

2.1
Introduction – The Household Appliance Industry

The household industry comprises a broad variety of large and small appliances, heating and climate control as well as security and safety devices. A more detailed product description is given in Chapter 1. The annual turnover figures of the industry are estimated at about 90 billion $ worldwide. East Asian countries represent the largest market, followed closely by Western Europe and the US. Of course, such figures can be nothing more than a rough guide, especially when the sources say nothing about the data, definitions and prerequisites these figures are based on.

The origins of the household appliance industry date back to the early decades of the last century, when "simple" tasks were transferred to household appliances. For example, in an early washing machine of the thirties, the water inlet and outlet as well as water motion and heating were controlled, while all other functions required were carried out manually. Refrigerators only provided the cooling power or the low temperature. In the forties of the last century, the first vacuum cleaners came on the market.

Over the last decades, however, requirements have changed, and the customer expects improved functionality as well as reduced and carefully controlled energy, water and detergent consumption.

Parallel to these customer-driven requirements, we have also seen a change in technology. Electronic control systems and microprocessors in particular have become more powerful, while prices have been falling.

As a result of both developments, the use of automatic control systems has strongly increased. This, in turn, has led to a more widespread use of sensors – not only the traditional low-cost sensors, but a whole host of new sensors that have been introduced – and even more are still being developed. Our book therefore wants to give an insight into the rapidly changing scene and explain what sensors are introduced into appliances for what purpose.

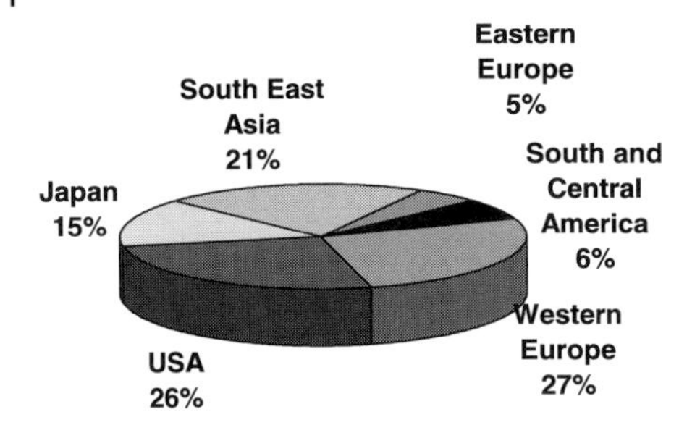

Fig. 2.1 Geographical distribution of the large and small appliance markets according to production quantity.

2.2 HA Industry Data – World Market Data

The household appliance field comprises several families of large household appliances, small appliances, heating and climate control devices and security instrumentation, as described in Chapter 1. The total turnover of all these appliances is estimated at over 90 billion $ worldwide. The worldwide production of larger appliances could be about 150 million units [1].

The Western European countries provide the biggest market, slightly bigger than the US, followed by the South Eastern countries and Japan, as depicted in Fig. 2.1.

A moderate growth of these production figures can be expected – depending on many socio-economic and technical factors. So the European appliance production is expected to grow by about 1 to 3% to about 65 million units in the next five years.

2.3 HA Industry Data – European Market Data

The European consumption of appliances is in the range of several million units per year, for example up to 50 million units per year for washing machines or refrigerators, as shown in Fig. 2.2 [2].

European production can only partly cover these needs, while the rest is imported. Nearly 70 million units of large appliances per year are produced in Europe.

The turnover of the household appliance industry is about 35 billion Euro, of which about 63% come from large appliances, 17% from small appliances, and 20% from heating, ventilation and air conditioning appliances.

A workforce of about 200,000 are directly employed by the European household appliance industry, and about a further 500,000 employees depend on it, working in upstream or downstream jobs [3].

The European production figures for large appliances are given in Fig. 2.3 [2]. They comprise:

- Automatic washing machines about 20 million units
- Refrigerators about 18 million units
- Microwave ovens, dishwashers and cookers, each between 6 and 7 million units
- Built-in ovens, and dryers, each somewhat more than 3 million units

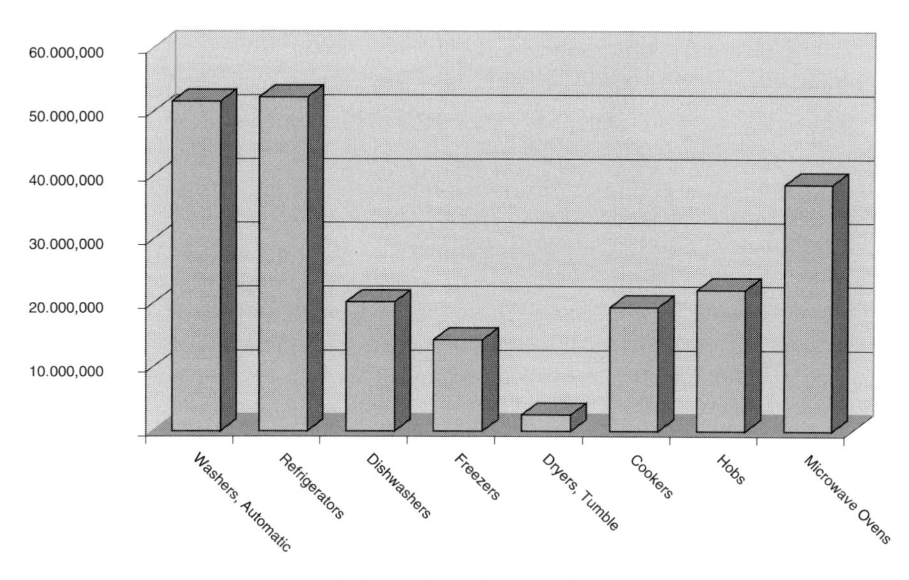

Fig. 2.2 European consumption of selected larger household appliances [2].

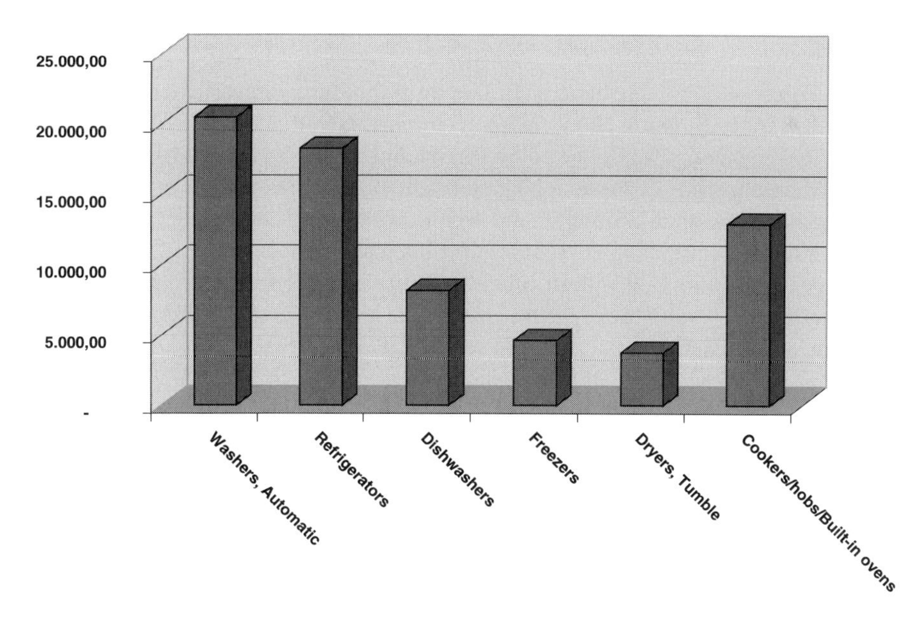

Fig. 2.3 European production figures for selected larger appliances [2].

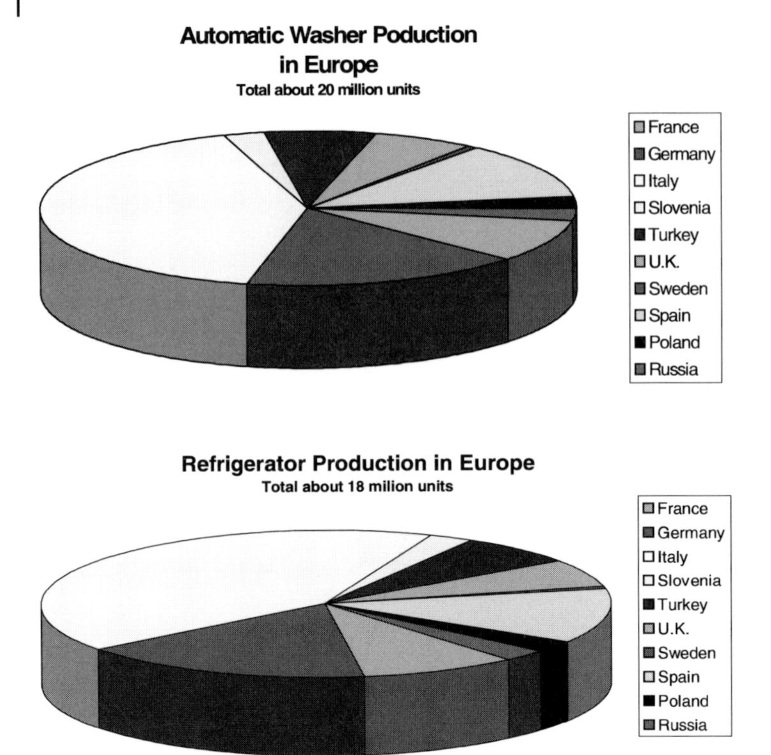

Fig. 2.4 Geographical distribution of the appliance production in Europe for washing machines (left) and for refrigerators (right).

The largest share of appliances in Europe is produced (measured in terms of quantities) in Italy (with about 38%), Germany (about 23%), the UK, France, Spain, Turkey (about 7%), followed by Russia and Slovenia (about 2%).

A moderate growth of these appliance production figures can be expected – depending on many socio-economic and technical factors as well as entrepreneurial activities among competing companies and brands.

The degree to which European households are equipped with appliances varies – the saturation level being highest in Central and Northern Europe, somewhat lower in Southern Europe and still lower in Eastern Europe. The average saturation level for some larger appliance families is shown in Fig. 2.5. For washing machines, cookers and refrigerators, a saturation of more than 90% has been achieved, around 50% for microwave ovens or freezers, and only 25% for dryers.

After a new product has been introduced, market behavior develops over time, and a higher degree of saturation will eventually be reached. The data given in Fig. 2.6. refer to Germany. The eighties saw the introduction of microwave ovens, whose market share has been steadily increasing to the present level of over 60%. By contrast, in the case of washing machines and refrigerators, which are found practically in every household, the saturation level remains more or less constant,

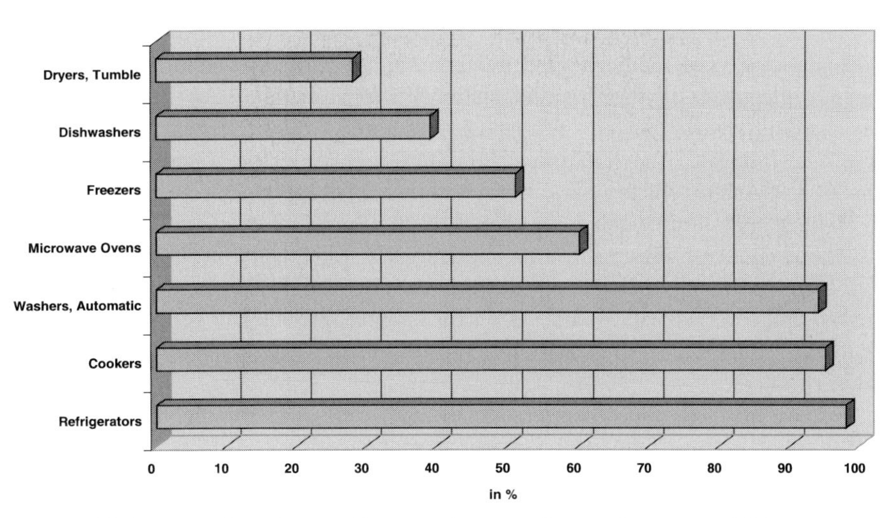

Fig. 2.5 Market saturation of selected large appliances in Europe, for the year 2000.

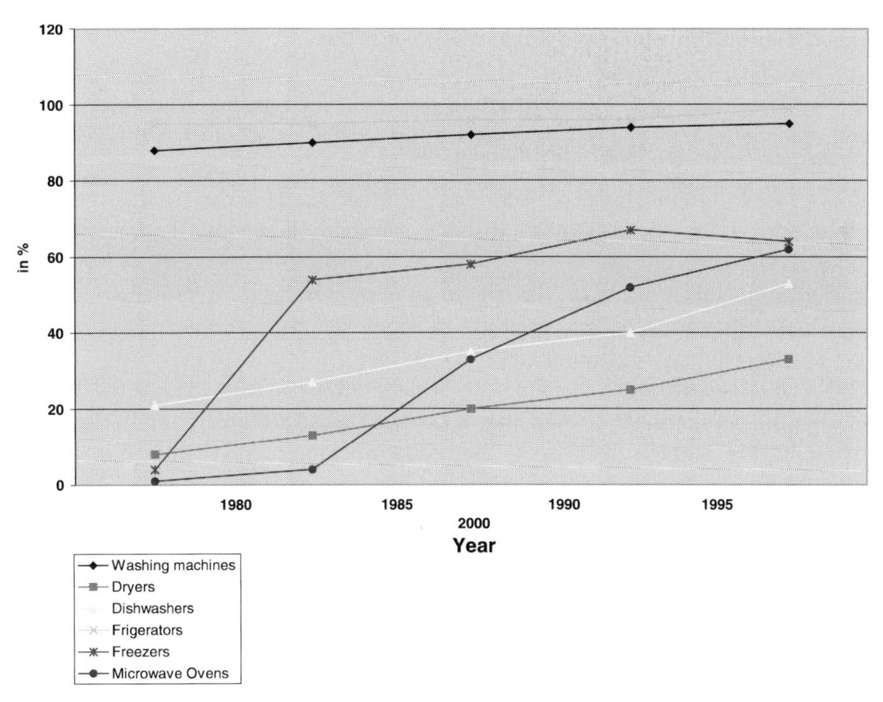

Fig. 2.6 Development of the market saturation for selected appliances [4]. Figures refer to Western Germany before 1990, and reunited Germany from 1990.

as business consists mainly in replacing existing appliances plus a small share of new households.

The appliance industry in Europe comprises about 280 companies, of which 90% are small and medium-sized independent enterprises. The larger companies include Electrolux (with a market share of approximately 20%), Bosch-Siemens Hausgeräte (about 16%) and Whirlpool (about 11%). These larger groups are followed by Merloni (Italy), Groupe Brandt, General Electric, Candy, Miele (Germany) and Arcelic (Turkey), and many more smaller appliance companies.

There is only limited scope for growth in the European household appliance industry, as could be shown for large appliances. While quality has been improved – e.g. improved energy efficiency – average real prices of the appliances have continued to drop over the past years. As a consequence, pressure to cut costs has been building up and led to the transfer of production facilities to countries with lower labor costs [3].

With the introduction of more sophisticated, consumer-friendly appliances, this trend could be overcome, and European competitiveness could be strengthened. The use of better control systems and the introduction of more and better sensors are a prerequisite for such a development, as we are going to point out in this book.

2.4
Sensors and MST in Household Appliances Market Data

The function of an appliance determines the choice of sensors. Thus, even in the early appliances, e.g. in the forties and fifties of the last century some relatively simple sensors have been used.

The most common control functions in these early appliances are the control of temperature, pressure, position or distance. Mechanical sensing devices were introduced for these purposes, such as bimetal temperature switches or liquid expansion temperature switches for ovens, washing machines, dishwashers, refrigerators, etc. Electromechanical pressure switches and potentiometric level sensors have also been introduced quite early.

When microelectronics and solid state devices developed over the last five to four decades, the development of solid-state sensors followed suit, resulting in the introduction of NTC and PTC resistors to monitor temperature, and first Reed relais and inductive sensors to determine position and distance, or tachometers for rotational measurements in washing machines and dish washers over the past two decades.

A description of sensors for appliances is given in Chapter 1. An additional overview of sensors in selected appliances is given in Tab. 2.1.

The sensor industry is particularly strong in Germany and the rest of Europe – about 2000 companies in Europe are producing or trading sensors, more than 600 companies in Germany, which are mostly highly specialized traditional small and medium-sized enterprises. They have developed a wide range of new devices which they sell to various industries, such as the process, manufacturing and ma-

Tab. 2.1 Examples of sensors in appliances used today [5, 6].

Appliance	*Monitored function*	*Type of sensors used today*
Washers	Temperature	NTC, Klixon
	Rotation	Tachometers
	Level	Pressure sensors
	Weight	Inductive sensors
	Position	Reed switches
	Unbalance	Tachometer, pressure sensors
Dryers	Temperature	NTC, Klixon
	Humidity	Conductivity sensing
Dish washer	Temperature	NTC
	Salt level	Reed contact + swimmer Density sensors
	Cleanser level	Reed contact
	Spray arm rotation	Magnetoresistive sensors
	Flow	Rotary water meters Pressure sensors
Freezers	Temperature	
Oven	Temperature	Expansion capillars, Pt 100
Microwave ovens	Temperature	NTC
	Humidity	Ceramic sensors
	Gases	
Hair dryers	Temperature	NTC
	Temperature (contactless)	IR Thermopiles
	Humidity	
	Air flow	
Toasters	Temperature	NTC
	Temperature (contactless)	IR Thermopiles
Irons	Temperature	Bimetal switch, NTC
	Humidity	
	Position, movements	Tilt sensors, acceleration sensors
Vacuum cleaners	Air flow	
	Particles	
Air cleaners	Gases	Ceramic Sensors
	Particles	

chine building, automotive, and environmental protection industries. Some of them, however, are exclusive suppliers to the household appliance industry.

The industry has its technological roots in the precision mechanics technology, in particular in Switzerland, in Germany and in some French regions (the former watch-making regions).

Over the past two or three decades, new technologies have been developed and introduced, often based on semiconductor technology. The trend in microelectronics toward miniaturization and increased complexity has also affected sensors.

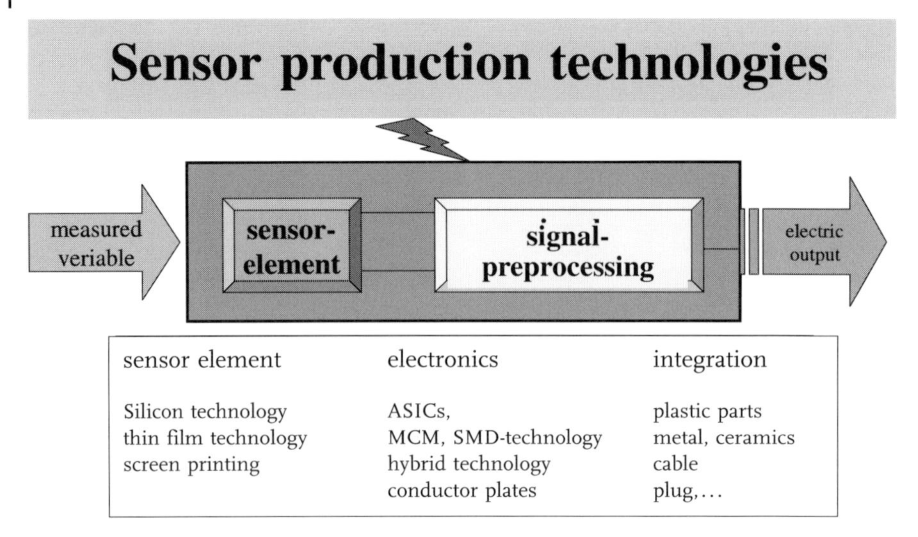

Fig. 2.7 Sensor production technologies needs a reliable and cost effective integration of technologies for the sensor element, the electronic data (pre) processing assembly.

The first silicon sensors were introduced in the fifties and sixties. They developed into semiconductor microsensors and in the past decade even to microsystems.

Furthermore, other solid-state sensors have been developed and introduced, ranging from thin film and screen-printed sensors to ceramic sensors and – latterly polymeric sensors.

Progress could be made in the development of sensors where it was possible to combine sensors and (micro-)electronic integrated circuits ("chips") in a common encapsulation with compatible technology. So today we have various types of sensors on the market, including microelectronics on printed circuit boards, SMD-like devices or electronics on ceramic hybrid circuits as well as flexible polymeric boards. In some cases fully integrated intelligent silicon sensors are used where some data processing and the sensor are combined on one chip – or sometimes two chips. The decision which of these technologies is suitable depends on technical requirements for sensors and data processing, size and weight of the end product and not least on cost in relation to the numbers of the item produced.

The latest development are micromechanical sensors. Their development began with the large-scale introduction of silicon micromachined pressure sensors to the automotive industry in the nineties, which entailed a massive price reduction. Then acceleration sensors for airbag firing, yaw rate sensors and more were introduced. Many devices are still being discovered. The next step is product evolution, with introduction times between a few years and over a decade, as shown in Tab. 2.2. Once customers in the industry have accepted a product, investment in large-scale production can go ahead. It helps to find more applications for the product. The time scale for the product evolution process varies from about five

Tab. 2.2 Product evolution time table of selected MEMS sensors [7].

Product	*Discovery*	*Product evolution*	*Cost reduction*	*Full commercialization*
Pressure sensors	1954–1960	1960–1975	1975–1990	After 1990
Acceleration sensors	1974–1985	1985–1990	1990–1998	After 1998
Gas sensors	1986–1994	1994–1998	1998–2005	2005?
Chemical Sensors	1980–1994	1994–1999	1999–2004	2004?
Yaw rate sensors	1982–1990	1990–1996	1996–2002	2002

years up to more than two decades, depending on acceptance and success as well as entrepreneurial initiative. Many of these reliable mass fabricated low cost devices were originally developed for the automotive industry and are now used in many other fields, e.g. the household industry.

We expect these micromachined sensors to become more and more important in the household industry, in many domestic applications of silicon pressure sensors, acceleration sensors, tilt sensors, infrared detectors and thermopiles, flow meters, as well as gas sensors and liquid constituent sensors.

Microsystems are also expected to be introduced in the near future, including for example artificial noses, fingerprint sensing systems, bar code readers, rf-tagging systems, microfluidic pumps and dosing systems, gas flow control systems, new flexible and low cost displays or electronic paper.

2.5 Conclusion

Several factors contribute to the total production costs of household appliances:

- Mechanical body construction between one quarter and half of the costs
- Electronics and power supply about one third
- Actuators about 20%
- Sensors 15% to more than 20%

As the domestic appliance industry caters for a mass market – the production volume of a single company may well be around a million units – interesting opportunities are opening up for sensors and microsystems producers, particularly, because silicon microsystems show their true potential especially when mass-produced. But there is also room for many other technologies in this interesting market, provided technology and cost are viable.

Many sensor – actuator control systems are already in use today, and many more will be introduced over the coming years, so that we can expect an increase in sensors, actuators and control systems, many of which will be microsystems, in the near future.

2.6 References

1. A. Lahrmann: Smart domestic appliances through innovations, 6th International Conference on Microsystems 98, Potsdam, Ed. H. Reichl, E. Obermeier, VDE-Verlag, Berlin (1998).
2. Appliance Magazine: Portrait of the European Appliance Industry, 2001.
3. Rainer Stamminger: Keynote for the components day at the Hometech, exhibition 28th February 2002.
4. Zahlenspiegel des deutschen Elektro-Hausgerätemarktes (2002), edited from ZVEI, Zentralverband Elektrotechnik und Elektronikindustrie, Hausgeräte-Fachverband, Frankfurt.
5. Data collected during the meetings of the European NEXUS User-Supplier Club 7 Household Appliances (see e.g. in *www.nexus-mems.com*).
6. G. Tschulena, mst news 2 (2001). G. Tschulena, Commercialization of Microsystems 2001, Oxford, UK.
7. Roger Grace: Commercialisation Issues for Microsystems, Proc. Microsystem Technologies, Düsseldorf, March 2001.

3
Appliances and Sensors

3.1
Home Laundry Appliance Manufacturers – Drivers of Change: Socioeconomics and Enablers

G. Wentzlaff, R. Herden and R. Stamminger

3.1.1
Introduction

To quote R. Olmedo et al. at the 38th International wfk-Detergency conference at Krefeld 1998, 60% of detergent products sold worldwide are still used for washing by hand. In the industrial countries, however, the use of a washing machine for everyday laundry has been pretty much standard over the last decades.

In developing countries as well as China and Russia, the use of domestic washing machines has been increasing constantly over the years in line with economic growth. All over the world doing the laundry is one of the least popular domestic chores; so consumers jump at the chance to make the tedious process easier.

It is therefore more than likely that the use of washing machines will continue to rise well into the next century, and we will probably see further development of the washing machine types currently available.

Over the years, two major trends have been driving washing machine development:

Trend 1 Environmentally friendly washing machines are in ever greater demand.

Trend 2 The availability of more sophisticated technology leads to the use of more and more control, monitoring and display devices in washing machines.

These trends are set to continue for at least another ten years, while the washing machine market is expanding into countries that have no significant home appliance industry. How the above-mentioned trends shape the current development of washing machines and what developments can be reasonably expected will be discussed in the following section. Attention should be paid at this point to those features of machines that are directly related to detergent.

3.1.2
Technical Outline

So far, we have only been talking about washing machines in general without differentiating between the various types. In terms of construction, the washing machines most frequently available on world markets can be divided into two groups, those with a horizontal drum axis and those with a vertical drum axis as shown in Fig. 3.1.

Machine sizes may of course vary with both types, depending on consumer requirements. Generally, it can be said that machines with a vertical drum axis usually have larger drum volumes and larger casing dimensions than those with a horizontal drum axis. Larger machines with vertical drum axis are practically exclusively the norm in the USA and in many Asian countries, whereas washing machines with a horizontal axis and an opening to the front are the more usual standard found in Europe.

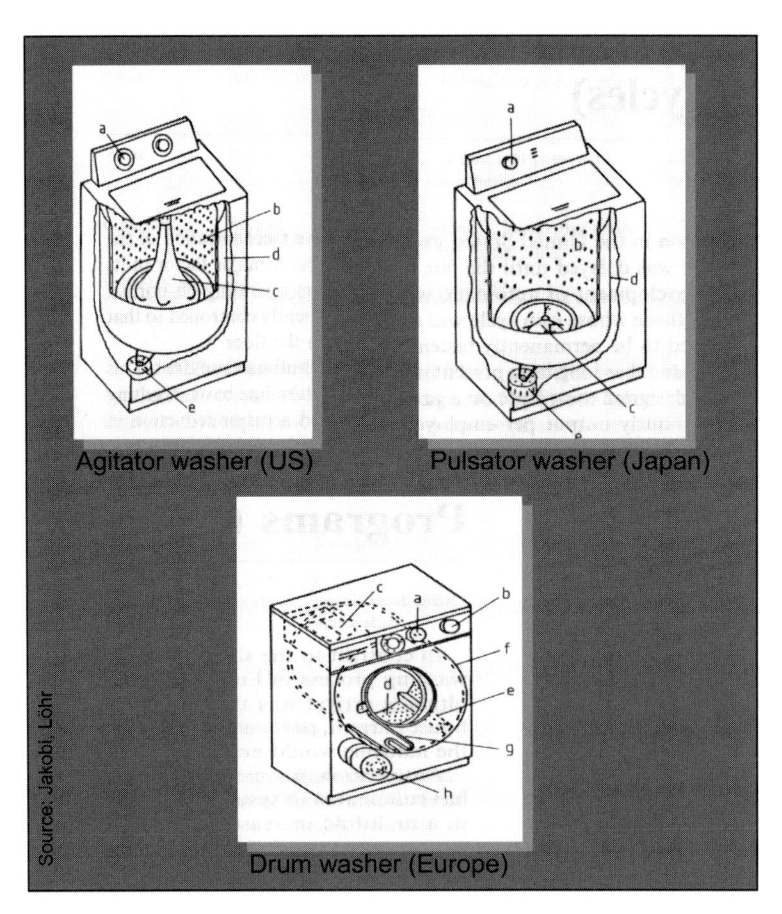

Fig. 3.1 Basic principle of washing machine construction.

For technical reasons, it is generally accepted that with given wash and rinse results, specific water and energy consumption in machines with a horizontal drum axis is less than in machines with a vertical drum axis. Very conservative consumer habits contrast with this. In other words, particularly in this area, consumers prefer familiar machine technology and are very difficult to convince of the advantages of other technologies. Whether the pressure from many quarters to save water and energy is great enough to persuade the consumer to accept other technologies therefore remains an open question.

These constructional differences set aside, it can be argued that the basic functions are very largely independent of the drum principle and can therefore be discussed in general here.

The basic technical functions of a washing machine consist of

1. Agitation
2. Water intake, water level
3. Temperature control
4. Detergent dispensing
5. Water and suds monitoring
6. Program sequence
7. Operation, display technology

3.1.3 Basic Functions

3.1.3.1 Agitation

Even today, particularly cheap and simple washing machines still have drive concepts that allow only one wash speed and one single spin speed and therefore do not require any electronic control. A motor of this type is shown in the left part of Fig. 3.2.

However, motor speed was also the first function in washing machines to be electronically controlled. The standard nowadays are controllable AC- or DC-motors. Depending on the textile type or wash program, such drives can be used to achieve optimal wash speeds, reversing rhythms and activation times. Usually a tachogenerator on the motor is used as a speed sensor. Such a smaller modern motor, connected to a small electronic control unit, is shown on the right hand side of Fig. 3.2.

European machines in particular are operated with very low water levels in order to save energy and water. It is therefore important to ensure that the laundry load is soaked in water or suds as quickly as possible so that dry textiles do not rub unnecessarily against the drum or the rubber seal. This is achieved in European machines by constructional features that improve water absorption, shown in Fig. 3.3, such as lifter bars, circulation (jet) systems or direct injection.

Here, too, electronics have come in handy. At the start of a program, additional agitation periods could be electronically introduced, which further enhance water absorption.

Fig. 3.2 Old-type double AC motor and new DC motor with electronic control unit.

Sensitive signal evaluation at the motor has a further advantage: If too much or the wrong kind of detergent has been added, excessive foam can be detected during washing. The increased load on the motor caused by the foam is registered and evaluated by modern electronics, and the wash process can then be adjusted accordingly to prevent foam overflow.

In electric laundry dryers, it is important from an energy-saving point of view to ensure that as much water as possible is extracted from the laundry beforehand. With increasing market penetration of laundry dryers, the importance of the maximum spin speed in washing machines also increases.

In order to achieve higher spin rates, it is paramount to recognize possible imbalances to prevent machine misalignment that would result in mechanical stress on the bearings as well as unnecessary noise development. The first step to avoid imbalance is a gentle run-up for a few seconds at the beginning, as shown in the upper panel of Fig. 3.4. Here the spin speed is continually raised in discrete and slow steps so that the laundry is distributed evenly around the drum.

In the lower panel of Fig. 3.4 you can see that if the current laundry load still develops an imbalance, this is registered and, depending on the machine features, either the run-up to spin speed is interrupted and a new attempt is made to reach spin speed, or the maximum spin speed is limited to ensure the machine always remains stable.

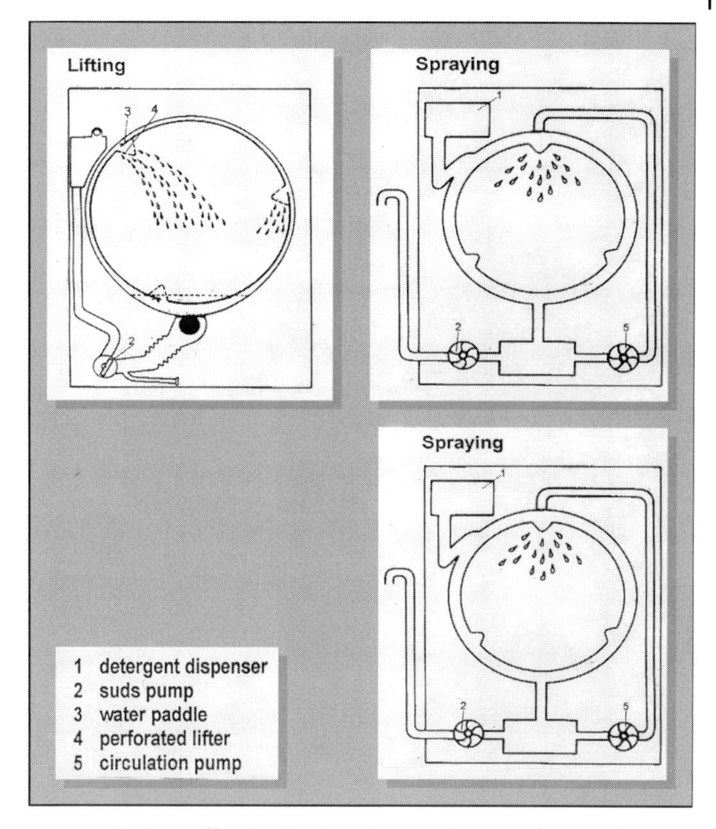

Fig. 3.3 Washing of low-level suds in horizontal axis washing machines.

Noise emission control is another area where the application of electronics goes hand in hand with environment protection. Special electronic circuits on DC motors ensure that the noise level can be reduced at the source rather than through costly insulation.

Looking at this first basic function clearly shows that electronics play a major part in lowering consumption figures and better machine management.

The way ahead lies in brushless direct drives or imbalance compensation systems that can only be realized with the application of electronics.

Brushless drives would avoid the noise caused by brush contacts on conventional motors. The solutions the motor industry has come up with so far have a major drawback – the noise developed by the required electronic unit itself, similar to the hum of a transformer, is still very loud.

Imbalance compensation systems not only reduce oscillation caused by a major shift in the center of gravity of the drum unit; they also make exact drum positioning possible as well as enabling immediate change of direction in the drum, which in turn gives more interesting processing technology options.

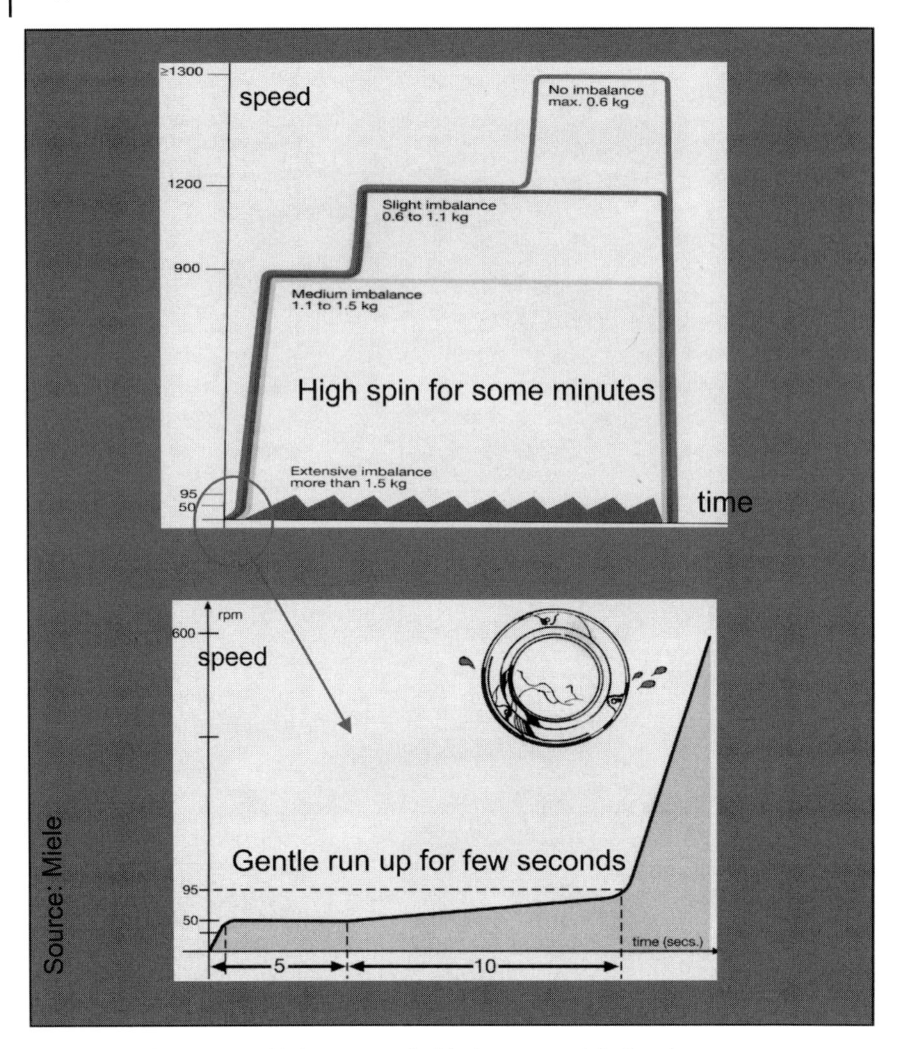

Fig. 3.4 Gentle run-up and balance-controlled high spin speed for laundry spinning.

3.1.3.2 **Water Intake, Water Level**

Water level measurement is carried out using the principle of pressure measurement and functions purely as a level limiter.

In the majority of cases electromechanical pressure measuring components are used as sensors as shown in Fig. 3.5 – the large item at the left-hand side. But here again electronics have been introduced. Fig. 3.5 demonstrates just how small a modern pressure sensor can be (right-hand corner).

Modern washing machines work with this type of pressure sensors with analogue output. In such machines the software can be used to set any desired water level.

Fig. 3.5 Examples of electromechanical and electronic pressure measuring devices.

Additionally, using the filling and absorption times established with the aid of electronics, the quantity of laundry or type of textile can be registered and the wash and rinse program can be modified as appropriate. The working principle as defined by water level in relation to process time is shown in the upper part of Fig. 3.6.

The lower part of the diagram shows that this modern system allows you to fine-tune water consumption and wash load. The upper line shows the old type half-load button version still used in some European machines where it is up to the user to decide if his actual load is half or full load. Sometimes he is wrong and the wash and consumption results are not as good as they could be.

With these or similar sensitive pressure sensors possible excess foaming can also be registered and an appropriate program modification activated, as already mentioned.

3.1.3.3 **Temperature Control**

Many Asian washing machines operate exclusively with cold water, although the cold-water temperature is not necessarily the 10–15 °C that can be expected in Europe. It may well be considerably higher. Therefore such machines are not fitted with a temperature sensor. Another machine type, particularly common in the USA, has a cold-water connection and an additional hot water connection but

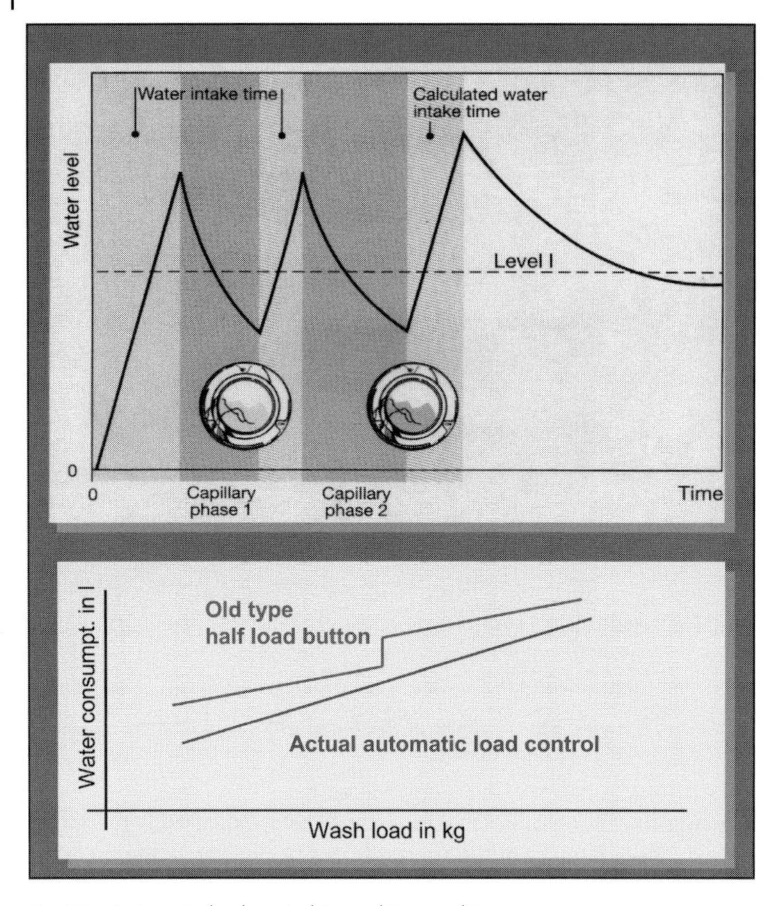

Fig. 3.6 Automatic load control in washing machines.

does not have its own heater element. Temperature measurement is frequently via a capillary thermostat functioning as a temperature limiter.

Fig. 3.7 shows a comparison between a capillary thermostat in the right part of the figure and the ever more frequently used NTC temperature sensor whose analogue electrical signal can easily be processed by an electronic control system. The NTC sensor type is increasingly used, particularly in modern European machines that always have their own heater element and sometimes also an additional hot water connection.

Let us now look at the ecological aspects linked to temperature regulation. While electronic water level regulation helps economize on water consumption, low wash temperatures provide an opportunity to save energy, particularly in connection with modern detergents. The increased use of various enzymes means that satisfying wash results can be achieved even at lower temperatures. With the resulting general trend to ever-lower temperatures, temperature control is, in our opinion, particularly important.

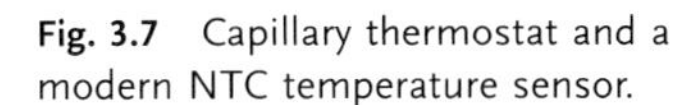

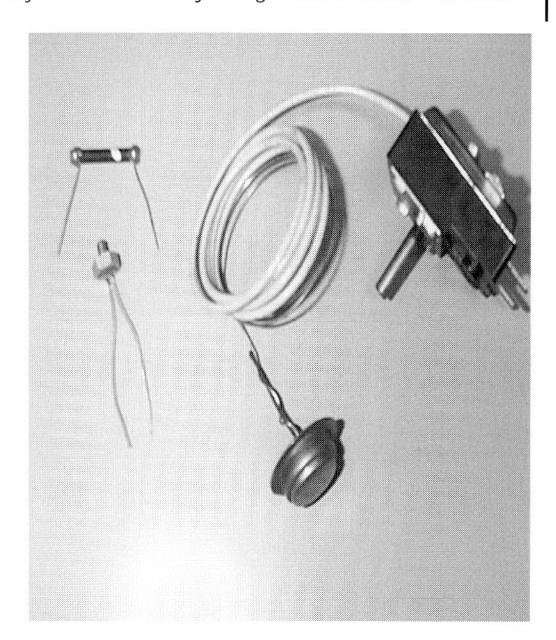

Fig. 3.7 Capillary thermostat and a modern NTC temperature sensor.

In cheap machines with simple capillary thermostats it is possible that due to the tolerances of this component the set temperature may vary by up to plus/minus 10 °C. This is shown in the upper part of Fig. 3.8. If the cold water temperature is relatively "high", the set lower temperature limit may be exceeded, and the heating may not be switched on at all. The washing process is then carried out with cold water with accordingly poor washing results.

By contrast, modern washing machines allow temperatures to be precisely controlled, virtually to the degree, through an NTC sensor. With appropriate monitoring, the set temperature can be maintained as long as required, as shown in Fig. 3.8. This means that even with low temperatures, the desired washing results can be ensured.

Another well-known application of accurate temperature control is known as the enzyme phase during heating, shown in the lower part of the diagram. Temperature-controlled suds cooling in cottons programs still significant in some European countries is also a good example. Thirdly, in the frequently used minimum iron programs, a temperature-controlled cooling process could avoid creasing and save a great deal of water compared to standard procedures.

3.1.3.4 **Detergent Dispensing**

Generally, detergent is added via special dispensers as shown in Fig. 3.9. In European machines there are usually 3 compartments, 2 for detergent for the pre- and main wash and one for fabric conditioner. In some European machines, one of these compartments can also be used separately to add bleach, as shown in

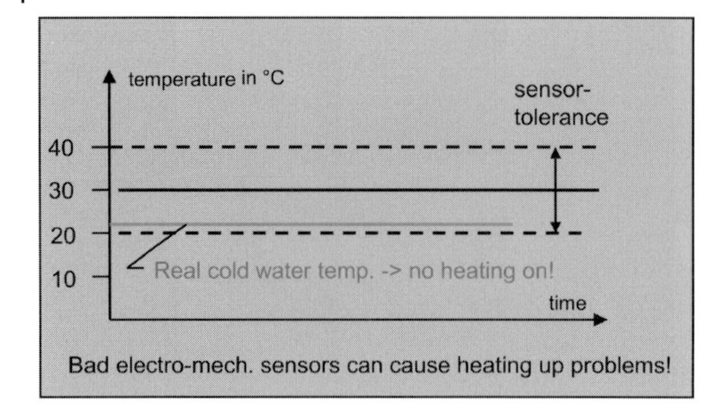

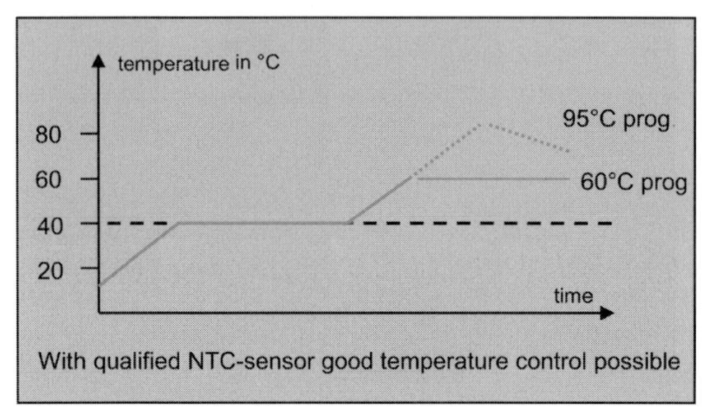

Fig. 3.8 Examples of temperature control.

Fig. 3.9. An additional separate compartment for water softener is only available in a very few European models. In conjunction with the correct detergent the wash process can be carried out fully automatically. So excellent washing results can be achieved without manually pre-treating the laundry or manually adding bleach, as it is common in some other countries.

Detergent dispensing is, of course, another application area for electronics. The solutions offered so far, however, have not met with market approval.

What is known as "Automatic Dosing System" which was introduced to the German market some years ago was a very complex solution of the problem and used a lot of electronics. The high purchasing costs due to the high technical requirements for 3 detergent components and a fabric conditioner, and problems in detergent logistics led to the abandonment of a then highly acclaimed system. Other systems for one detergent only that were clearly simpler technically have not yet been optimized for practical consumer purposes.

Looking chronologically at the basic functions of agitation, water level and temperature control, they have all been automated one after the other. The only manual procedure remaining is still detergent dispensing. In our opinion, what is called for is a common development field for both the detergent and washing machine industries, as they share the purpose of making laundry washing as easy as

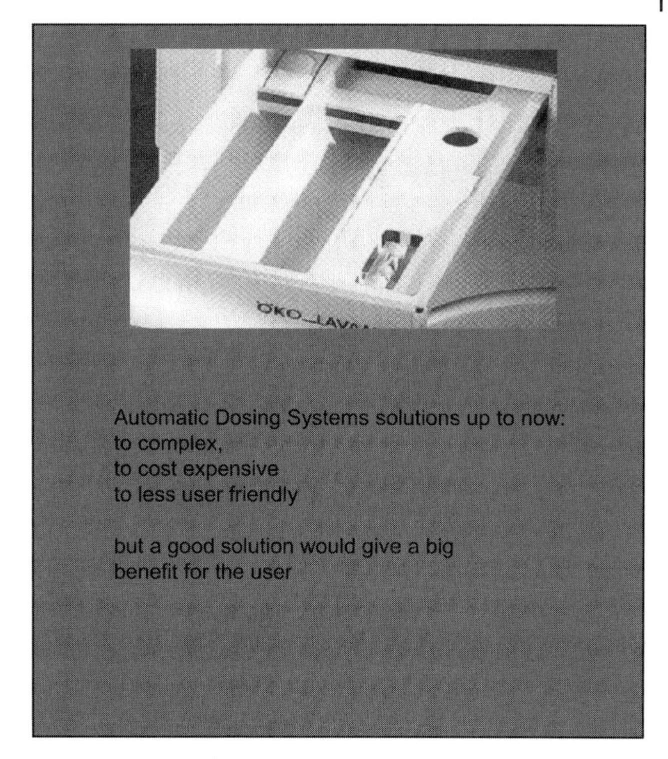

Fig. 3.9 Detergent dispensing system.

possible for the end user. The washing machine industry seems to be very willing to go along this route.

Recent suggestions to solve the problem of monitoring and dispensing detergent quantities involve technically simple solutions, such as making dispensing instructions machine-readable, for example via bar codes or chip cards. As an alternative, some cheap and easily identifiable additive – possibly fluorescent – could be added to detergents. Such standardized marking should be selected so that it can be detected simply and cheaply in the washing machine.

Naturally, it has also been suggested that textile care instructions could be electronically transferred to the washing machine. However, with frequent misuse and errors within the current wash-care labeling system, such methods should be treated with caution.

3.1.3.5 Water and Suds Monitoring

Water quality and suds monitoring is not possible without the use of electronics and has therefore only been introduced to some domestic washing machines over the last few years. The greatest improvements in this area are still to come, as the relevant components continue to come down in price.

As detergent dispensing instructions are often given in relation to the degree of water hardness, it seems reasonable to display the actual water hardness automatically or to adjust the wash and rinse processes as appropriate. It would therefore be really useful to be able to measure water hardness, which could be done by sensors measuring electrical conductivity. Some European machines have been fitted with such sensors, but so far, their measurements have not been clear enough, since any salt dissolved in the water will affect the results. Moreover, the temperature dependence and the durability of the sensors is not satisfactory. A further obstacle preventing large-scale introduction of this technology is the high cost of components.

It is quite obvious that the suds should also be measured. In recent years sensors have been fitted that directly measure the clouding of suds in Japanese machines with vertical drum axes as well as in European machines with horizontal axes. This uses relatively inexpensive infrared sensors which register the weakening of a beam of infrared light.

Fig. 3.10 gives an overview of major measuring parameters, using the European machine type as an example. Turbidity in the suds mainly depends on the type of detergent as well as the amount of dirt and fluff in relation to the amount of water added, as shown in the upper part of the diagram.

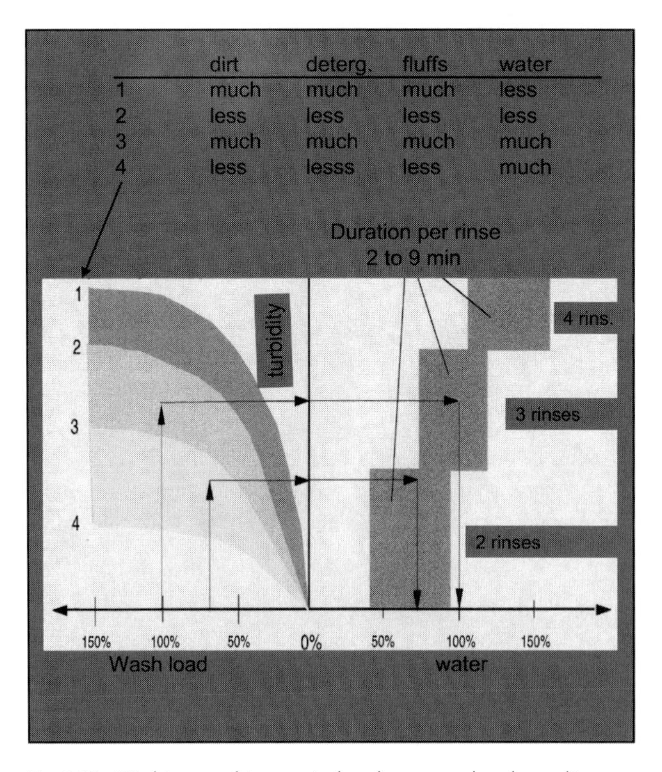

Fig. 3.10 Washing machine control and measured suds quality.

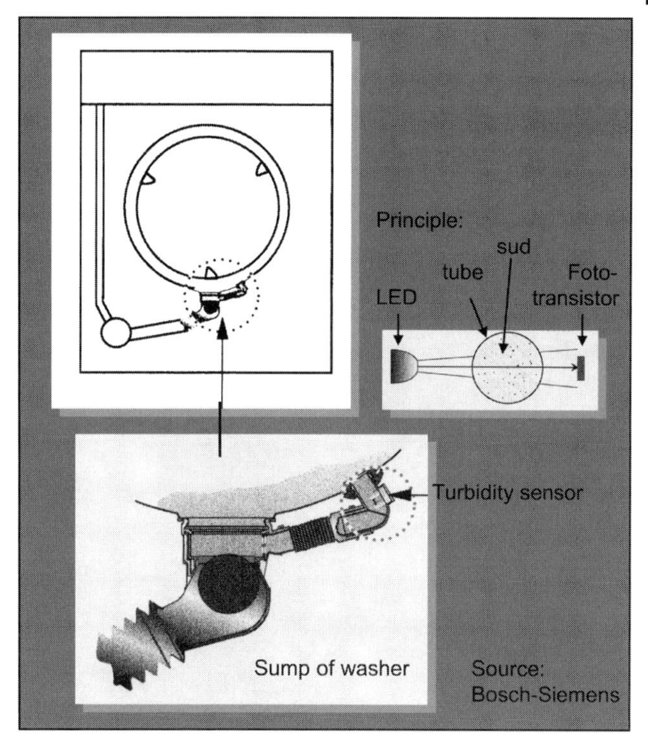

Fig. 3.11 Operation principle of suds quality measurement by turbidity sensing.

The left panel shows the possible range of the input parameters, while the right panel shows possible machine responses after signal evaluation. Depending on the combination of registered signals, the wash process is varied in duration and the number of rinse cycles.

Fig. 3.11 shows the position of the sensor in a bypass pipe in the bottom area of the drum. This ensures that the sensor is always immersed in the suds and the pipe can always drain at the end of the program, which helps to avoid the pipe clogging up with calcareous deposits.

It is known from laboratory tests that surface tension measurement can provide reliable information regarding existing detergent concentration. Work is being carried out in various institutes on such sensors for the commercial sector. However, for use in domestic washing machines, only sensors that are extremely inexpensive, maintenance-free and durable are suitable. How much of a breakthrough can be achieved here in the future remains to be seen.

Furthermore, special detergent electrodes are currently undergoing tests, but so far, they have only registered anionic surfactants. It is well known that modern detergents consist of anionic as well as cationic surfactants, which complicates monitoring. It would, of course, be extremely helpful to have information about the contents of the detergent used in order to optimize control of the washing and rinsing process.

The ever-greater use of a wide range of enzymes in detergents to improve washing results is another complication in the monitoring process, particularly as they occur in very low concentrations. Solutions for applications in domestic washing machines are not currently known.

3.1.3.6 **Program Sequence**

These are the most common applications of electronics in a program sequence:

- Measuring a variety of functions
- Control and regulation processes
- Evaluation and control via microprocessors
- Conventional signal processing
- Fuzzy algorithms
- Neural networks

With the abundance of wash process signals mentioned here it is fair to say that this information can only be evaluated effectively with a fully electronic program control. From the actual wash requirements and the information registered, an optimal process sequence is worked out and then set automatically, taking into account important parameters such as wash and spin speeds, program duration, water levels and number of rinses.

Evaluation in the microprocessor may be carried out conventionally, in accordance with fuzzy logic algorithms or even on the basis of so called neural networks. To what extent such terms can be advertised to the end user as a type of quality criterion remains to be seen. For consumers these differences are largely of no consequence as they always receive clean, hygienic, problem-free laundry for which only the absolutely essential quantities of the required resources of water, energy and chemicals have been used.

3.1.3.7 **Operation and Display Technology**

The applications for electronics discussed so far are generally not apparent to the consumer they are effectively invisible. However, during operation, the use of electronics becomes obvious to the consumer, not least due to the ever-improving display technology.

Japanese washing machines in particular frequently operate with large 7-segment or LCD displays. In Europe as well such technology, although cost-intensive, is being introduced in more and more machines. Until now it would be used to display the set spin speed, wash temperature or program duration. Fig. 3.12 shows a modern fascia with some additional features.

The display of the expected residual moisture level of the laundry in percent and the percentage of the actual load related to the maximum possible load is relatively new. As this is highly textile-dependent, additional measurements carried out by ever more critical consumers will almost inevitably show variations. New and actually inappropriate consumer complaints could be the result.

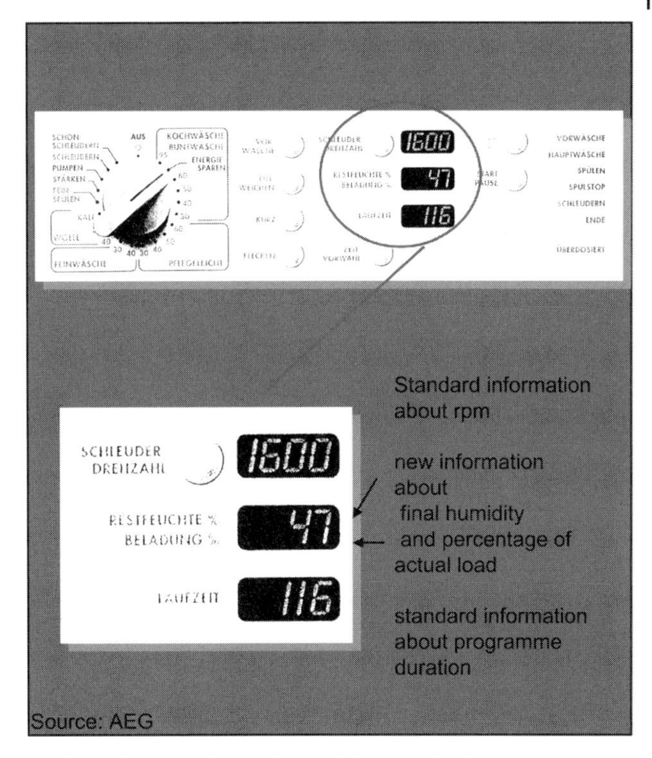

Fig. 3.12 Example of a fascia panel (Source: AEG).

Fully electronic control also allows machines to be updated, thus extending their lifespan. See Fig. 3.13.

If machine or detergent manufacturer's research laboratories develop new improved wash programs, these can very easily be transferred retrospectively to machines in situ using existing update technology. Fig. 3.13 shows how a service technician could use a special infrared interface. Other manufacturers achieve the same result using a conventional PC interface.

The retrospective implementation in machines already in use means that new process developments can be introduced much more quickly to many more machines and hence to a larger customer base, e.g. for a new special detergent. Until now, market penetration of new technology depended on the purchase of new machines, which slowed down the process considerably, given the long lifespan of modern washing machines.

These interfaces are also of great assistance when carrying out repairs. The service technician can obtain detailed information about the machine status via an appropriate diagnostic program and then carry out targeted repair.

Complex operating panels as shown above will in future certainly be reduced to a multiline display with fewer setting elements as already used in oven control.

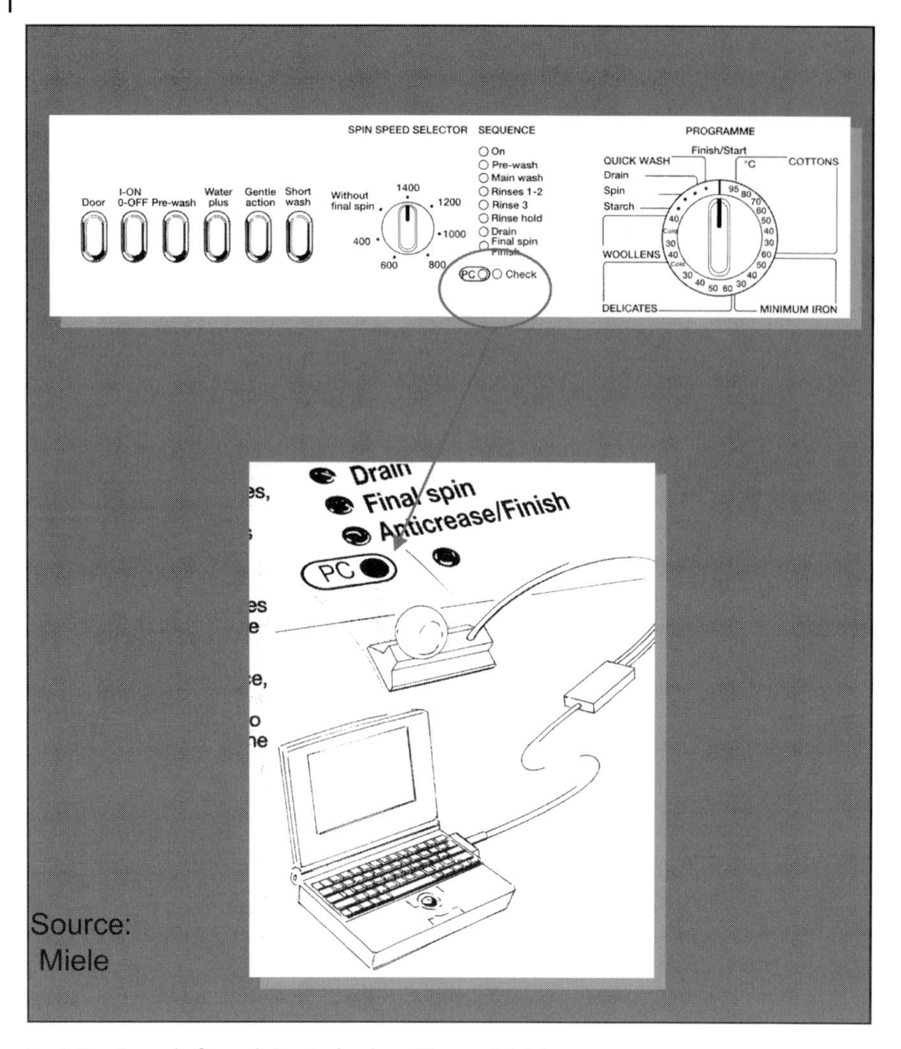

Fig. 3.13 Example for updating technology (Source: Miele).

For clarification, the user can also check all information for correct program setting through such a display in his/her habitual language. This operating concept could reduce number of control knobs and buttons drastically. Having program information clearly laid out at the start of each program would also be an improvement.

Another thought for the future, which is inconceivable without the use of electronics, is the so-called intelligent home concept, for example the linking of all domestic appliances to a house bus system. The consumer can then keep tabs on the status of each appliance through a central control PC, a television screen, or from practically any point in the world. The appliances could thus to a certain extent be operated by remote control.

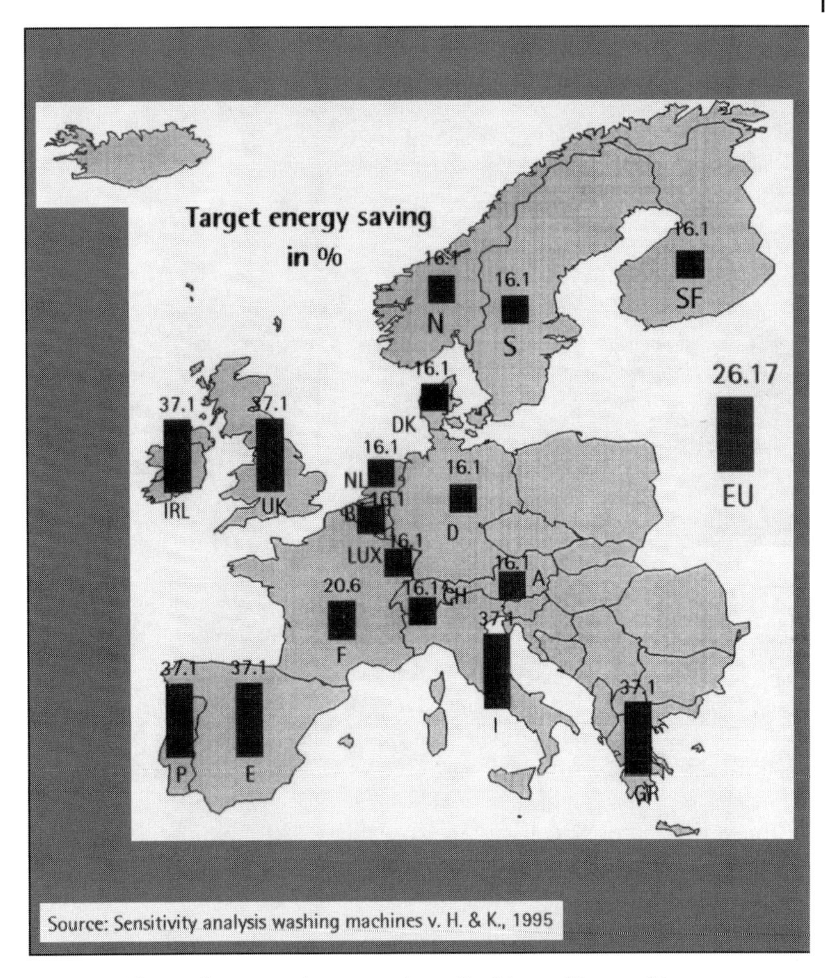

Fig. 3.14 Volume of energy savings yet to be realized by washing machines.

If a fault occurs in a machine, the service department could carry out a remote diagnosis and dispatch a technician with the correct parts to carry out necessary repairs.

3.1.4
Summary

We would have liked to finish by giving an overview of the worldwide water and energy saving potential, but unfortunately, those figures are not available, so we have to limit ourselves to Europe instead.

Fig. 3.14 contains the results of a study carried out by a neutral agent on behalf of a Commission DG XVII of the European Union. This was to establish what

o Increase in the needs for low consumption values with at least the same washing results.

o Further development of existing machine concepts through increased use of electronics
- for all basic functions (agitation, water level,temperature control, water and suds monitoring
- as well as for automatic programme modification.

o Improved operating and display technology also by the use of electronics.

o Wash programmes can be modified retrospectively, e.g. to match new detergent developments

o Electronics may also be used as an aid to the still open question of simple automatic detergent dispensing.

Fig. 3.15 Summary

savings could be achieved if all European washing machines operated with the most modern technology available. The energy saving volume per country is between 16 and 37% and is thus much higher than the savings volumes that could be achieved through further improvements of today's best machines. As an average for Europe, the study expects 26%.

The various examples discussed have shown that electronics play their part in all basic functions of state-of-the-art washing machines and that it would be difficult to imagine a future without them. Above all, electronic control is crucial if washing machines are to be operated economically. Fig. 3.15 gives a final summary of the most important points.

(1) We expect an increase in the needs for low consumption values with at least the same washing results.
(2) Further development of existing machine concepts will come through increased use of electronics (for all basic functions, such as agitation, water level, temperature control, water and suds monitoring, as well as for automatic program modification)
(3) Operating and display technology will also be improved by the use of electronics
(4) It will be possible to modify wash programs retrospectively in existing washing machines, e.g. to match new detergent developments. Finally electronics may be the answer to the so far unsolved problem of simple automatic detergent dispensing.

3.1.5
Acknowledgements

Finally, I would like to thank my former colleagues and partners, Mr. Rudolf Herden and Mr. Hellhake, both Miele, Gütersloh, Germany, Dr. Rainer Stamminger, Electrolux Europe, Nuremberg, Germany, and Ms. Gundula Czyzewski, Bosch-Siemens Hausgeräte GmbH, Berlin, Germany, who have given a lot of help to prepare this report.

3.2
Intelligent Combustion

H. Janssen, H.-W. Etzkorn, and S. Rusche

3.2.1
The Situation

As member states of the European Union have been releasing new regulations to protect the environment by reducing pollutant emissions, the heating equipment industry as well as their customers is well aware of the need for a new sensor-controlled burner concept. The answer is continuous innovation in small burner technology. Burners with radiant surfaces, for example, reduce temperatures in the reaction zone of the flame, which in turn, reduces NO_x emissions. Most modern furnaces use fully premixed burners. The stabilisation of the flame becomes more difficult, because a rise of the fluid flow yields a decrease of the burning velocity.

In order to reduce energy consumption, new low-energy houses have been built, using improved insulation. These houses require furnaces with a wider heat load range. For the low heat demand of these homes, the furnaces had to become smaller, while they also had to be capable of producing high amounts of hot water in a short time for sanitary needs. All these demands lead to modern small heater systems, which became extremely sensitive to varying boundary conditions. Additionally, the liberalization of the EU gas market leads to varying gas qualities from different sources in the public gas supply. As a result, gas burners have to deal with a wider range of gas compositions.

Traditionally, controls of small gas boilers have been using the differential pressure of the streaming air to adjust the gas flow or they mechanically combine the change of the gas flow rate and the air supply. Both types do not take into account that changing gas qualities have a major impact on the required air flow for an optimized combustion. The resulting incomplete combustion leads to high emissions of carbon monoxide and other polluting gases. Another aspect concerning the gas quality are ignition problems with low caloric gases.

Therefore sensors and controlling elements, which are able to control the combustion reliably, are becoming more and more important. A number of different sensor appliances for gas detection are available on the market. Nevertheless, the suitability of a sensor strongly depends on the type of burner because several different principles of measurement can be combined in various ways in sensors to indicate the quality of combustion.

There are three main methods of monitoring the combustion status, using sensors. Firstly temperature, flow rate and composition of the fuel gas (to calculate its calorific value and the air supply needed) can be measured before gas and air are mixed for combustion. Secondly, it is possible to monitor the reaction zone. The number of ions or of some uncharged radicals can give information about the conditions under which the reaction takes place. The easiest way to obtain this information is by measuring the quantities of some products in the flue gas. If the O_2-volume-fraction is known, for instance, it is possible to conclude whether the gas or the air supply has to be adjusted. The state of the combustion can be determined by sensing the amount of CO_2, NO_x and CO in the exhaust gas. The correlation between changes measured by a chosen probe and the variation in the characteristics of combustion should facilitate a reliable control system. Furthermore, low cost, high reliability, long-term stability and high sensitivity are all aspects to be taken into account when choosing suitable sensors.

This chapter will give an overview of combustion control methods for small domestic appliances. From the explanation of the working principles of the most important sensors to the description of their application these contents will give useful hints to any manufacturer of gas appliances. Some boiler prototypes with sensor-supported control and safety devices have already been built for research purposes in several laboratories. These new burner concepts have been tested with very positive results. The new controls are capable of reducing emissions and improving the ignition behaviour by adjusting the gas supply to the actual air flow or vice versa. The new appliances are suitable for use with different gas types without any adjustments during the installation procedure. Simultaneously, the possible heat load range can be widened without increasing the emission of pollutants while the control system keeps the gas-to-air ratio always at an optimum level. Boilers equipped in this way will soon represent the state of the art.

3.2.2
Possible Measurements of Gases Characterising the State of Combustion

To obtain hygienic combustion, it is essential to adjust the equivalence ratio Φ to an ideal value. This value characterises the ratio of the fuel quantity needed for a stoichiometric combustion to the fuel quantity supplied. In most of the common gas appliances, the air supply slightly exceeds the amount of air needed for complete stoichiometric combustion. The exact value for the surplus of air – often referred to as "lambda (λ)" – depends on the configuration of the burner system in question.

Generally, the volume fraction of single flue gas compounds for a known combustion system, an adjusted equivalence ratio and a given composition of the fuel gas supplied can be regarded as fixed. In order to adjust the air and gas supply, several different strategies have been developed. The most important strategies to predict correlations between the concentration of flue gas compounds and an ideally adjusted combustion are explained below.

At present, the volume fraction of O_2 in the flue gas is already being used in some cases in the field as a value for adjusting the excess air ratio in combustion processes. A control system based on monitoring the O_2 fraction in the flue gas can help control compliance with a desired value.

The correlation between O_2 fraction and stoichiometric value λ is given with the equation

$$\lambda = 1 + \frac{\dot{V}_{\min, flue, dry.}}{L_{\min}} \cdot \frac{[O_2]}{0{,}21 - [O_2]}$$

where $\dot{V}_{\min, flue, dry.}$ is the minimum total dry flue gas flow rate, $L_{\min}$ the minimum required amount of air and $[O_2]$ the volume fraction of oxygen in the flue gas.

Another possibility to adjust the combustion air flow to the supplied fuel gas flow is to maintain a constant level of CO_2-content in the flue gas. Thus, a stoichiometric value can be adjusted for any type of fuel gas, which varies only slightly from the optimum value.

The excess air ratio λ can be obtained with

$$\lambda = 1 + \frac{\dot{V}_{\min, flue, dry.}}{L_{\min}} \cdot \left(\frac{[CO_{2,\max}]}{[CO_2]} - 1 \right).$$

The air and gas supply of domestic appliances is usually adapted using the fraction of CO_2 in the flue gas. The changes in the minimum required amount of air and in the total flue gas flow rate cannot be taken into account, but are negligibly small. This also applies to the maximum CO_2 fraction for the combustion of common natural gas, which differs by less than 1%. Therefore the variations in equivalence ratio Φ remain within tolerable limits.

In condensing appliances there is a chance that, due the water solubility of carbon dioxide, small CO_2 fractions are assimilated by the condensate and cannot be measured in the flue gas. However, compared to the deviations from the maximum CO_2 fraction caused by variations in fuel gas composition, this is a negligible measurement corruption.

Furthermore factors such as stoichiometric value, heat load and design of the burner as well as the combustion chamber have a significant impact on the emission of pollutant gases. Depending on the reaction of a combustion system to a changing equivalence ratio decisions can be made how to minimize the pollutant emissions by adapting the flow rate of air or gas. A combustion control system based on monitoring the CO fraction in the flue gas could thus be considered.

Generally speaking, CO is produced when air is in short supply during the combustion process. It can also be the result of an interrupted reaction, caused by a too short retention period in regions with sufficiently high temperature. This can occur when a too high air excess results in a very high flow velocity, thus resulting in difficulties concerning flame stabilization. Therefore a necessary manipulation can be initiated once the CO emission exceeds certain limits.

These examples show that there are in fact correlations between the concentration of certain substances in the flue gas and the equivalence ratio of the combustion. Integrating sensors into gas appliances would be a good way of improving their safety, reliability and efficiency.

3.2.3 Suitable Sensors for Controlling Combustion Processes

Various requirements have to be met for the use of sensors in domestic gas appliances. First of all, the temperature conditions at the probe must be chosen carefully. Secondly cross-sensitivities, accuracy, measuring range, size of the sensor and its response time are to be taken into account, and finally production and maintenance costs have to be low.

Most sensors show cross-sensitivities, i.e they react to substances other than the targeted flue gas compound. Without detailed knowledge of the combustion system and its behavior under changing boundary conditions, it would thus be preferable to use selective sensors for single substances. In a laboratory situation there is a whole range of suitable technologies, such as chemiluminescence, paramagnetism or mass spectroscopy. Unfortunately, the instruments required are much too expensive to be integrated into gas appliances under serial manufacturing conditions. Low-priced, highly precise and long-term stable sensors are hard enough to find, and even harder if they have to cope with problems like humidity in the flue gas and cross-sensitivity.

The following table shows the most important functional principles that are able to detect gases in mixtures frequently found in residential gas appliances.

3.2.3.1 Capacitive Measuring

Capacitive Measuring is based on detecting changes in the dielectric constant as a result of changing gas concentration. These sensors are mostly used for measuring humidity, but can also detect CO_2. One major drawback of these systems is

Tab. 3.1 Functional principles for detecting single gases in mixtures.

Functional principle	*Detectable gases*
Change of capacity	CO_2, H_2O
Acoustic principles	Binary gaseous mixtures
Optical principles	CO_2, CO, CH_4, NO, NO_2, SO_2
Measurement of heat conduction	H_2, CH_4
Pellistors	Combustible gases
Electrochemical cells with liquid electrolyte	O_2, CO, NO, NO_2, SO_2, H_2, and others
Electrochemical cells with solid electrolyte	O_2, H_2, CO
Semiconductors	O_2, H_2, CO, CH_4, NO_2, Cl_2

the fact that they are made of organic polymers with a very short life span – less than a year, which prohibits their use in domestic appliances.

3.2.3.2 **Acoustic Principles**

Sensors based on acoustic principles are suitable for in-situ-measurements of gas concentration in industrial processes. They can even be used for aggressive media [1]. The velocity of sound – in the following formula marked as "VOS" – is taken as the characteristic criterion.

$$VOS = \sqrt{\frac{dp}{d\rho}} = \sqrt{\frac{RT}{M} \cdot \kappa}$$

Up to now, sensors using this parameter have not been taken into consideration, as they are generally not selective. Water for instance is traceable in air because its velocity of sound is significantly higher. The VOS of carbon dioxide is just around 1/3 the VOS of air. In a mixture of air, water and CO_2 none of the compounds can be quantified. As the VOS of carbon monoxide is similar to that of air, CO cannot be quantifiedby this method either. Hence, sensors based on acoustic principles cannot be taken into consideration; neither for the measurement of single species in a flue gas flow nor for the identification of fuel gases.

3.2.3.3 **Optical Principles**

Optical principles are based on the fact that technical gases have distinct absorption spectra in different wavelength ranges of electromagnetic radiation. The widespread "infrared spectral photometrics" uses the fact, that certain gases absorb infrared radiation in a characteristic manner. O_2 and N_2 are "IR-inactive" and therefore other compounds in air or flue gas can be easily detected. This technique has a very high selectivity for single compounds and shows only a few cross-sensitivities.

Optical methods are especially useful for the selective detection of CO and CO_2 concentrations. In low-priced sensors, a simple miniature light bulb is used as IR-source. The radiation emitted enters an absorption chamber, through which the flue gas is pumped. An added interference filter lets only the absorption spectra of the target gas pass. The IR detector determines the reduction of the light intensity, which is then transformed into an electrical signal. The correlation between the source intensity and the received intensity is given in the Lambert-Beer equation.

$$I(l) = I_0 \cdot \exp(-a \cdot c \cdot l)$$

Here, I is the received radiation intensity, I_0 is the source intensity, a is the absorption coefficient, c stands for the concentration of the gas to be measured and the length of the radiation pathway filled with gas is called l. Fig. 3.16 shows an optical sensor for detecting CO_2.

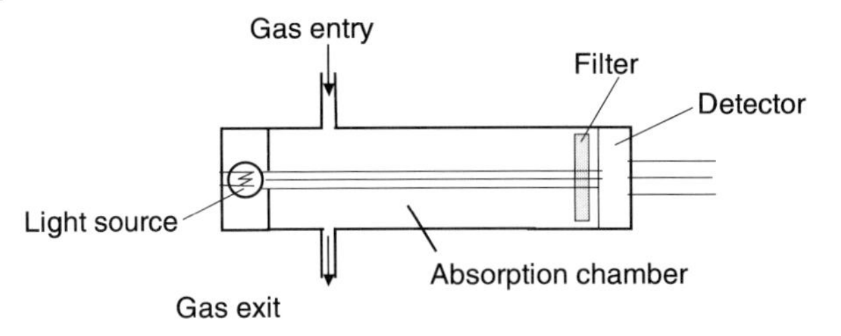

Fig. 3.16 Schematic view of an optical sensor for the detection of CO_2.

As CO absorbs radiation in a similar way to CO_2, a control system could be developed, based on the use of light sources with suitable wavelengths or appropriate filters to measure CO.

3.2.3.4 Measurement of Thermal Conductivity

Another characteristic of gases, which can be determined with little expense, is their thermal conductivity. At constant temperature and pressure this depends only on their molar mass, see Tab. 3.2.

The thermal conductivity of methane is about twice as high as that of any other flammable compound of natural gas. Sensors for determining the methane number use this effect, and the principle is already in use for gas engines [2], as their performance depends heavily on the methane number.

Yet thermal conductivity alone is not sufficient for the characterization of gaseous mixtures. Given a fixed air ratio and temperature, thermal conductivity of flue gases, resulting from the combustion of different fuels, does not vary by more than 1%. A changing air ratio has a smaller effect than a small rise in the flue gas temperature. Therefore the thermal conductivity alone is not suitable as reference value. Further information is required to identify fuel gases.

3.2.3.5 Heat of Reaction

Sensors following the principle of "heat of reaction", so-called "Pellistors", see Fig. 3.17, serve as indicators for flammable gases. With the catalytic conversion of fuel gases on the surface of a heated sensor, its electrical resistance changes proportionally to the concentration of the gas.

Tab. 3.2 Thermal conductivity of pure gases (at $t=0\,°C$).

Gas	CH_4	C_2H_6	C_3H_8	C_4H_{10}	H_2	N_2	O_2	CO_2
Thermal Conductivity k in [10^{-3} W/mK]	30,06	18,0	15,1	13,8	175,4	23,8	24,2	17,01

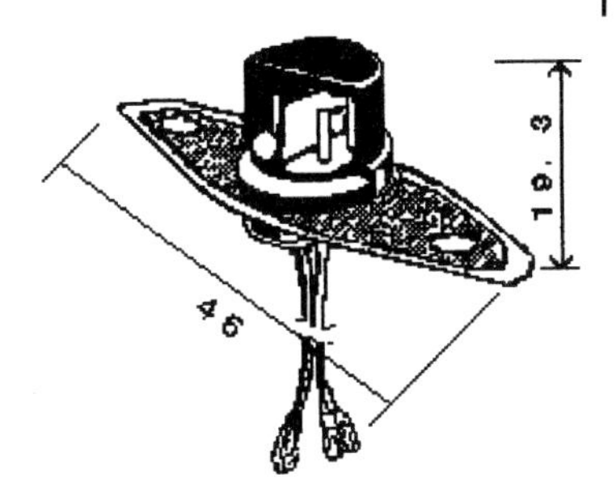

Fig. 3.17 Cross-section of a pellistor.

One temperature-sensitive resistor as compensator and another one as detector are integrated into adjoining strings of a Wheatstone bridge circuit; the voltage can be measured. Since both resistors are exposed to the test gas flow, disturbances caused by changes in temperature and humidity are compensated.

Currently, pellistors are often used as guarding sensors in rooms where there is a risk of flammable gases leaking and causing explosion. Pellistors react to concentrations far below the explosion limits. As these pellistors have been specifically developed for this purpose, nearly all that are currently available work at an ambient temperature of below 50 °C.

Current developments are aimed at reducing the lower detection limit and at extending the upper ambient temperature limit. The chance to integrate sensors of this type into commercial gas boilers increases with positive results of these efforts.

3.2.3.6 **Electrochemical Cells with Liquid Electrolyte**

Electrochemical sensors with a liquid electrolyte are widely used for the detection of corrosive or toxic gases in the workplace. Portable monitors are used in short time measurements of exhaust gases as well. These sensors work amperometrically – an external voltage supply is connected with the electrode on both sides of the measuring cell.

The test gas, arriving at the measuring electrode (cathode) either by diffusion or by pumping, is electrochemically converted. The resulting ions pass the electrolyte and are discharged at the anode; the measurable voltage is proportional to the partial pressure of the test gas.

However, the system has several drawbacks – its lifespan is very short, and measuring can only be carried out intermittently. The range of the ambient temperature is also limited. Progress towards continuous measuring have not been observed, so it is not very likely that the system will be applied to measuring fuel gases.

3.2.3.7 **Electrochemical Cells with Solid Electrolyte**

Electrochemical cells where ions dissipate through a solid electrolyte are mainly used for measuring oxygen concentrations. The most common sensors based on this principle are the λ-probe for monitoring combustion engines in motor vehi-

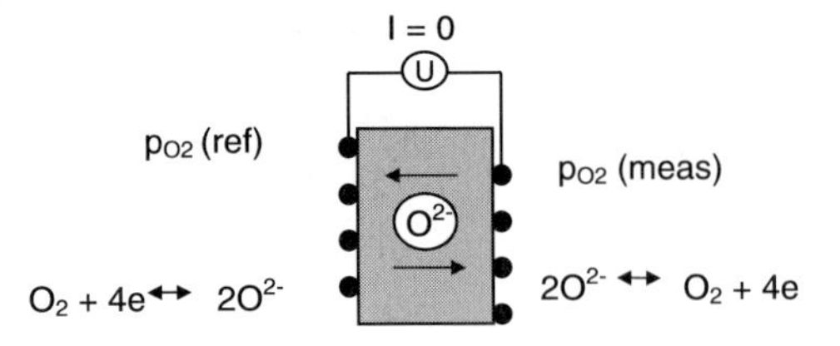

Fig. 3.18 Scheme of a potentiometric O_2-probe.

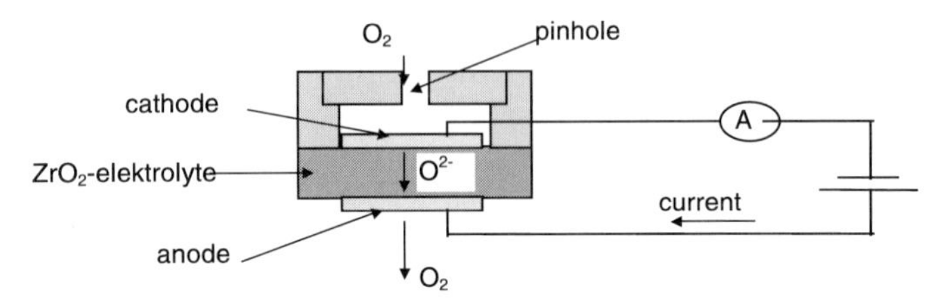

Fig. 3.19 Schematic view of a saturation current probe.

cles. The solid electrolyte with attached electrodes (see Fig. 3.18) separates the test gas from the reference gas. The electrode is usually made of catalytic material, in order to be able to ionise oxygen easily. With sufficient heating (working temperature of the electrolyte 650–800 °C) the oxygen ions from the reference gas pass through the electrolyte towards the opposite electrode.

Besides these potentiometric sensors there are also amperometric sensors using the principle of ion conductive solid electrolytes. In addition to the heating voltage those sensors are also equipped with a second voltage supply, inducing a current, which varies depending on the concentration of the test gas. Fig. 3.19 shows a schematic view of these so-called saturating current probe.

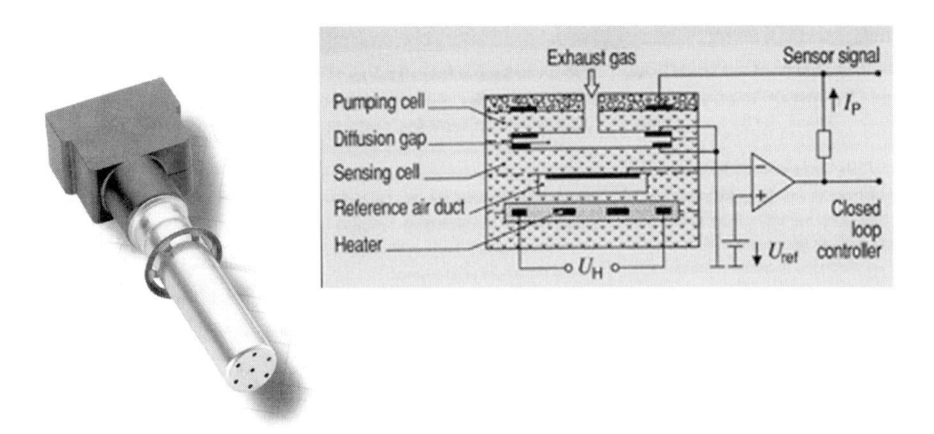

Fig. 3.20 Schematic view of an O_2-probe with combined potentiometric and amperometric functional principle [4].

There are also some types of sensors that use the amperometric as well as the potentiometric principle. These seem to have an even higher potential for the application in surveillance and control of small performance combustion systems. With the knowledge of the strength and weakness of either principle, these sensors have been developed specifically for use in small systems. One example of these is given in Fig. 3.20.

3.2.3.8 **Semiconductors**

Semiconductive sensors are based on the changing electrical conductivity of the ion-conducting material, which depends on the concentration of reducing or oxidizing compounds in the flue gas. This functional principle makes use of three effects: Charge transport via crystal borders, through crystals or along their surface. Together with the given ambient conditions these effects determine the electrical conductivity.

If the ambient contains reducing gases, they accumulate at the surface where they react with oxygen. This results in a reduced electrical resistance by lowering the potential barrier between single crystals. With rising temperature the oxygen gaps inside the semiconductor diffuse towards the inner parts of the material, thus raising the conductivity.

The choice of suitable materials depends on the desired sensitivity for the test gas and the other compounds of the ambient gas. In the past, Sensors based on SnO_2 have been widely used. Later developments are based on the use of Ga_2O_3, which is more resistant to changes in humidity of the ambient because of its higher operating temperature. This sensor type is also affected by chemisorption, but at temperatures above 650 °C, this effect can be considerably reduced. Furthermore Ga_2O_3, compared to SnO_2, shows a much better stability of the signals and a higher reproducibility [5]. More is described in Chapter 5.3. The material can be used for detecting CH_4, C_xH_y, H_2, CO, alcohol and aerosols in the ambient as well as in flue gases.

Problems with crystal borders can be overcome using a high operation temperature. The charges are no longer limited in their movement by the crystal borders, only by the crystal lattice itself. Thus, electrical conductivity depends directly on changes in the charge carrier distribution that can be controlled and reproduced. The ceramic Ga_2O_3 is not completely selective; the sensitivity for different test gases can be adjusted by choosing an adequate operation temperature. Nevertheless, the signal of the sensor is always a mixed signal, composed of the compounds given above; changing the operation temperature lowers or raises the signal for the one or the other compound. Since the resistance of the sensor is highly dependent on temperature, only small changes in the operation temperature can be made.

Ga_2O_3-sensors with operating temperatures around 600 °C can be used for detecting reducing gases. The functional principle is based on the chemisorption of the reducing gases. At temperatures between 850 °C and 1000 °C, the volume effect can be observed, responsible for the oxygen sensitivity of the sensor. At lower temperatures, the effect becomes smaller and smaller; below 700 °C it is not traceable anymore.

The Japanese company FIS Inc. has developed a λ-probe with a semiconductive cell made of $BaSnO_3$. The functional principle is based on the change in conductivity of the probe. The signal is generated by surface reactions with the local atmosphere and also is sensitive for the intermediate products and the free radicals resulting form combustion [6].

Platinum wires ($\varnothing = 100$ µm) attach the semiconductive cell to a heat-resistant Al_2O_3 housing. The electrical resistance of the cell mainly depends on the temperature and the oxygen concentration of the ambient. The recommended temperature range lies between 700 °C and 900 °C. As these temperatures cannot be reached by electrical heating equipment, the probe has to be placed inside the combustion chamber. According to the manufacturer $BaSnO_3$ is much more stable than the commonly used SnO_2. Nevertheless sub-stoichiometric combustion combined with temperatures above 1000 °C leads to damages and therefore has to be avoided.

The well-known step function of λ-probes as they are commonly mounted in combustion engines of motor vehicles is much smaller with this probe and is shifted to higher air ratios. The assumed reason are reactions with intermediate combustion products, free radicals and carbon monoxide on the sensor surface whose intensity increases with lowering air ratios down to stoichiometric level. A solution for the set up based on this sensor is presented further below.

Most probably a commercial utilization of these semiconductive sensors in domestic appliances can be expected for the short term.

3.2.3.9 **Additional Possibilities**

Several methods have been investigated to find correlations between physical properties of fuel gas mixtures and the excess air ratio to optimize the combustion procedure. In spite of the varying composition of natural gas it is said to be possible to control a heater system by measurements of the dynamic viscosity of the gas [7]. One explanation could be the correlation between Wobbe number and viscosity: With increasing Wobbe numbers the viscosity decreases, and if the Wobbe number of a gas is known, the excess air ratio can be adjusted, resulting in an open loop control.

Another method is the indirect measurement of gas concentrations in a mixture: To detect a flame in household gas appliances it is usual to record the ionization current, which reveals more than just the existence of a flame. The junction between the ionization of a reacting gas and the stoichiometric value λ of this combustion is one very promising possibility of controlling the combustion [8].

As indicated in Fig. 3.21, the ionization current shows a maximum near the stoichiometric level of combustion. By introducing a set point for the ionization current the excess air can be adjusted in order to achieve optimum combustion conditions.

Apart from the primary air ratio, the ionization current depends on various other factors, such as differences in heat load, fuel gas composition, voltage supply of the probe, design of the burner, the position of the electrodes and also different temperatures of solids in the combustion chamber. All these disturbing fac-

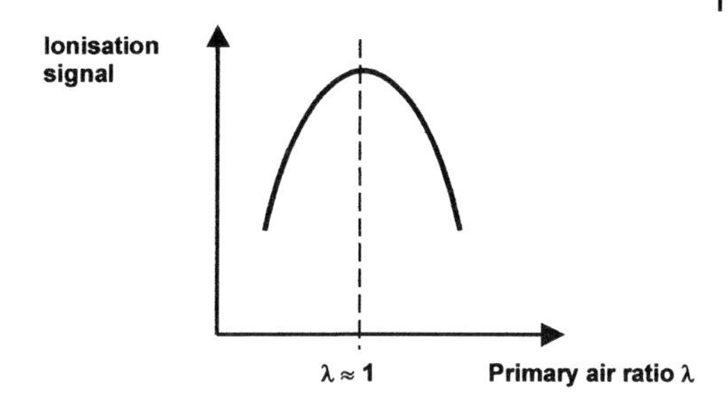

Fig. 3.21 Characteristic relation of the ionization signal and the primary air ratio.

tors may vary even for two appliances of the same type. Thus there is a lot of investigation to be done until a series of appliances using this feature can be established on the market.

3.2.4 Some Exemplary Solutions

The use of sensors for surveillance of industrial combustion processes is state of the art. Especially optical sensors detect the existence of a flame. With decreasing furnace performance the amount of installed sensors declines. In residential appliances there are nearly no sensors installed because the costs for both, sensors and actuators, have to be balanced with the technical profit. The efforts for setting up combustion controls are very ambitious but in many cases not successful. First a distinction has to be made between sealed boilers and those that are open towards the room in which they are installed. The resulting controls cannot be interchanged between these two groups of appliances.

3.2.4.1 Combustion Control of a Boiler Open to the Room

In order to control the air supply of what is known as an atmospheric appliance, the main challenge lies in finding a low-cost and robust actuator to achieve an optimal air ratio. Numerous methods are mentioned in the relevant literature and in patent specifications. Unfortunately, most of these concepts are far too costly and elaborate to be used in serial production. Despite these difficulties there have been some promising approaches.

Some concepts for regulating the air ratio in an atmospheric appliance involve local intervention at specific parts of the air intake. In a wider approach, the total flow can be adjusted. Fig. 3.22 shows the airflow in a sealed atmospheric appliance with a flue gas fan [9].

In fuel-lean premixed burners, the primary air ratio determines the quality of combustion; changing the rotational speed of the flue gas fan has also some influence. An ionization probe is used to determine the quality of combustion. A dedicated system, developed at GWI, provides accurate detection and analysis of the ionization signal. This includes a metering device, which is provided with a rectangular supply voltage, thus warranting very accurate ionization signals.

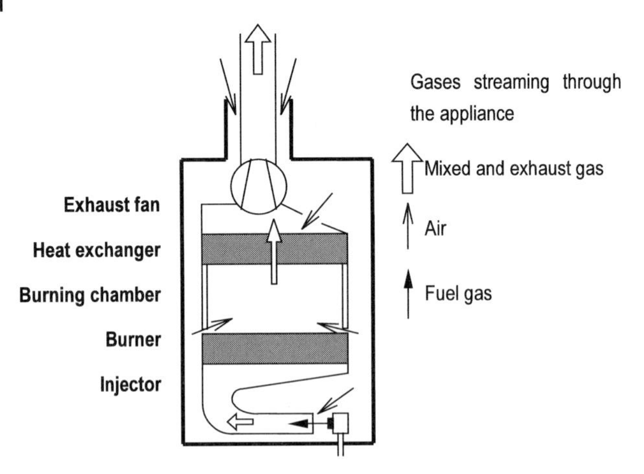

Fig. 3.22 Schematic view of an atmospheric appliance with a flue gas fan.

In a comprehensive test program the characteristic functions of the ionization current and the primary air ratio as well as of the primary air ratio and the fan supply voltage have been gathered. A closed loop control has been designed on this basis, which also includes appliance start-up and calibration as well as a suitable, dynamic set point and actuator controls.

This results eventually in a closed loop control allowing an adjustment of the primary air ratio and thus providing constantly low pollutant emissions regardless of the fuel gas composition and the heat load of the appliance.

3.2.4.2 **Surveillance of a Fan-assisted Sealed Boiler Using a Semiconducting Element**

During investigations to analyze the behavior of measuring instruments that can reliably monitor the combustion process, a sensor was tested, consisting of semiconducting $BaSnO_3$, which is heated by the flame (see above). Its electrical resistance changes with the temperature and the O_2- and the CO-content in the combustion chamber (see Fig. 3.23). With an air ratio of less than $\lambda=1.1$ the resistance decreases and starts oscillating at an irregular frequency (see: Fig. 3.24) [10].

The sensor needs an ambient temperature of 700 to 900 °C over a longer period of time to provide stable and reliable signals (see: Fig. 3.25). Therefore, it is necessary to find a position in the combustion chamber, the temperature range can be maintained at all times (i.e. with variations in burner loads and excess air ratios).

The sensor was integrated in a standard boiler available all over Europe. The only conditions to be fulfilled for the application of this sensor are:

- the gas and air supply must be fully premixed,
- the avoidance of long-term high concentrations of reducing gases near the sensor,
- the use of a circuit with a power dissipation of the sensor of less than 10 mW.

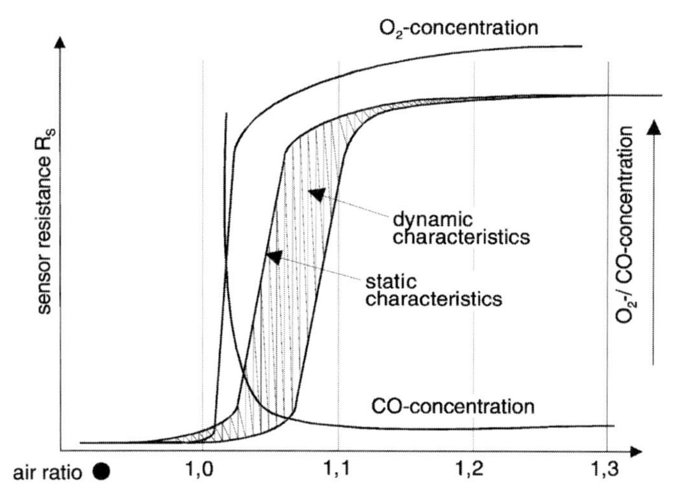

Fig. 3.23 Sensor resistance in relation to the CO-/O_2-concentration [6].

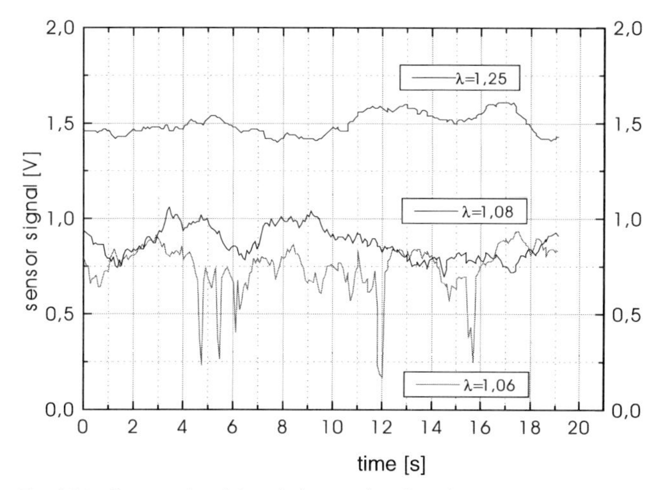

Fig. 3.24 Sensor signal in relation to the air ratio.

In the selected boiler, comprehensive measurements helped find a position where the temperature varied only within the above-mentioned range at all times, contributing to a reliable long-term stability of the $BaSnO_3$-element.

All sensors' dependencies on the CO-, the O_2-concentration and on the temperature have to be taken into account due to the fact that the use of only one of these parameters would lead to false readings during optimization of combustion.

The control of combustion is based on regular readings of the resistance and its change over time. The gas valve opens with a heat load-dependent velocity, so the air excess ratio decreases. While the change exceeds a certain limit and the resis-

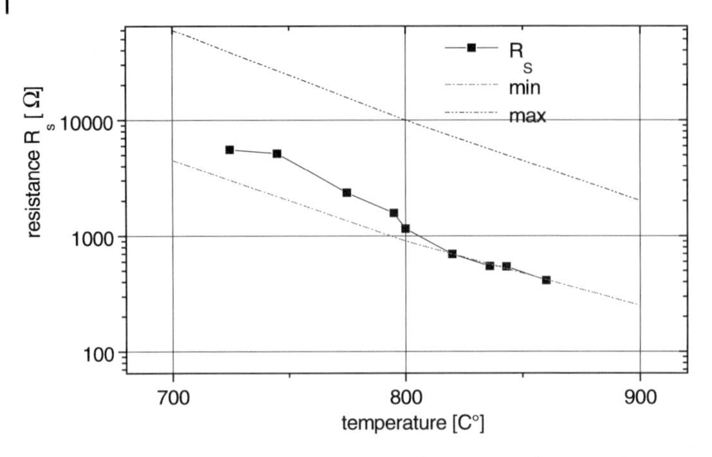

Fig. 3.25 Sensor resistance at an air ratio of $\lambda = 1.2$ in relation to the temperature.

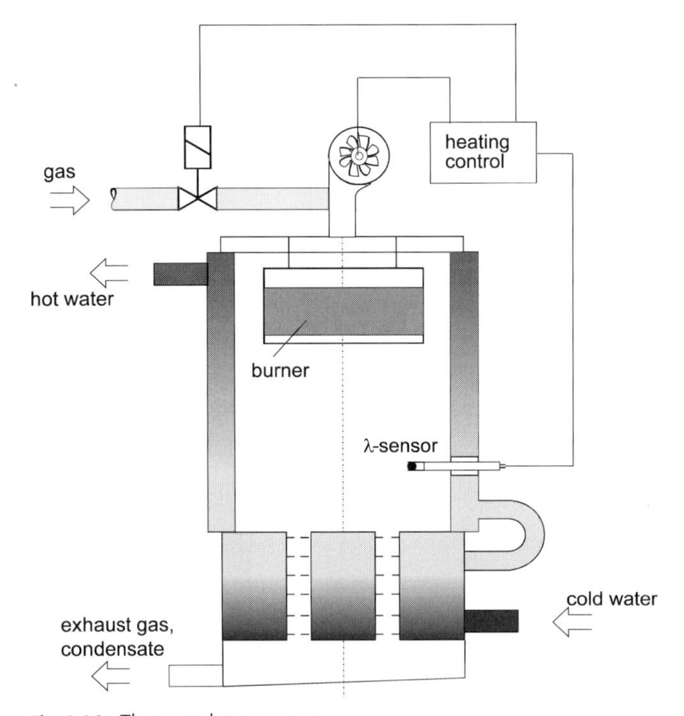

Fig. 3.26 The complete concept.

tance is in a defined area a signal will be created by both values that forces the gas valve to close in a defined step size. So the regulation is highly dynamic, the valve never rests in a tight position for more than one second.

In addition there had to be created a new furnace control device. Now a burner control exists which uses the sensor's signals to adjust the gas-air ratio under various circumstances with a very low emission of polluting gases (CO and NO_x). The new appliance is suitable for different gas types without any adjustments, not even at the first start after installation.

This new burner concept (see: Fig. 3.26) was tested with very positive results. The new control reduces emissions and improves the ignition behavior by adjusting the gas supply to the actual air volume. At the same time, the performance of the burner system can be adapted within a wide heat load range without increasing the emission of pollutants, as the sensor keeps the gas-to-air ratio always constant.

3.2.5 References

1 Zipser, L.; Labude, J.: Akustische Gasanalyse. tm – Technisches Messen, 11/1991 bzw. 12/1991, R. Oldenbourg Verlag.
2 Schollmeyer, H.-J., Korsmeier, W.; Wolf, D.: Anpassung des Motorbetriebs an wechselnde Gasbeschaffenheiten. gwf – Gas, Erdgas 135 (1994) Nr. 3
3 Sensor product information. Nemoto & Co. Ltd., Tokyo, Japan: 1997
4 The new oxygen sensor for exhaust gas measurement in gas- and oil burners. Product information, Robert Bosch GmbH, Karlsruhe: 2001
5 Bernhard, K.; Fleischer, M. und Meixner, H.: New Materials for High-Temperature Gas Sensors. Proceedings Sensor 95, Nürnberg, 1995
6 FIS Metal Oxide Lambda Sensor LS-01 for COMBUSTION CONTROLS. Technical Information – 11.2, Product Information, FIS Inc., Osaka, Japan, Febr. 1998
7 Wawrzinek, K.; Trimis, T.; Pickenäcker, K.; Löhr, S.: Wobbeindexmessung für eine luftzahlgeregelte Verbrennung. gwa 80 (2000) Nr. 12
8 Herrs, M.; Merker, R.; Naumann, R.; Nolte, H.: Verbrennungsregelung mittels Flammensignal. gwf – Gas, Erdgas 139 (1998) Nr. 1
9 Kostrzewa, G.; Rusche, S.; Feldpausch-Jägers, S.; Janssen, H.: Intelligente Verbrennung – Abschlussbericht. Gaswärme-Institut e.V. Essen, Essen. 2000
10 Rusche, S.; Kostrzewa, G.: Intelligent Combustion: A Gas boiler with a new control and safety device using the signals of a semiconductor-sensor. Proceedings Eurogas 99, Bochum: 1999.

3.3
Condition Monitoring for Intelligent Household Appliances

J. Goschnick and R. Körber

3.3.1
Introduction

Human beings have always been using their noses to trace chemicals in their environment, whether they had to decide about food quality, monitor cooking processes in the kitchen, judge air quality or detect fires – to give just a few examples. Our olfactory sense is nothing else but a sensitive gas detector 'wired' to that powerful computer with its enormous storage capacity that we call brain. However, while classical differential gas analysis tries to describe a mixture, ideally identifying the presence and concentration of every single chemical component, biogenic olfaction treats the usually complex atmospheric gas ensembles as one single overall odor with certain characteristics and a defined intensity.

An odor usually consists of a wide range of components that make up its characteristics. The concentration relationships of these components remain more or less constant. The characteristic odor is then memorized like a fingerprint and, again in contrast to classical analysis, is assigned to the source or the cause of the odor. We learn how rotten fruit typically smells or that a characteristic (burning) odor comes from smoldering electrical cable insulation. The olfactory analysis then takes advantage of the stored odor impressions. When we perceive an odor, we compare it to the stored odor fingerprints. Once we have recognized an already known odor, we can immediately draw conclusions about, for instance, the freshness of food or the condition of a process. If all comparisons fail, the odor is classified as unknown, which might be of some significance, as it could indicate that something untoward is happening.

Although smell perception is integral, any change in the composition is detected, provided the nose is sensitive to the gas type in question. No information is lost. Variations in the concentration of the gas ensemble that do not affect the concentration ratio will be noticed as variations in the intensity of the same odor, whereas a change in the compound spectrum or a change in the concentration ratio will change the character of the odor. Moreover, the primary integral characterization is not the end of the olfactory analysis: By subsequent data processing in our brain, the integral perception of a multicomponent odor can be broken down into a limited number of components and their concentrations. Thus, odor fractions or gas concentrations in a mixture can even be quantified simultaneously. It seems that Mother Nature has found a very efficient alternative to classical analysis. It is effective when it comes to analyzing complex gas mixtures in everyday life, helping to spot e.g. a malfunctioning process or poor quality food with minimal technical effort.

The enormous significance of olfaction for the interaction of humans with their environment is evident. According to their type and condition, most objects or processes give off a typical smell, caused by fluids and solid materials containing a characteristic spectrum of volatile components that are released into the surroundings. This is known as an odor signature. The olfactory sense, combined with our brainpower, allows us to detect and analyze the odor instantaneously, providing real-time information about the chemical conditions in the environment. This is then being used for decisions if and how to react. The capability of on-line remote chemical sensing is a precious addition to the range of human senses that help us respond intelligently to our ever-changing environment.

Obviously, a device that would provide artificial olfaction by continuously measuring and analyzing gases could play an important role in household appliances. Such an Electronic Nose (EN) could for example be used for simple warning purposes in applications such as fire or pollution alarms (e.g. should exhaust gas enter a room through an open window). However, even more could be gained by integrating the EN into household devices in order to feed regulatory loops which enable intelligent reactions of the device. One example might be a smart extractor hood which automatically starts and adjusts ventilation when kitchen fumes are detected and could switch off the stove as a last resort when a smell of burning is sensed. Similarly, cooking, frying or baking could be automated by monitoring the process and regulating the hob and oven controls. This would ensure that energy is saved and burning is avoided. In toasters, a more sophisticated device that switches off when the toasted bread produces the desired flavor could replace time-controlled heating and avoid burning. Another area of use is air quality monitoring to control the ventilation of rooms, based not only on temperature, humidity and the CO_2 concentration, but also on the actual odor content caused by the people present as well as the outgassing materials of the furniture.

In all these applications, the crucial criterion is the ability of the analytical device to distinguish gas ensembles as odors. Only very rarely is the detection of individual gases required. On the contrary, the spectrum of detectable gases should be as broad as possible. A third criterion is a high enough sensitivity to detect gas components even at a ppb level. Surprisingly, accurate quantification over a broad range is seldom required. Hence, a sensitive Electronic Nose (EN) with a broadband sensitivity spectrum can be expected to have a wide range of applications in the household. However, production costs of the complete device (including the sensor system, its packaging and the electronic circuitry) have to be kept within the usually very narrow cost margins for mass production. Moreover, only little energy and space (compared to the monitored object, e.g. the kitchen stove) should be required. Long-term stability is mandatory, and no maintenance should be necessary. All these requirements can only be met by microsystem technology. In order to demonstrate the feasibility and the application potential of EN systems with properties especially suited for consumer products, microsystems and their fabrication technology are developed at the Institute for Instrumental Analysis of the Karlsruhe Research Center in collaboration with industry.

3.3.2
High-Level Integrated Gas Sensor Microarrays

Gas analysis can be carried out with a vast variety of techniques. Conventional gas-analytical instrumentation (e.g. infrared spectroscopy, mass spectrometry or gas-chromatography) provides a high standard of odor recognition, but price, energy consumption and size are totally incompatible with mass production. Only gas analytical sensor systems with manufacturing costs of clearly less than 50 € (including electronics) are likely to be applied in consumer goods. For more than a decade now, small and simple gas sensors that provide only one output signal have been commercially available. However, these allow only one component to be quantified and do not pick up a range or mixture of gases. Moreover, these sensors usually suffer more or less from cross-sensitivity, i.e. they do not respond to just one particular target gas, but also show some reaction to other gases. Therefore, a single output sensor is never enough, even if only one target gas has to be detected.

This is why usually several gas sensors, each with a different sensitivity spectrum, are combined to form what is known as a sensor array. This array delivers signal patterns characteristic of the gases to which it has been exposed and allow the user to distinguish between individual gases or gas ensembles. To what extent they can be distinguished largely depends on the selectivity range of the sensors used in the array. What was a major drawback in single sensors, cross-sensitivity, turns out to be an advantage in an array setting, as low selectivity now allows the array to respond to a wide range of gases.

The first ENs, based on this technology, were launched on the market in the early 90s. These conventional ENs, however, still dominating the market, are equipped with arrays assembled from separate sensors. Individually manufactured sensors equipped with sockets are placed in plugs on a carrier plate of several

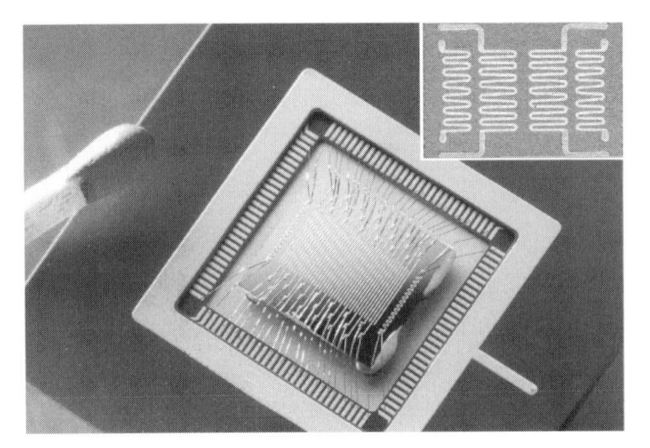

Fig. 3.27 KAMINA gas sensor chip with gradient microarray mounted in its housing.

square centimeters in size. Clearly, high manufacturing costs are the consequence. Moreover, a sophisticated gas sampling system splitting the analyte gas into identical fractions for each sensor, adds to the high production costs. Not only their price of more than 10,000 € , but also the sheer size of these instruments and their energy consumption make them unsuitable for consumer products.

The answer lies in microsystem design and fabrication, applied to a relatively simple arrangement of gas sensors (the higher the integration the better). Combined with microelectronics, it is perfectly suited for the mass production of EN modules. Microsystems are usually produced in batches and will meet demand at low cost. Additionally, small size, low energy consumption and long-term stability can be achieved. Of course, not only consumer applications will benefit from the microsystem approach, since the improvements are also relevant to instruments used in industrial applications, medical care or in environmental monitoring.

KAMINA, the Karlsruhe Micronose (**Ka**rlsruher **Mi**kro**na**se), is based on a novel type of gas sensor array developed at the Karlsruhe Research Center, the so-called gradient microarray, one of which is displayed in Fig. 3.27. The simple construction and the high-level integration of the sensor elements help reduce production costs, while also providing advantages for gas analysis. The Micronose exploits the fact that electrical conductivity in semi-conducting metal oxides, e.g. tin oxide, depends on the ambient gas composition. Four good reasons can be given to choose variance in conductivity as a measuring principle the extremely high conductivity response to changes in the gas composition, the broad spectrum of detectable gases (only passive gases such as noble gases and nitrogen give no response), the chemical robustness of metal oxides even at high temperatures and the easy fabrication of microstructured metal oxide films. The influence of gases on the electrical properties of certain metal oxides has been well known for many years [2]. The effect of gases on the electrical conductivity is caused by the adsorption of gases at the surface of the metal oxides, which is often followed by a catalytic oxidation (organic gases, H_2, NO, CO) increasing the electron density and hence the conductivity within the surface layer [3]. In other cases, the mere adsorption causes a charge transfer from the substrate to the adsorption complex which, e.g. for NO_2, results in an immobilization of electrons at the surface and reduced conductivity as a consequence.

The special structure of the gas sensor array (Fig. 3.28) is a technological novelty: In contrast to conventional macro-arrays and other gas sensor microsystems, one single monolithic metal oxide film alone forms the basis of the whole array. This film is divided into sensor segments by parallel electrode strips that measure electrical conductivity in each segment. A simple gradient technique is used to vary the selectivity of gas detection from one sensor segment to the next. An ultra-thin gas-permeable membrane layer that covers the metal oxide film increases in thickness across the array. Additionally, a controlled temperature gradient is maintained across the array by four platinum heating meanders located at the back of the chip (see Fig. 3.27). As temperature-dependent diffusion through the membrane and temperature-dependent reactions of the gases with the metal ox-

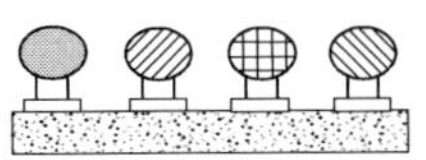

Conventional macro-structure
Separately housed sensors mounted on large substrates - differentiation by different metal oxides

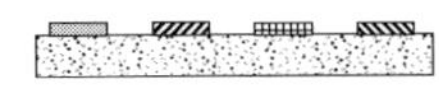

Microarray with separate sensor layers
Separate thin-film layers; sensor differentiation by different metal oxides

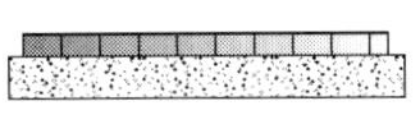

Microarray as a segmented metal oxide film
High integration by simple division of a monolithic metal oxide film into sensor segments - sensor differentiation by temperature and membrane gradient

Fig. 3.28 Gas sensor array structures.

ide interface are specific functions of the various gases, these gradient variations modify the selectivity from one sensor segment to the other. Exposure to single gases or gas ensembles (such as odors) causes a change in the electrical conductivity of the segments. A set of gradually different conductivities from sensor strip to sensor strip occurs, depending on the nature and concentration of the gas components in the gas ensemble to which the microarray is exposed. Accordingly, this characteristic conductivity pattern reflects the types and quantities of the ambient gases. Hence, this gradient microarray can be applied to realize a sensitive EN system, as gas discrimination and quantification can be achieved by a straightforward evaluation of the conductivity pattern.

A small serial production has been set up at the Institute for Instrumental Analysis to develop and demonstrate the fabrication of the microsystem. The production can be subdivided into four phases: The wafer-based formation of the fundamental structure, the packaging stage including separation, housing assembly and contact formation of the chips, the deposition of the gradient membrane and the final annealing treatment [4, 5].

In a first step, high frequency sputtering is used to deposit the gas-detecting metal oxide film on silicon wafers that have been oxidized on both sides and on the platinum metal structures, i.e. the electrode pattern including two meander-shaped thermoresistor strips at the front and the four heating meanders at the back. Usually the operational temperature is set between 250 °C and 300 °C, controlled by the two temperature sensors of the front side. The temperature gradient is formed by different settings of the power input for the four independent heating elements. A simple shadow mask technique is used to perform the lateral structuring. Currently, tin dioxide (SnO_2) and tungsten trioxide (WO_3) are used as gas detecting materials. The second phase starts with the singularization of the microarray chips, which are mounted on a carrier and the electrical contacts are made. The subsequent third phase is the deposition of the gradient membrane made of SiO_2, which is only a few nanometers thick and covers the whole array structure, including the metal oxide film and the electrode strips. Ion Beam-Assisted Deposition (IBAD) has been chosen for fabrication because this method makes it easy to produce lateral variation in the thickness of the membrane, which varies across the array by about 10 nm, giving each sensor segment a distinctive membrane thickness – leading to a variation in the selectivity of gas detection in the individual sensor elements [6]. The main advantage of the gradient microarray over arrays made of separate sensors (regardless whether macro- or microstructure) lies in the simplification of the fabrication process. The entire basic structure of the microarray requires no more than three production steps. In contrast to macroarrays that have individually mounted sensors, targets and sputter masks do not have to be changed because the array uses one single metal oxide detector film instead of a whole range of them. Consequently, very low production costs can be achieved, and the low number of required processing steps allows high yields. In a recent joint project with industrial partners the fabrication process for routine mass production was worked out whereby the production costs of the chip were estimated to be lower than 5 Euro/chip in high volume production [4].

3.3.3 Gas Analytical Performance of the Gradient Microarray

Exposures of gradient microarrays to well-controlled model gas atmospheres have been performed with a vast variety of test gases to investigate the response of gradient microarrays. In order to examine the stationary as well as the dynamic behavior of the microarray chips, these were periodically exposed to humid air containing certain concentrations of the test gas in alternation with clean air of the same humidity (see Fig. 3.29). Both types of microarray chips (equipped with SnO_2 or WO_3) gave a quick and sensitive response to the changing composition at the test pulse transitions. Response times of less than 5 seconds have been achieved. Nearly all gases, simple inorganic gases such as H_2, CO, NH_3, H_2S as well as organic gases were detectable at concentrations of even less than 1 ppm. SnO_2 showed a higher sensitivity to hydrocarbons, while WO_3 was somewhat better in detecting gases bearing nitrogen or sulfur. However, a platinum-doped SnO_2 chip (SP-chip) turned out to be the better solution for an overall screening of indoor air, as the sensitivity deficit of WO_3 for hydrocarbons is much higher than the relative sensitivity shortcoming of the SnO_2 chip for nitrogen or sulfur-bearing gases.

Moreover, differentiation of the 38 segments of an SP-chip brought about by the temperature and membrane thickness gradient has proved to provide high quality gas discrimination power. An analysis of the signal patterns obtained for some gases present in the early stages of fires is shown in Fig. 3.30. Prior to the

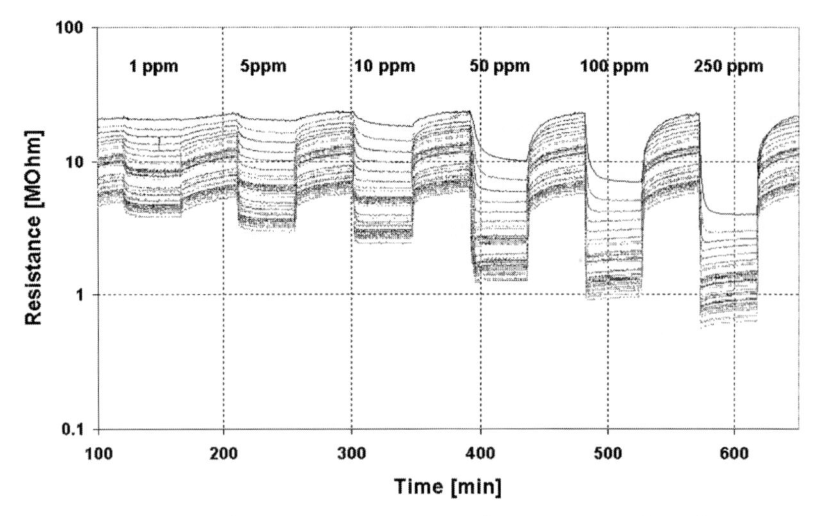

Fig. 3.29 Resistances of the 38 sensor segments of a SnO_2-microarray equipped with SiO_2-membrane during alternating exposure to carbon monoxide contaminated humid air and clean air with the same relative humidity of 60%. CO gas pulses of 1 to 250 ppm were used to test the microarray. The model atmospheres were kept at room temperature. The operating temperature range of the microarray was 250–300 °C.

signal pattern analysis, the signals of the sensor segments were normalized to the median of all sensor segments in order to remove the quantitative information. The median is chosen to represent the mean of all sensor segments rather than the arithmetic average because of its inherent higher stability towards changes in the signal pattern and mathematical operations.

The applied Linear Discrimination Analysis (LDA) basically performs a projection from the original signal space of 38 dimensions down to the three dimensions shown by presenting the maximum of differences between the data of the different gases while the data belonging to repeated measurements of the same gas are minimized [9]. The pattern components of the resulting LDA are linear combinations of the original sensor signals. Each position in the LDA plot represents a certain signal pattern. Large distances in the plot indicate a low similarity of the corresponding signal patterns, while near data points correspond to signal patterns of high similarity. The ability of the gas sensor microarray to distinguish between the model gases is evident when looking at the well-separated data clusters of the different gases compared to the scatter within the clusters comprising data of repeated measurements of the same gas. Moreover, although the data were gathered from two different concentration samples (10 and 100 ppm) the homogeneity of the data clusters demonstrates that the median normalization is appropriate to remove the concentration dependence.

However, in addition to its low production costs and excellent gas analytical features, the gradient microarray chip offers further advantages over the classical arrays consisting of separate gas sensors of different chemistry. Since all sensor ele-

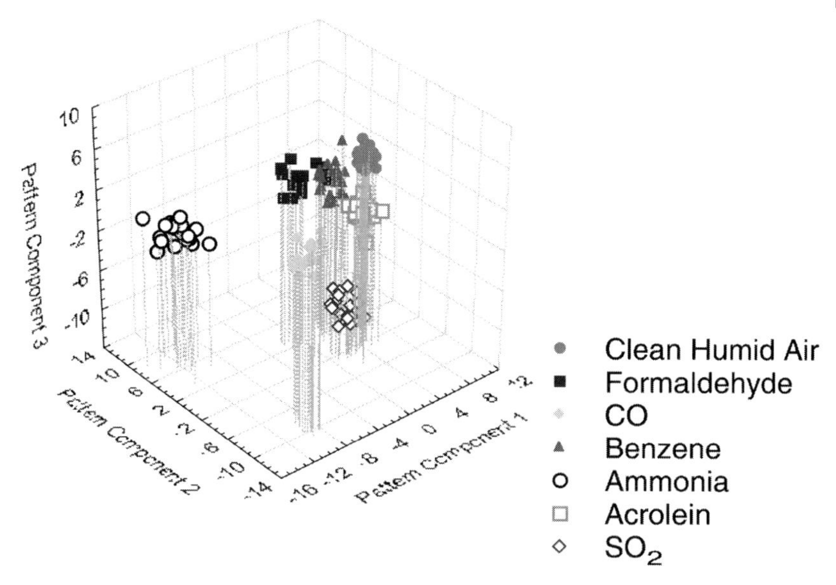

Fig. 3.30 LDA analysis of signal patterns of typical fire gases. The data were obtained using a SnO_2 microarray equipped with SiO_2 membrane (T = 250–300 °C). Data were used of 10 and 100 ppm model gas pulses and were taken 2 and 30 minutes after pulse start.

ments use the same gas-sensitive material, the influence of metal oxide aging is considerably reduced. In contrast to conventional sensor arrays made of sensors of different metal oxides and/or dopants, the sensor segments are chemically identical. While conventional arrays suffer from individual sensor aging due to their different chemical compositions, the segmented metal oxide array behaves differently. If aging occurs, all sensor segments drift consistently. The signal relations remain stable. Thus, gas recognition is rarely affected by the drift, as opposed to measurements by conventional arrays, where aging of individual sensors changes the characteristics of gas patterns.

Moreover, the gradual variation of the sensing properties permits checking the reliability of the sensor signals. Unlike conventional gas sensor arrays where the relation between sensor signals usually cannot be predicted, the sensor signals of a gradient array must reflect in their line-up the gradual difference between neighboring sensors. Thus, the sensor signal of every element has to be close to the average of its two neighbors. Deviations from this correlation indicate a fault. Furthermore, this relation permits smoothing the signal train of the gradient array. This cuts out the noise from the data, clarifying the detection results and saving data recording time.

3.3.4
Application Examples

3.3.4.1 Controlling Frying Processes

The standard gradient microarray chip equipped with 38 sensor segments of platinum-doped SnO_2 and coated with a SiO_2 membrane a few nm thick was installed in the current KAMINA operating unit, a demonstrator device the size of a beverage can. The unit merely serves as a tool to test the sensor system and its sampling periphery in practical scenarios (see Fig. 3.31). A simple fan in the head of the KAMINA demonstrator device facilitates air sampling. However, this device is not intended for mass products. An operating unit devised for consumer applications is under way, which contains the electronics and the microarray chip on one board, half the size of a credit card (see Fig. 3.32). The microarray chip was operated at a surface temperature ranging from 250 to 300 °C.

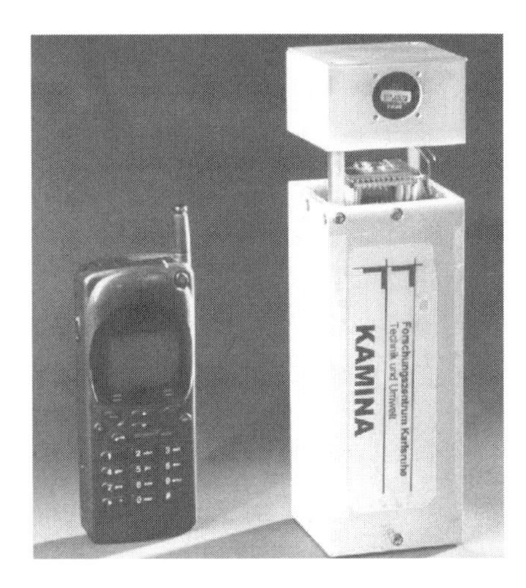

Fig. 3.31 The KAMINA demonstrator. The lifted device head shows the gradient microarray which receives ambient air by a fan. The complete microprocessor-controlled electronics is contained in the lower part of the device.

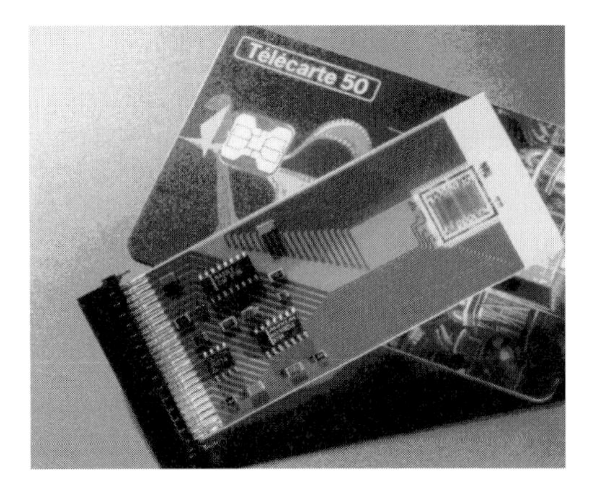

Fig. 3.32 A complete electronic nose based on gradient microarray technology does not even take up the space of a credit card and is thus applicable in intelligent user products.

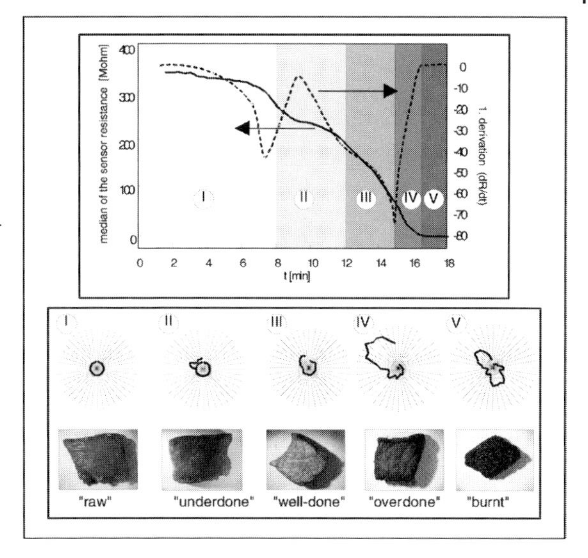

Fig. 3.33 Analysis of the vapor produced when frying beef in oil. Top: The average (median) resistance vs. time during frying of meat pieces. Below: The meat during its different frying phases and the corresponding conductivity patterns of the microarray, normalized to the initial status of raw meat in cold oil. The microarray was installed behind a Teflon membrane about 10 cm above the meat.

A KAMINA unit was installed 10 cm above a pan containing pieces of meat in oil, its gas inlet protected by a Teflon membrane, to monitor the aroma of the meat during the frying process. Fig. 3.33 shows the progressing stages. They can be clearly distinguished by looking at the signal patterns. Differentiation between the stages "under-done", "well-done" and "over-done" is possible. Here, KAMINA could not only help save energy if the heaters were adjusted accordingly, but also make the frying process easier to reproduce. The signal patterns can clearly be used to define the final point of frying for switching off the heaters. Finally, safety could be improved by switching off the stove immediately, as soon as the slightest smell of burning is detected.

3.3.4.2 **Indoor Air Monitoring**

Applying an electronic nose for continuous monitoring of indoor air provides a wide range of information: the level of air comfort (level of odor nuisances), the presence of unhealthy gas components and hazardous potential (for instance the presence of inflammable gases). In order to evaluate the practical performance of the gradient microarray, its sampling arrangement and data processing procedures have been investigated in two application scenarios, one checking its efficiency for fire prevention and the other for general air quality control.

One important issue concerning fire prevention is the early detection of overheated wire insulation, a common source of smoldering fires. Therefore model experiments were performed to examine how far early indications of overheated insulation could be recognized by the gradient microarray and its simple sampling arrangement. The tests were carried out in a closed box with KAMINA placed next to a cable overheated by a current overload. The experiments were performed using various different wire insulation materials (e.g. kapton, teflon and ethylene tetra fluorine ethylene [EFTE]). The wires were about 30 cm away from the inlet port of KAMINA. Although the current was high enough to finally melt or ignite the insulation, it was kept sufficiently low in order to prevent early smoke emission. The time between switching on the current and initial smoke emission was about 10 min. In order to check the discrimination power of the microarray chip for different gas ensembles released by the overheated insulation and possible interfering gases, vapors of several solvents (e. g. toluene, xylene, acetone, isopropanol, ethanol) were admitted to the test chamber by placing tissues soaked in solvent inside the test box.

Fig. 3.34 shows the results of a Linear Discriminant Analysis (LDA) of the signal patterns obtained in different exposure scenarios. Prior to the actual pattern analysis the resistances of the sensor segments were normalized to the median of all sensor segments in order to eliminate quantitative information that would be useless in this application, as the gas transport away from the hot cable occurs in an uncontrolled and discontinuous way. Furthermore, most of the signal noise was eliminated by a preceding Principal Component Analysis (PCA), reducing the number of variables to the ones that significantly contribute to the variance of the data set. Two different kinds of LDA were carried out. In the top plot the data of each insulation and solvent vapor were treated as separate classes with individual signal patterns. In this case, the LDA algorithm performs a projection from the original signal space down to the two dimensions shown which exhibit the maximum of differences between the signal patterns of all different cable insulations and vapors. The resulting LDA clearly shows that the differences between the signal patterns of the overheated insulation materials and clean air are significant enough to detect overheating of cables, and it is even possible to distinguish between different types of insulation. This is also true for the vapors that are clearly distinguishable by their signal pattern. Moreover, the signal patterns of the solvents are well separated from those of the overheated insulation materials. Hence, the signal patterns obtained by solvent vapors cannot be misinterpreted as an indication for cable smoldering. Furthermore, the excellent discrimination between different insulation materials performed by the gradient microarray can even be exploited to distinguish cables from each other by intentionally marking them with an odor.

However, variety of signal patterns of different insulation materials or different solvent vapors does not interfere with the main task of detecting smoldering cables. If the LDA algorithm is instructed to classify the signal patterns into three categories only, clean air, overheated insulation and presence of solvent vapors, a well-separated three-class system can also be established, as the bottom LDA diagram in Fig. 3.34 shows. Hence, the results demonstrate the potential of the gradient microarray to act as an intelligent early fire detector, which first of all detects overheated wires by their typical gas release long before smoke emission can be observed. In a second step, cables with different insulation materials can be discriminated by their smell. The advantage over conventional smoke detection is evident – more information is provided earlier.

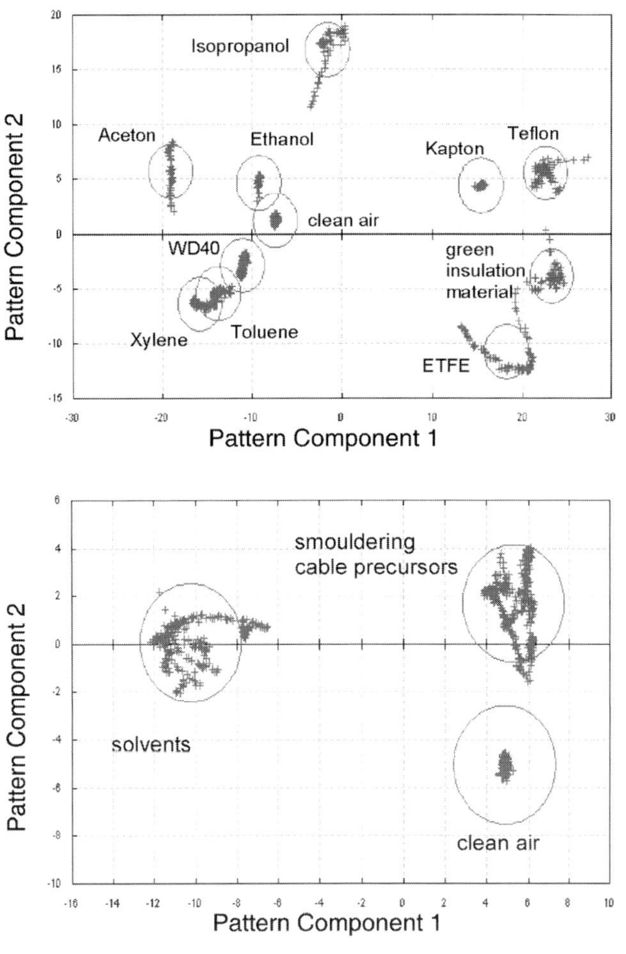

Fig. 3.34 LDA of the microarray signal patterns obtained in practical KAMINA tests to detect gaseous precursors of smoldering fires through overheated cable insulation. Prior to the LDA the measured resistances of the sensor segments were normalized to the median of all sensor segments and the number of variables was reduced by a Principal Component Analysis (PCA) to cutback the noise. Clear distinction is achieved between clean air and solvent vapors as well as possible interfering components and overheated wire insulation materials. Top plot: Clean air, solvent vapors and the different insulation materials were individually classified. Bottom plot: LDA obtained from the same data set with classification into three categories only: smoldering cable precursors, solvent vapors and clean air. The circles describe a confidence range of 95%. (EFTE: ethylene tetra fluorine ethylene).

When it comes to indoor air quality, interaction of people with their gaseous surroundings is the most important factor. In rooms with poor air exchange, the presence of people causes a steady deterioration in air quality. Air quality may be understood here and henceforth as a numeric measurement for the content of unhealthy and odorous components (air comfort). Additionally, building materials and furniture are also gas sources contributing to the air inventory, but they usually don't cause short-term air deterioration, whereas people do. This is why KAMINA was tested for its response to air quality changes caused by the presence of people – after all, this is what air conditioning mainly has to deal with.

The KAMINA chip indeed proved sensitive enough to detect human breath and perspiration. Fig. 3.35 shows one out of a series of measurements of changes in room air quality due to people present in the room. In this case 18 people gathered in a meeting room. Soon after they entered, the median resistance began to drop considerably.

The decrease of the sensor resistances indicated oxidizable organic components that dominated the gas ensemble, released into the air by perspiration and breathing. It is possible that gas release from clothing also contributed to that effect. About two hours after the people had entered the room, the gas emission rate and the low air exchange rate (leakage through closed door and windows) reached a kind of equilibrium, keeping the pollution level constant. Temporarily increasing the exchange of air by opening the window increased the median resistance, indicating a slight temporary improvement of the air quality level. Only after the end of the meeting, when the people had left the room and a window was opened for longer, the fresh air caused a substantial improvement of the air quality.

In order to investigate the discrimination power of the gradient microarray regarding the changes in air composition, a signal pattern analysis was performed

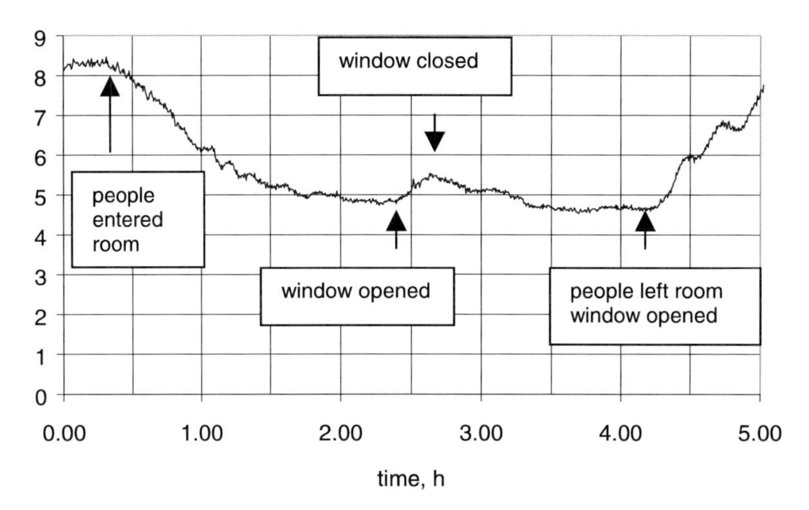

Fig. 3.35 Median of the resistances belonging to the 38 sensor segments of the gradient micro-array during a meeting of 18 people in a room of 258 m^3 without ventilation and with the windows closed for most of the time. The height of the ceiling is 2.80 m. The KAMINA was operated at the back of the room at a height of 1.0 m above the floor. Soon after people entered the room, the median resistance decreased significantly, due to the continuous degradation of the air quality. About 2 hours later, a window was opened for a period of 15 min, resulting in a slight recovery of air quality. However, the latter improved significantly only after people left the room and windows were opened for a longer period.

with LDA, basically following the same procedure as in the case of overheated wire insulation. Apparently, the changes in air quality indicated by the median are more than just a quantitative variation in the same gas ensemble. To eliminate transient conditions, approximate stationary data, characteristic for certain air situations were selected from the complete data set for LDA analysis. The LDA clearly separates different signal patterns, demonstrating the capability of KAMINA to distinguish different air compositions, depending on the presence of people in the room and the air exchange conditions. The LDA shown in the top plot of Fig. 3.36 gives four different states of air quality: the air in the empty room with opened windows, the air in the empty room with closed windows (taken in the morning), the air in the same room with people present and the air in the empty room with an isopropanol (0.1 ml)-soaked tissue placed at a height of 1.0 m above the floor. In this sequence the level of the median resistance decreases. This is plausible because the opened windows allow fresh air to flow into the room, diluting the gases released by building materials and furniture which also occur in the air of the empty room. When the window is closed, the levels of these pollutant gas components rise because the air exchange has been disrupted. People entering the room with closed windows further reduce the air quality. These three levels of air quality can all occur during normal usage. The corresponding signal patterns are collected in the LDA, indicating only moderate differences within the air inventory. By contrast, the isopropanol-contaminated air, in this case simulating the situation of an unnoticed accident with a release of unhealthy gases, shows a significantly different signal pattern at a location different from the area of the signal patterns occurring during common usage of the room. Hence, the deviation of the air inventory from its normal state can easily be detected. Of course, the origin of the abnormal air quality can only be identified if the signal pattern is known to the data evaluation system and a comparison can be made. For a number of accidents, the situation can, indeed, be rehearsed in a model experiment (cigarette smoke, presence of glues and solvents, natural gas leaks, incoming polluted air from outdoors, fire detection). However, even if the observed signal pattern is unknown, people in a room can be warned by an acoustic or visual signal that something out of the ordinary has happened.

If we leave aside the accident scenario and concentrate on everyday air quality analysis, even more subtle differences in the air inventory can be picked up. The LDA then makes it possible to distinguish the slight difference in the empty room between day and night. Moreover, the smell of smoke coming from a pipe being smoked in the hallway outside the investigated room was still detectable, although the doors were closed. The small amount of air exchange through the sparse openings of the closed door was apparently enough to cause a smoke odor in the meeting room to be detected by KAMINA.

This is a simple example for the capability of KAMINA to be used as a sensitive fire warning device. Unlike conventional devices for that purpose, which work on the basis of optical aerosol particle detection, the gas analytical KAMINA will not give false alarms caused by dust or soot particles.

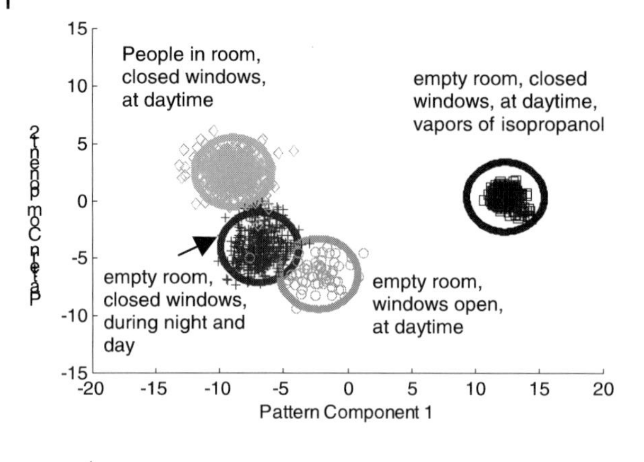

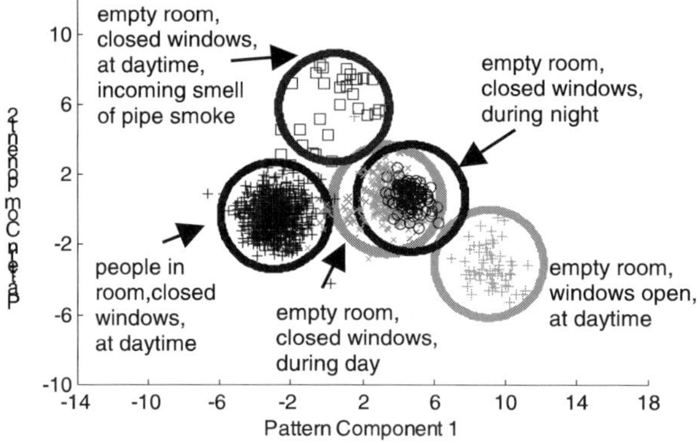

Fig. 3.36 LDA of the microarray signal patterns measured in a meeting room during night and day. The room temperature was about 22 °C in all cases. Data were collected at two meetings a week apart and in the empty room with an isopropanol source at a distance of 0.8 m from KAMINA. The resistances of the sensor segments are normalized by reference and by median. Clearly the air of the empty room with closed and opened windows can be distinguished from the lower air quality caused by people. However, these differences are smaller compared to the presence of a pollutant isopropanol which has been released to the air by a soaked tissue. Bottom plot: LDA that only accounts for the normal range of air variation, permitting within the air inventory a more subtle discrimination. Additionally, the smell of a pipe smoked on the hallway is clearly distinguished and a slight difference between the empty room by night and day is found.

3.3.5
Summary and Outlook

The variety of household appliances give plenty of opportunities to use an electronic nose for condition monitoring, as many of them give off their particular signature of components. Most of these appliances will be found in the kitchen, and food storage as well as preparation are prime targets for the application of electronic noses. Earlier, we gave the example of monitoring a meat-frying process, which gave us some idea of the detecting potential of KAMINA with its unique gradient microarray. The detection of burning can of course also be used in other non-food applications, such as spin-drying, ironing and hair-drying.

Our preliminary results show that the indoor climate can also be closely monitored by the gradient microarray of KAMINA. Moreover, the analysis of signal patterns also seems to help already recognize a fire in an incentive state. An extensive collection of data over a longer period of time is, however, necessary in order to obtain a profound statistical basis of data. Clearly, many further tests have to be performed in various room situations, e.g. different room sizes, different building materials, different furniture and rooms of various usage to find out how the instrument and its data-processing algorithms should be designed to suit practical application conditions. Such investigations are under way.

The technology of ENs in general is to a large extent still in an infant state. However, worldwide R&D on these systems have revealed an immense applicability of such systems that has hardly been exploited yet, but this technology will not be widely applied unless analytical potential combined with high reliability can be achieved at a very low price, using very little energy and taking up minimal space. This means in concrete terms that for consumer applications, the production cost must not exceed 50 $, energy consumption should be no more than 1 Watt and the device no bigger than a matchbox. Only if these requirements are met will the EN find its way into consumer applications.

The R&D work at the Forschungszentrum Karlsruhe has shown that a microsystem based on a sensor array structure of great simplicity and of high integration level allows EN systems to be built combining low fabrication costs and a high level of gas-analytical performance. The developed gradient microarrays can be expected to meet the consumer product requirements of price, small dimensions and low energy consumption under mass production conditions. The work with such microarrays has only just begun, leaving plenty of scope for future improvements. Similarly to the way that microelectronic devices have continuously improved their price/performance ratio by shrinking structures, leading to an enormous surge in microelectronics applications, the application spectrum of EN devices will benefit from the microsystem approach, as higher levels of integration will be achieved.

3.3.6
References

1 J.W. Gardner and W.E. Gardner; Insight 12(1997) p 865

2 T. Seiyama, A. Kato, K. Fukiishi, M. Nagatini; Anal. Chem. 34 (1962) p. 1502.

3 W. Göpel, K.-D. Schierbaum in "Sensors", Vol.2 "Chemical and Biochemical Sensors"; eds. W. Göpel, T.A. Jones, M. Kleitz, J. Lundström and T. Seiyama; Verlag VCH, Weinheim, New York 1991, p 429

4 J. Goschnick; Microelectronic Engineering, 57–58 (2001) 693–704

5 M. Bruns, J. Fuchs and J. Goschnick; Final Report on the Joint Project „Wirtschaftliche Produktionstechnik für oxidische Mehrschichtsysteme mit lateraler Mikrostruktur am Beispiel eines Mikrogas-

sensorsystems (PROXI)"; Report of the Forschungszentrum Karlsruhe, FZKA 6523, 2001 (in german)

6 J. Goschnick, M. Frietsch, T. Schneider; J. Surface and Coatings Technology, 108–109 (1998) p 292

3.4 Sensor Examples in Small Appliances

T. bij de Leij

Philips Domestic Appliances and Personal Care (Philips DAP) is one of the product divisions of Royal Philips Electronics, alongside Consumer Electronics, Lighting and Medical Systems. In this article we will focus on sensors that are needed within the DAP division. Philips DAP has four Business Units, each with its own specific field of interest: Shaving & Grooming, Body Beauty & Health, Food & Beverage and Home Environment Care.

Apart from the four Business Units, Philips DAP has two Corporate Centers, four Centers of Competence (centers where development and production take place) and about 40 National Sales Organizations, which are organized in regions. DAP employs some 10,000 people worldwide. The DAP division markets about 400 different products, selling around 70 million units a year. Some of the appliances they make and sell are very straightforward, while others have been given

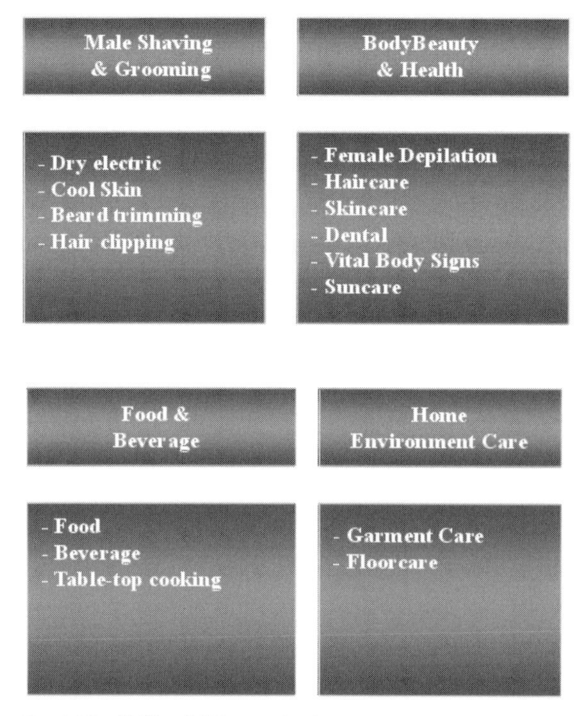

Fig. 3.37 Philips DAP organisation structure.

more specific added features to make them stand out among all the other products on the shelves, and this is where the various sensors come in.

3.4.1 Reinventing Appliances

To boost sales, it is vital to keep on improving products and adding attractive features. DAP has the ambition to add value by offering consumers superior solutions that fulfil their needs with regard to personal wellbeing and home management. This means that the Centers of Competence need to (keep the) lead in innovation strategies and have to establish the manufacturing infrastructure to realize these strategies.

Over the years, DAP has proven itself with numerous innovations such as the Quadra Action and the Cool Skin in Shaving&Grooming and the Cellesse and the Satin-Ice in Body Beauty. Philips DAP has also forged a few strategic marketing alliances, for example with Beiersdorf for the development and production of the additive for the Cool Skin shaver. Last year Philips DAP acquired the Optiva Corporation, a major US producer of electric toothbrushes, and became market leader in oral care products in North America.

3.4.2 Facts and Figures

Over the past years, the annual revenues of Philips DAP have been rising steadily, underlining the importance of this division. These last five years an overall growth of almost 28% was generated, resulting 2,130 million euros in sales. It is the combined effort of marketers, developers, designers and manufacturers that has given Philips DAP a solid market position in the field of domestic appliances.

However impressive these results may be, there is no time to lean back and enjoy the success. Keeping the favor of today‘s fickle customers is a hard job, requiring frequent innovations as advantageous selling points. The appliances DAP markets are meant to improve the quality of life for consumers. However, functionality alone is not enough. The appearance of a product plays an increasingly important role. That is why Philips DAP puts a lot of effort into the look and feel of a product. This has not gone unnoticed, for its designers won 47 awards over the past few years.

3.4.3 Sensors in DAP Appliances

Sensors are used to add something extra to the functioning of an appliance. Some recent examples of this use of sensors are the Natura hairdryer, the Cucina toaster and the Senseo Crema coffeemaker.

The Natura hairdryer is equipped with an infrared sensor that monitors the temperature of the hair without making direct contact with it. It is equipped with

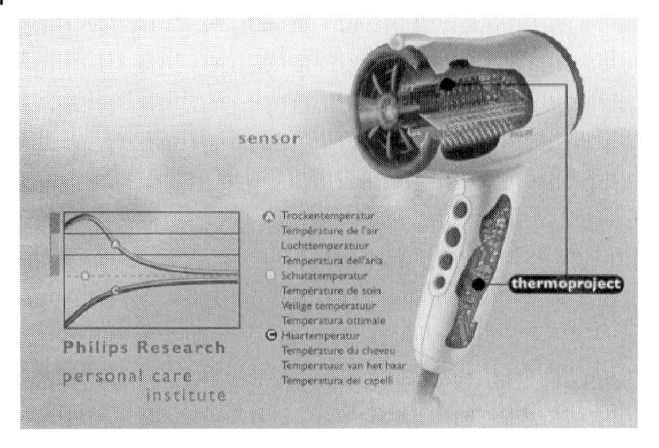

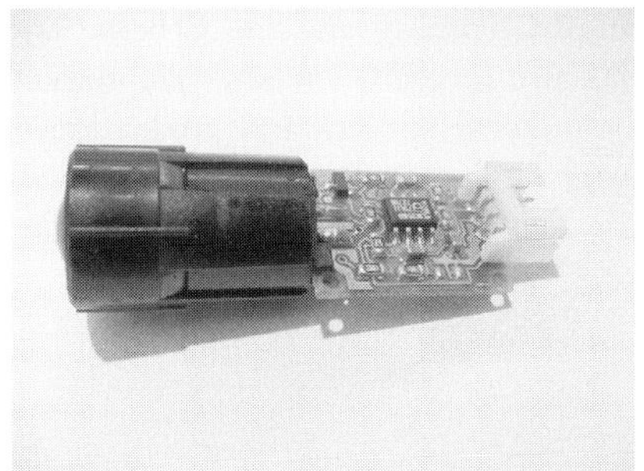

Fig. 3.38 Natura hairdryer with contactless temperature sensor.

Cucina toaster

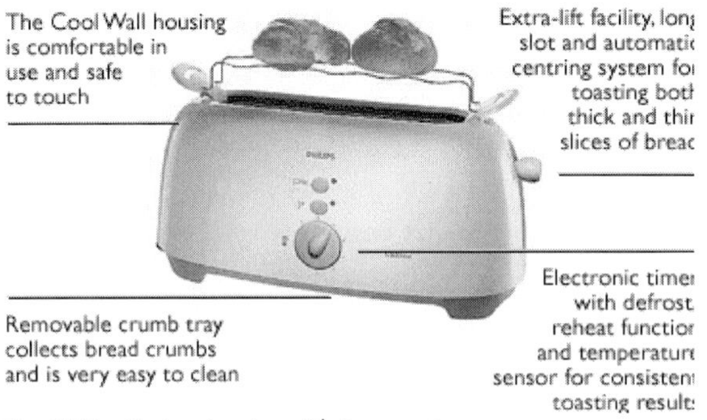

Fig. 3.39 Cucina toaster with temperature sensor.

Fig. 3.40 Senseo Crema with temperature sensor.

sensor electronics that check and control the temperature of the airflow, thus preventing the hair from being damaged by overheating.

This process is also applied in the Cucina toaster. The toaster is fitted with an NTC sensor that monitors the temperature in the toasting chamber and adjusts the toasting time, if necessary.

Of course, the first slice of bread needs longer to become golden brown and crispy than the ones that follow, because the toaster will have heated up properly by then. This also goes for the Senseo Crema, the latest sensation in coffee brewing.

This machine forces hot water through a coffee pad, which results in a cup of coffee with an espresso-like creamy finish. A sensor monitors the temperature of the water. It keeps the water 'standby', ready to make a new cup of coffee, until the machine is turned off.

3.4.4 Sensor Criteria

Most sensors in DAP products, like the ones mentioned earlier, are used for temperature monitoring. Some sensors are in direct contact with the fluids they monitor, others are monitoring in a contactless way. Partly due to European Committee regulations more and more contactless sensors are needed to meet safety and hygiene regulations. It is not only these regulations that make Philips DAP crave for more sophisticated sensors. DAP also needs them to realize its ambition to create more personalized appliances. A sensor can enable an appliance to recognize the user and adjust itself to his or her preferred settings.

Many aspects related to sensors influence the 'time to market' of newly developed appliances. The easier it is to add a sensor to a product, the faster it can be released. Apart from having to meet certain functional requirements, a sensor

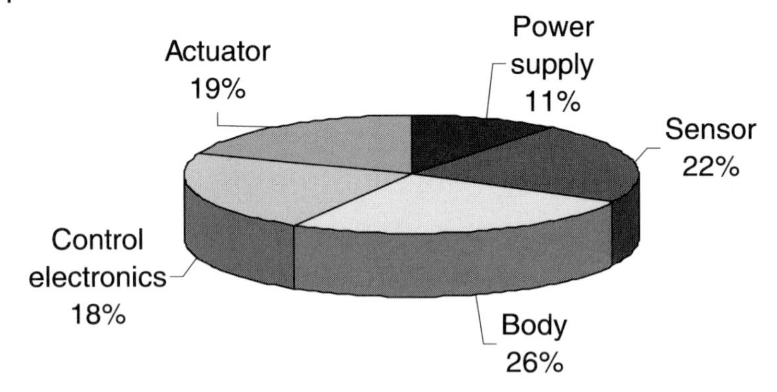

Fig. 3.41 Component cost price distribution of an average DAP appliance.

also needs to be easy to apply in a product. The sensor should be easy to incorporate during assembly and the interface with the control electronics should be rather simple. As most DAP appliances are directly connected to the mains, it also is essential that safety is supplied through the sensor.

There is one requirement, however, a sensor does not have to meet and that is a long life span. There is little point in using a sensor that outlasts the appliance. Therefore sensors produced with less expensive techniques and materials can be used, resulting in a lower cost price of the appliance. As shown in the diagram, a relatively large part of the cost price can be attributed to the sensor.

When a sensor is used in an appliance, additional components like a low-voltage power supply and control electronics are required. As these components have a big influence on the actual cost price of the product, it is very important that the cost price of the sensor is reduced to the bare minimum, since a lower cost price makes the product more competitive.

3.5 Infrared Ear Thermometers

B. Kraus

3.5.1 Introduction

Measuring body temperature is important for the detection of disease and assessment of the response to treatments. The first thermometer was developed by Galileo in 1603. Thermometers for measuring body temperature have been in use since about 1870. The first measurements taken were axillary, and later oral and rectal measuring methods were introduced. The working principle of those thermometers, the expansion of matter by temperature increase, is still used for body temperature measurement in mercury-in-glass thermometers. Electronic thermometers, first introduced in the 1970s, usually employ a thermistor sensor and give digital temperature readouts. They eliminate the risk of glass breakage and require shorter measuring times. The traditional measurement sites however remain the same.

In clinical settings core temperature measurements, including pulmonary artery and esophagus measurements, are often required. In 1959 Benzinger [1] first proposed the human tympanic membrane as the ideal site for core temperature measurements. The tympanic membrane is ideal, because it is located near the carotid artery and shares its blood supply with the hypothalamus, which controls body temperature. First temperature measurements in the ear were performed with thermistor sensors in direct contact with the tympanic membrane. The invasiveness of this method limited its use mainly to anaesthetized patients.

The first infrared ear thermometer (IRET) for non-contact temperature measurement was introduced in the U. S. hospital market in 1986 (IMS, First Temp 2000A) [2]. The main advantages of IRET are the very short measuring time and the convenient handling (see Fig. 3.42). Compared to the other traditional measurement sites there are additional advantages. Axillary temperatures are simply skin temperatures and therefore are affected by sweating, ambient air temperatures and by alterations in perfusion. Orally temperatures can be influenced by the way the thermometer is placed, by breathing, recent eating, drinking or speaking. Rectal temperatures are thermally decoupled from the central blood circulation and lag behind the core temperature, and therefore may be too slow to reveal fast temperature changes, e.g. after medication.

The first clinical IRET used thermopile sensors to achieve non-contact temperature measurement in the ear. In 1991 a tympanic thermometer for home use was first introduced to the consumer market (Thermoscan HM 1). It utilized a pyroelectric sensor which requires the use of a suitable mechanical shutter or chopper mechanism, since it is only sensitive to temperature changes [3]. The main advantage of the pyroelectric sensor unit was its lower cost. However, prices for thermo-

Fig. 3.42 Body temperature measurement with an infrared ear thermometer.

piles have gone down since and thus their advantages prevailed. Today nearly all non-contact ear thermometers on the market use thermopile sensors.

3.5.2 Sensor Unit

The sensor unit of an IRET usually consists of an infrared sensor, in most cases a thermopile sensor in a TO-5 or TO-46 housing, a gold plated barrel, which reflects the infrared radiation from the ear to the sensor and reduces the sensitivity of the sensor to ambient temperature changes (see Fig. 3.43).

A thermopile sensor generates an output voltage that depends on the temperature difference between its hot and cold contacts. For infrared temperature measurement, the hot contacts are normally thermally insulated and placed on a thin membrane, whereas the cold contacts are thermally connected to the metal housing. Infrared radiation, which is absorbed by the hot contacts of the thermopile, causes a temperature difference between hot and cold contacts. The resulting output voltage is a measure for the temperature difference between radiation source and cold contacts of the thermopile sensor. It is therefore necessary to measure also the temperature of the cold contacts by an additional ambient temperature sensor in order to determine the temperature of the radiation source.

Unfortunately, the output voltage of the thermopile not only depends on the temperature of the radiation source but is also affected by changes of the ambient temperature. The ambient temperature of the thermopile housing is influenced for example by moving the thermometer from a colder room to a warmer one or by warming up the measurement tip of the thermometer due to contact with the ear. This causes a temperature gradient inside the sensor housing. If the cap of the housing is warmer than the base plate, additional infrared radiation from the cap is emitted to the thermopile and leads to an incorrect output signal. A second effect is caused by the temperature gradient within the gas filling of the sensor

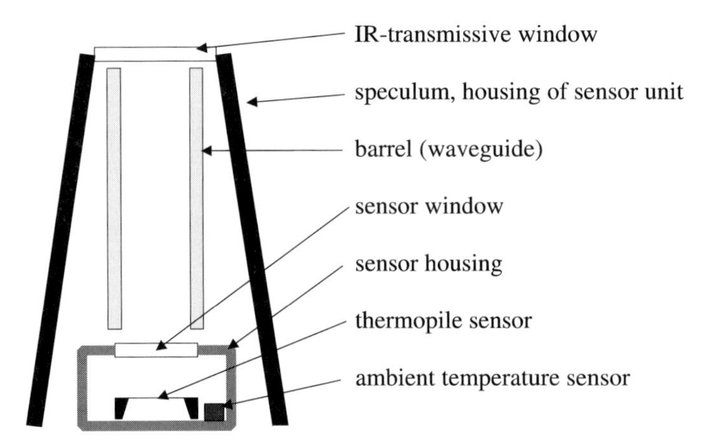

Fig. 3.43 Schematic of a sensor unit of an infrared ear thermometer.

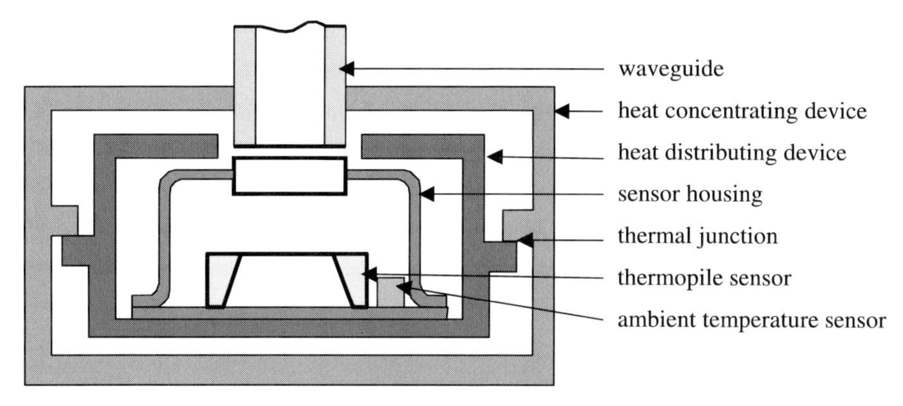

Fig. 3.44 Principal schematic of the thermal coupling arrangement in the sensor module.

housing. While the temperature of the surrounding gas influences the thermally isolated hot contacts, the cold contacts are not affected because of their higher thermal mass. These two effects can easily distort measurements by more than 1 °C.

To avoid these errors, the Thermoscan IRT 3000 thermometer from Braun uses a sensor unit [4] which is schematically shown in Figures 3.44 and 3.45. The thermopile sensor in its housing is surrounded by a heat concentration device and a heat-distributing device. The infrared radiation from the ear is directed to the thermopile via the waveguide. On heating or cooling of the waveguide, or the outer heat concentration device, temperature gradients occur. Part of the heat is transferred to the inner heat-distributing device through the thermal junction and onward simultaneously to the bottom and the cap of the sensor housing. This ensures that no temperature differences occur between bottom and cap of the sensor housing.

Fig. 3.46 shows a cross-section of the sensor head used in the IRT 3000. Heat-concentrating and heat-distributing devices consist of four die-cast zinc parts.

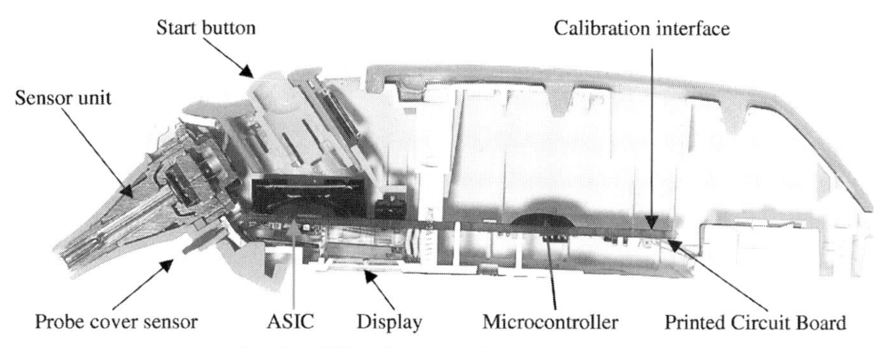

Fig. 3.45 Cross section of an IRT 3000 infrared ear thermometer.

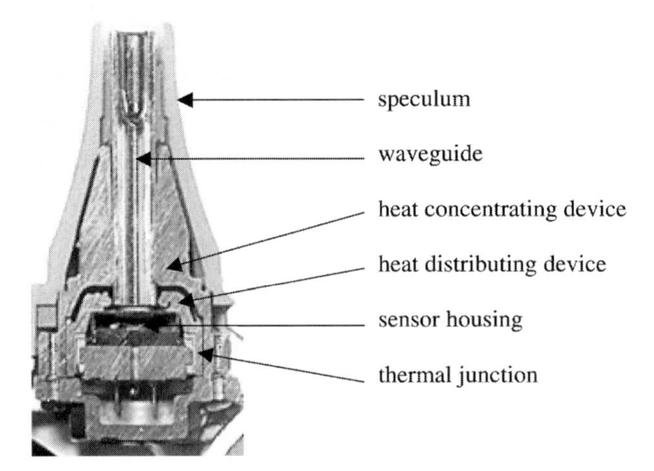

Fig. 3.46 Cross section of the sensor unit of an IRT 3000 infrared ear thermometer.

Three of these parts, and the gold plated waveguide, are pressed one into each other for good thermal contact.

3.5.3
Electronics and Temperature Calculation

Fig. 3.47 shows the schematic of the electronics of the IRT 3000, that is typical for infrared ear thermometers. The output signals of the thermopile and the ambient temperature sensor are switched by a multiplexer to an amplifier and are converted by an analogue to digital converter (ADC). The microcontroller then calculates and displays the measured body temperature. The output signals of standard thermopile sensors are in the range of 10 μV per °C temperature difference between ear and sensor. The required accuracy of 0.1 °C therefore demands electronic noise, zero drift and resolution of below 1 μV. To obtain these values, an application specific integrated circuit (ASIC) with integrated offset compensation low noise chopper amplifier and high resolution ADC has been developed (see Fig. 3.47).

The ambient temperature sensor in the IRT 3000 is a KTY type spreading resistance sensor. The resistance of the sensor can be written as a second order polynomial function

$$R = R_0[1 + \alpha(T_a - T_0) + \beta(T_a - T_0)^2] \tag{1}$$

where R_0 and α are sensitivity and temperature coefficients of the sensor, β can be treated as constant, T=25 °C and T_a is the ambient temperature to be measured. In order to simplify the calculation, the microcontroller calculates the ambient temperature by a third order polynomial equation:

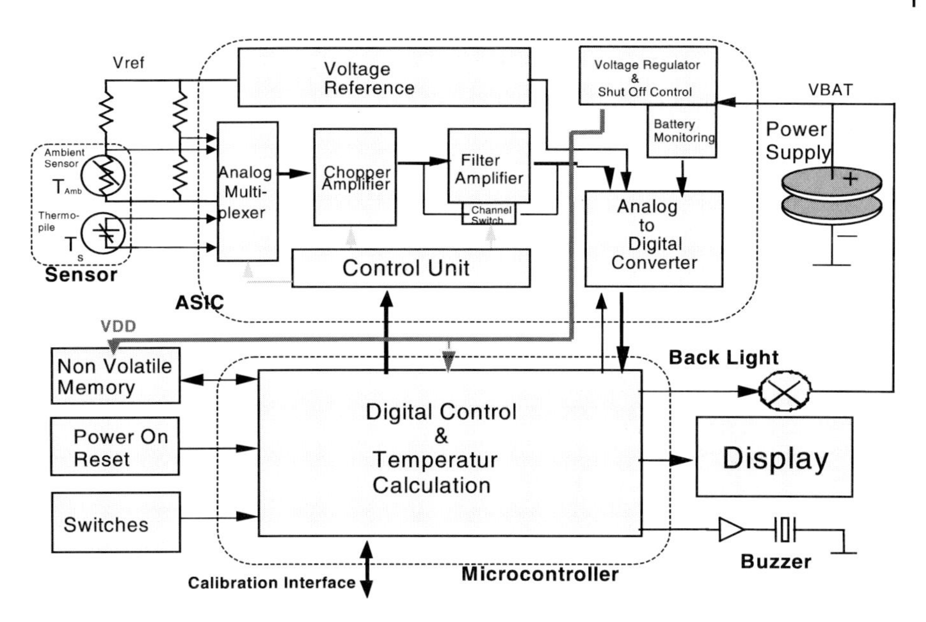

Fig. 3.47 Schematic of the electronics of the IRT 3000 ear thermometer.

$$T_a = a_0 + a_1 Z_R + a_2 Z_R^2 + a_3 Z_R^3 \tag{2}$$

where a_0, a_1, a_2 and a_3 are polynomial coefficients and Z_R is the conversion result of the microcontroller for the ambient temperature.

Using Planck's law, the output voltage of the thermopile sensor can be written as

$$U = S(T_a)[L(T_b) - L(T_a)] \tag{3}$$

Where $S(T_a)$ denotes the sensitivity of the thermopile, T_b is the temperature of the body to be measured and $L(T)$ is Planck's integral:

$$L(T) = k_1 \int_0^\infty \varepsilon(\lambda)\tau(\lambda)s(\lambda)\frac{1}{\lambda^5}\frac{1}{e^{\frac{k_2}{\lambda T}} - 1}d\lambda \tag{4}$$

with λ Wavelength
k_1, k_2 Constant factors
$\varepsilon(\lambda)$ Emissivity of the body to be measured
$\tau(\lambda)$ Transmission of the optical components (including probe cover)
$s(\lambda)$ Wavelength dependent part of sensor sensitivity

The measurement target, tympanic membrane with surrounding tissue, can be treated as a blackbody radiation source with an emissivity of $\varepsilon(\lambda) \sim 0.98 \ldots 1$. The emissivity is mainly independent of skin color, cerumen or hair in the ear canal.

In the ideal case without wavelength dependencies, i.e. $\varepsilon(\lambda) = \tau(\lambda) = s(\lambda) = 1$, the integration of Planck's law results in the Boltzmann formula

$$Z_b = \sqrt[4]{T_a^4 + U/S(T_a)} \tag{5}$$

In the non ideal case (all cases) the integral is approximated or solved numerically.

In the IRT 3000 ear thermometer the integral $L(T)$ and also the reverse function $L^{-1}(T) = T(L)$ are approximated by polynomial functions.

$$L(T_a) = b_0 + b_1 T_a + b_2 T_a^2 + b_3 T_a^3 \tag{6}$$

$$T_b(L) = c_0 + c_1 L + c_2 L^2 + c_2 L^3 \tag{7}$$

with the polynomial coefficients b_i and c_i. These coefficients depend on the parameters $\varepsilon(\lambda), \tau(\lambda)$ and $s(\lambda)$ and are stored in the non-volatile EEPROM memory during calibration. They are different for the consumer models IRT 3000 and the professional model PRO 3000 due to the different window materials (silicon for PRO 3000 and polyethylene for IRT 3000).

The body temperature can then be calculated by transposing equation 3 to

$$T_b = L^{-1}\left[\frac{U}{S(T_a)} + L(T_a)\right] \tag{8}$$

The thermopile output signal is converted to a digital value Z_U. To take into account the temperature coefficient of the thermopile, the value U/S is written as:

$$\frac{U}{S} = Z_U S_0[1 + S_1(T_a - T_0)] \tag{9}$$

Then the body temperature is calculated in the following steps [5]:

$$L = Z_U S_0[1 + S_1(T_a - T_0)] + b_0 + b_1 T - a + b_2 T_a^2 + b_3 T_a^3 \tag{10}$$

$$T_b = c_0 + c_1 L + c_2 L^2 + c_3 L^3 \tag{11}$$

3.5.4
Calibration

For the calibration of most infrared ear thermometers the sensitivities S_0 and R_0 and the temperature coefficients S_1 and a for both sensors have to be determined. Typically a two-step calibration is performed. In the first step the ambient sensor is calibrated by immersing it into two different temperature controlled baths. In the second step the thermopile sensor is calibrated by measuring the output signal while placing it before two different blackbody radiation sources.

For the IRT 3000 a simultaneous calibration concept [6], that is shown schematically in Fig. 3.48, has been developed. At two different ambient temperatures two measurements at different blackbody temperatures are performed with completely assembled thermometers connected to external computers. This results in four independent sets of output values of both ambient and thermopile sensors (see Eq. 6 and 7):

$$L(T_{bi}) - L(T_{ai}) = Z_{Ui} S_0 [1 + S_1 (T_{ai} - T_0)] \tag{12}$$

This equation system with the unknown variables $T_{a1} = T_{a2}$, $T_{a3} = T_{a4}$, S_0 and S_1 can be solved analytically. The ambient temperatures are thus calculated indirectly by radiation temperature measurements. The ambient temperature of the calibration chamber or the thermometer need not to be measured.

Insertion of the ambient temperatures in Eq. 1 results R_0 in and α. Since the ambient temperature is calculated from a polynomial funtion (eq. 2) it is necessary to determine the polynomial coefficients by a least square fit regression over the ambient temperature range 0–50 °C. Together with S_0, S_1, b_i and c_i these values are written by the calibration computer to the non volatile memory of the microcontroller. As a calibration check two additional blackbody readings are performed at a third ambient temperature (see Fig. 3.48).

The main advantages of the simultaneous calibration of both sensors are the contactless measurements, so that no contamination with fluid can occur, and the possibility to calibrate fully assembled devices without the risk of decalibration due to subsequent assembly steps.

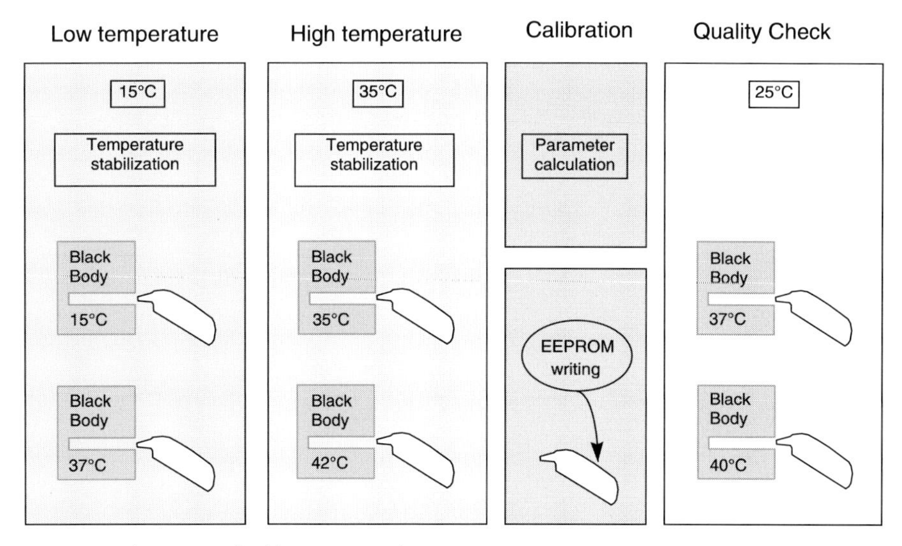

Fig. 3.48 Schematics of calibration procedure.

3.5.5
Conclusion

The infrared ear thermometer is a major step in the development of thermometers for body temperature measurements. Compared to traditional mercury-in-glass or electronic contact thermometers it is more convenient, safer and faster. During its 10 years in the consumer market it has been gradually replacing conventional thermometers, especially for temperature measuring in children.

The development of low-cost infrared sensors was the prerequisite for the development of infrared ear thermometers. The demands on high accuracy, high quality and low price can only be fulfilled by modern sensor technology, highly integrated analog and digital electronics, advanced calculation and calibration concepts and highly automated mass production. Further research and development, especially in the fields of sensor technology and adapted electronics, will further improve infrared ear thermometers so that their advantages will further increase.

3.5.6
References

1 T.H. Benzinger, "On physical regulation and the sense of temperature in man", Proc. Natl. Acad. Sci. 45, pp 645–649 (1959).
2 G.J. O'Hara, D.B. Philips, "Method and apparatus for measuring internal body temperature utilizing infrared emissions". U.S. Patent 4602642 (1986).
3 J. Fraden, "The development of Thermoscan instant thermometer", Clinical Paediatrics, Supplement 1991, 18–22
4 F. Beerwerth, B. Kraus, K. Honnefeller "Measuring tip for a radiation thermometer", U.S. Patent 6152595 (2000)
5 B. Kraus, M. Kaiser, "Radiation thermometer and method of computing the temperature", U.S. Patent 6149298 (2000)
6 B. Kraus et al., "Method of calibrating a radiation thermometer", U.S. Patent 6065866 (2000)

4 Sensorics for Detergency

W. Buchmeier, M. Dreja, W. von Rybinski, P. Schmiedel, and T. Weiss

To meet customer as well as ecological demands on modern detergency application and to further develop existing products, an exact knowledge of the washing process is essential. Washing, cleansing and tumble-drying are processes that involve many interfacial effects. Important steps during washing and cleaning are for example dissolution of the detergent, wetting of the substrate to be cleaned, complexation and ion exchange, removal of stains on the substrate, stabilization of the soil in the washing liquor and modification of the substrate e.g. by softeners.

Since nearly all products for washing and dishwashing are complex mixtures of different ingredients, an in-situ access to the relevant data is of high interest. This can be achieved by using on-line sensors in the specific applications. Washing and dishwashing machines equipped with on-line sensors will be discussed as an example.

Another way to achieve high detergency performance while keeping environmental impacts low is by making 'intelligent' products with built-in sensor functions. In these applications, the products can respond to external stimuli and fulfil their task. The inherent sensoric principles of two recent innovations, detergent tablets and dishwashing tablets with built-in rinse aid, will be discussed.

It becomes clear that research and development in the household and detergency area depend on a profound knowledge of physico-chemical parameters. Thus, an overview will be given of selected physico-chemical methods used in detergency research, most of them based on sensor technologies.

4.1 Introduction

Compared to the first fully automatic front loader washing machines that appeared in Europe in the early 60s of the last century, tremendous changes in performance, wash temperature, water consumption and specifically energy consumption have taken place. An additional push in this direction came with the introduction of the European Energy Label in 1996 together with voluntary commitments of the washing machine industry to further energy savings.

When looking at Sinner's famous circle that describes wash performance as the result of four components – temperature, time, mechanical and chemical action, it becomes clear that with decreasing wash temperature and time, chemistry has to become all the more sophisticated.

Accordingly, modern laundry detergents consist of up to 25 different constituents, which can be roughly classified as surfactants, builders, bleaching agents, and auxiliary agents. Each individual component has its own very specific function in the washing process. At the threshold to the new century, new developments in the appliance technology seem to show that new characteristics such as fuzzy program controls as well as some specific sensors will increasingly determine the washing process. While modern household appliances become more efficient in terms of performance, energy and water consumption, they are also becoming more and more convenient, and it is these convenience aspects that have also been recently determining the development of modern detergents. They, too, are becoming more sophisticated as for example in terms of shape, disintegration, solubility or controlled release of specific ingredients.

Thus, the challenge of the future lies in the development of intelligent household appliances with sensory controls, combined with intelligent detergents. In the following article, a rough overview will be given of the different types of household laundry products. Important aspects of production, composition and their use are reviewed. Then, the specific functions of the different ingredients are explained more in detail. The third section concerns the physico-chemical parameters which are important in detergency, followed by aspects of their on-line determination in product development and control. In the fourth section, state-of-the-art sensors for household appliances are reviewed and finally, product concepts of the market for detergents with a built-in sensoric function will be discussed.

4.2
Household Laundry Products [1]

Laundry detergents currently on the market in various parts of the world can be classified into the following product categories:

- Heavy-duty detergents
- Specialty detergents
- Laundry aids

The category of heavy-duty detergents includes those detergent products suited to all types of laundry and to all wash temperatures. They are offered in the shape of conventional powders, extrudates, tablets, bars, and liquids and pastes, making up some 65% of the world production of detergents. In 1998, compact powders worldwide accounted for 24% of total detergent powder production. Depending on quality, product concept, manufacturing process, local standards, local regulations, and voluntary agreements, great differences are found from one formulation to another. For example, detergent manufacturers market laundry detergents with and without phosphate in various regions and countries for ecological and regulatory reasons.

The world production of laundry detergents amounted to 21.6×10^6 t in 1998.

4.2.1 Conventional Powder Heavy-Duty Detergents

Significant differences in composition exist among powder heavy-duty detergents around the world. Recommended detergent dosages in Europe generally lie between 3 and 8 g of detergent per liter of wash water, whereas values of 0.5–1.5 g/L are usual in the Americas and Eastern and Southern Asia. There are a number of reasons for the latter being so much lower: for example, differences in washing machines, softer water, higher bath ratios, separate addition of bleach, e.g., hypochlorite solution or sodium percarbonate, and different detergent formulations. Detergents in North America, Japan, and Brazil are designed for agitator and impeller type washing machines. These detergents lack foam suppressors. They sometimes even contain foam boosters instead.

The rise in popularity of ever more colored and easy-care fabrics in the last 30 years, along with the trend toward energy conservation, has resulted in a decline in the once common European practice of washing at 95 °C, while the 30–60 °C wash temperatures have significantly gained favor. Detergent manufacturers have responded to these changes by increasing the content of those surfactants that are noted for their effectiveness at low temperature washing as well as by adding various enzymes, bleach activators, soil repellents, and phosphonates. As such, detergents have become more and more multifunctional over the last decades, particularly those for use at low washing temperatures.

4.2.2 Compact and Supercompact Heavy-Duty Detergents

The global introduction of compact and supercompact powder detergents in the late 1980s and early 1990s has had a major impact on detergent categories of large markets such as Japan, Europe, and the USA, as well as on composition and manufacturing processes of heavy-duty detergents [2]. As they have to be very effective at a much reduced recommended dosage, compact detergents have markedly higher contents of surfactants, active oxygen for bleach, bleach activators, and enzymes than conventional detergents. On the other hand, the total level of alkaline builders has been reduced and the conventional filler/processing aid sodium sulfate has been largely omitted. The apparent density of most compact detergents is above 0.75 kg/L. Most of the first-generation compact detergents were produced by downstream compaction of tower powder.

Heavy-duty detergents featuring bulk densities of from 0.8 to 1.0 kg/L are called supercompact detergents or second-generation compact detergents. The prevailing manufacturing methods used are wet granulation/compounding and extrusion.

Supercompact detergents have made their way worldwide since 1992 and are meanwhile state of the art in many countries. The market shares vary significantly from country to country and from region to region. In the USA compact powders accounted for 80% of total detergent powders, whereas their share in Japan was even 92% in the total powder category in 1998.

Sodium triphosphate has been increasingly losing its importance as a key ingredient in detergents in Europe, the USA, and some East Asian countries since 1985. It has been replaced mainly by zeolite A, special silicates, sodium carbonate, specific polycarboxylates, and citrate. The widespread use of supercompact detergents has been the key driving force behind the increasing market shares of non-phosphate powder heavy-duty detergents.

4.2.3
Extruded Heavy Duty Detergents

Extruded detergents are an innovative and unconventional form of second-generation compact detergents. The large-scale extrusion process for detergents was applied for the first time in 1992. Extruded detergents have since then been marketed under the registered trademark of Megaperls by Henkel throughout Europe. Extruded detergents feature spherical particles of uniform size (for example 1.4 mm) in contrast to conventional spray-dried detergent powders and granulated supercompact detergents that are characterized by a broad distribution in particle size. The density of each extruded particle is approximately 1.4 kg/L. Extruded detergents have as such one of the highest densities achieved in detergent manufacturing.

Apart from being an extraordinary basis for manufacturing supercompact products, extruded detergents feature additional advantages such as complete absence of dust particles, very high homogeneity, no segregation of particles, and excellent free-flowing characteristics. Extruded detergents allow anionic surfactant contents of more than 20% along with very high densities.

4.2.4
Heavy-duty Detergent Tablets

Another form of non-liquid heavy-duty laundry detergents are detergent tablets. Apart from very few regional distributions in the past, detergent tablets were first introduced on a large scale in Europe in late 1997. They have been conquering the European market ever since and holding a share of 10% in the heavy-duty detergent category by mid-2000.

Consumer relevant characteristics of tablets are easiness of dispensing and convenience in handling, i.e., no dosing and dispensing aids are needed. Other advantages include precise dosing and smaller packages compared to powder products, due to their concentration.

Tablets are the most compact form of non-liquid detergents. Their density ranges from 1.0 to 1.3 kg/L, providing further benefits such as lower packaging volume, easiness of transportation and storage, and reduced shelf space. Detergent tablets therefore belong to the supercompact detergent category. Fast disintegration is an essential property in the tablets.

Heavy-duty detergent tablets are also required to be sufficiently hard to be handled during packaging, transportation, and in-home use. A balance has to be struck between conflicting desired properties. Usually, laundry detergent tablets weigh 35 to 45 g with diameters ranging between 40 and 45 mm. The most common recommended dosage in Europe is 2 tablets per wash cycle. One- and two-phase (colored) tablets are available on the market. Two-phase tablets allow the separation (in each one phase) of certain detergent ingredients that otherwise might adversely affect each other during storage, for example enzymes and activated bleach.

The main difference in composition between heavy-duty detergent tablets and other supercompact detergents lies in the use of disintegrants.

4.2.5 Color Heavy-Duty Detergents

In 1991, a new category of heavy-duty detergents appeared on the West European market, for which the name color detergents was coined. So far, this type of compact detergent has mainly been successful on the Western European market. Color detergents differ from conventional European heavy-duty detergents in that they contain neither bleach nor optical brighteners and feature specific dye transfer inhibitors such as poly-(*N*-vinylpyrrolidone) or poly-(vinyl pyridine-*N*-oxide). Color detergents are recommended and used for washing colored laundry to prevent discoloring.

4.2.6 Liquid Heavy-duty Detergents

Liquid detergents have been common in the United States since the 1970s for a variety of reasons. The USA is the largest single market of liquid heavy-duty detergents (HDL) in the world. The market share of these products was some 50% in 1999. Until 1987, such products were insignificant in Europe [3], although liquid heavy-duty detergents had been available on the European market since 1981. In 1998, liquid heavy-duty detergents represented 12% of the market in Europe. Their worldwide share is 14% of total detergent production.

Liquid heavy-duty detergents are distinctive because of their usually high surfactant content (up to ca. 50%). They rarely contain builders such as zeolite or triphosphate and are generally devoid of bleaching agents, because they cannot retain active oxygen which would also be incompatible with enzymes during storage. HDL is most effective when it comes to removing greasy and oily soil, especially at wash temperatures $<60\,^{\circ}C$.

The physical appearance of liquid laundry detergents remained unchanged for a long time. They were on the market as low viscous (250–300 mPa · s) water-thin products with Newtonian flow behavior. This changed with the launch of detergent gels in Europe, which were pourable heavy-duty detergent concentrates featuring very specific rheological behavior [4]. These gel detergents have much higher viscosities than conventional HDL (ca. 1500–3000 mPa/s). and were introduced in almost all European markets in 1997. Because of their high viscosity and rheological properties gels show improved detergency performance upon pretreating of stains as compared with conventional HDL [4]. Due to the continuous increase in the use of liquid concentrates and the reduction of the recommended dosage, it has been possible to reduce package sizes without decreasing the number of wash loads per package. This has resulted in large savings on packaging material and transport costs.

4.2.7 Specialty Detergents

Specialty detergents play a relatively minor role in the United States, but they are quite important in Europe. Specialty detergents are products developed for washing specific types of laundry. Such detergents are generally used in washing machines but also for hand washing. They usually require the use of special washing programs (e.g., a wool cycle to prevent felting or a gentle cycle to prevent creasing). In these specialty detergents, fabric care has a higher priority than in heavy-duty detergents that simply prioritize the removal of soil and stains.

The range of products include special detergents for:

- delicate and colored fabrics
- woolens
- curtains
- washing by hand

Many of these products are appropriate for either machine or hand washing. Most of them are marketed as compact detergents and have an average bulk density of 0.4–0.5 kg/L.

Detergents specially designed for delicate and colored laundry contain neither bleach nor fluorescent whitening agents which may adversely affect the sensitive dyes used in some of these fabrics. These detergents are particularly useful for laundry colored with dyes sensitive to oxidation or for pastel fabrics that otherwise might experience color shifts if treated with fluorescent whitening agents. Most products today contain cellulases, which helps fabrics to look new longer and keep colors bright. Dye transfer inhibitors are added to some specialty detergents.

Detergents for wool are primarily intended for use in washing machines. Particular care must be taken to prevent damage to the sensitive fibers that constitute natural wool, including applying low temperature washing, short washing times, high bath ratios, and avoiding vigorous mechanical action.

High foaming detergents for washing by hand are intended for washing small amounts of laundry in the sink or in a bowl. Due to the increased number of household washing machines, the use of specialty handwashing detergents has been gradually declining in Europe through the last decades. Their share was some 3% in the entire detergent category in 1999.

Liquid specialty detergents have been on the market in Europe for a long time. Some liquid specialty detergents are intended for hand washing. However, most specialty liquid products are intended for washing machine application, e.g., detergents for woolens. These may be free of anionic surfactants, in which case they usually contain mixtures of cationic and nonionic surface-active agents. The cationic agents act as fabric softeners to help keep wool soft and fluffy.

4.3 Detergent Compositions [1]

4.3.1 Surfactants

Surfactants constitute the most important group of detergent components and are present in all types of detergents. Generally, surfactants are water-soluble surface-active agents that consist of a hydrophobic portion (usually a long alkyl chain) attached to hydrophilic or solubility-enhancing functional groups.

A surfactant can be grouped in one of the four classes – anionic, nonionic, cationic and amphoteric surfactants, depending on what charge is present in the chain-carrying hydrophilic portion of the molecule after dissociation in aqueous solution. Tab. 4.1 shows examples of surfactants most commonly used for detergents.

Tab. 4.1 Examples of the most common surfactants used for detergents formulations.

Anionic surfactants	*Nonionic surfactants*	*Cationic surfactants*	*Amphoteric surfactants*
Alkylbenzenesulfonate (LAS and TBS)	Alkyl polyglycol ethers, Alcohol Ethoxylates (AE)	Dialkyldimethyl-ammonium Chlorides	Alkylbetains Alkylsulfobetains
Secondary alkane-sulfonates (SAS) α-Olefinsulfonates (AOS)	Alkylphenol polyglycol ethers, alkylphenol ethoxylates (APE)	Imidazolinium Salts Alkyldimethylbenzyl-ammonium Chlorides	
α-Sulfo fatty acid esters (MES)	Fatty acid alkanol amides (FAA)	Esterquats (EQ)	
Alkyl sulfates (AS) Alkyl ether sulfates (AES)	Alkylamine oxides N-Methyl glucamides (NMG)		
	Alkylpolyglucosides (APG)		

The structure of the hydrophobic residue also has a significant effect on surfactant properties. Surfactants with little branching in their alkyl chains generally show a good cleaning effect but relatively poor wetting characteristics, whereas more highly branched surfactants are good wetting agents but have unsatisfactory detergency performance. Surfactants suitable for detergent use are expected to demonstrate the following characteristics [5, 6]:

- Specific adsorption
- Soil removal
- Low sensitivity to water hardness
- Dispersion properties
- Soil antiredeposition capability
- High solubility
- Wetting power
- Desirable foam characteristics
- Neutral odor
- Low intrinsic color
- Sufficient storage stability
- Good handling characteristics
- Low toxicity to humans
- Favorable environmental behavior
- Assured raw material supply
- Competitive costs
- Chemical stability

4.3.2
Builders

Detergent builders play a central role in the course of the washing process [7–9]. Their function is largely that of supporting detergent action and of water softening, i.e., eliminating calcium and magnesium ions, which arise from the water and from soil.

The category of builders consists predominantly of several types of materials – specific precipitating alkaline materials such as sodium carbonate and sodium silicate; complexing agents like sodium triphosphate or nitrilotriacetic acid (NTA); and ion exchangers, such as water-soluble polycarboxylic acids and zeolites (e.g., zeolite A).

Detergent builders must fulfill a number of criteria [9]:

- Elimination of alkaline-earth ions originating from water, textiles, soil
- Soil and stain removal
- Multiple wash cycle performance
- Handling properties
- Human toxicological safety assurance
- Environmental properties
- Economy

4.3.3 Bleaches

The term bleach can be taken in the widest sense to include the induction of any change toward a lighter shade in the color of an object. Physically, this implies an increase in the reflectance of visible light at the expense of absorption.

Chemical bleaching is used to remove colored non-washable soils and stains adhering to fibers and is accomplished by oxidative or reductive decomposition of chromophoric systems. Only oxidative bleaches are used in laundry products to a great extent.

Two procedures have attained major importance in oxidative bleaching during the washing and rinsing processes – peroxide bleaching and hypochlorite bleaching. Their use is more or less widespread, largely depending on laundering habits in the various regions on the globe.

The dominant bleaches in Europe and many other regions of the world are of the *peroxide* variety. The usual sources of hydrogen peroxide are inorganic peroxides and peroxohydrates. The most frequently encountered source is *sodium perborate tetrahydrate. Sodium perborate monohydrate and sodium percarbonate* have been increasingly used at the expense of sodium perborate tetrahydrate.

Hypochlorite is used for bleaching in many regions of the world where laundry habits, such as washing in cold water, would make sodium perborate less effective.

To achieve satisfactory bleaching with sodium perborate and sodium percarbonate at temperatures $<60\,^{\circ}C$, so-called *bleach activators* are commonly utilized. These are mainly acylating agents incorporated in laundry products. When present in a wash liquor of pH value 9–12, these activators preferentially react with hydrogen peroxide (perhydrolyze) to form organic peroxy acids in situ. As a result of their higher oxidation potentials relative to hydrogen peroxide, these intermediates demonstrate effective low-temperature bleaching properties. Among the wide variety of bleach activators investigated [10–11], only the following compounds have been incorporated on a large scale in detergents worldwide:

- Tetraacetyl ethylene diamine (TAED) [1–13]
- Sodium *p*-isononanoyloxybenzene sulfonate (iso-NOBS) [14]

4.3.4 Auxiliary Ingredients

Surfactants, builders, and bleaches are quantitatively the major components of modern detergents; the auxiliary agents discussed in this section are introduced only in small amounts, each to accomplish its own specific purpose. Their absence from current detergent formulations is difficult to imagine.

4.3.4.1 Enzymes

The effectiveness of proteolytic, amylolytic, and lipolytic detergent enzymes is based on enzymatic hydrolysis of peptide, glucosidic, or ester linkages. The mainstay of the market has been the *protease* types.

Especially in the US and European markets, *amylases* have been added to detergents along with proteases since 1973 to capitalize on the activity of the amylases toward starch-containing soils. From different amylases available, only α-amylases are used for detergents. They are able to catalyze the hydrolysis of the amylose and amylopectin fractions of starch, i.e., cleavage of the α-1,4-glycosidic bonds of the starch chain [15]. This facilitates the removal of starch-based stains by the detergent.

In this context, *lipases, which, due to their substrate-specificity, facilitate the removal of triglyceride-containing soils,* are also worth mentioning. Their activity is highly dependent on temperature and concentration. Studies of model greasy soils have shown that removal of solid fats at very low temperatures (e.g., 20 °C) is largely due to the action of added lipases [16]. The removal of triglyceride-based fatty stains is mostly evident after several washings and drying cycles. At present lipases are contained as further enzymes in premium branded detergents.

Cellulases are capable of degrading the structure of damaged (amorphous) cellulose fibrils, which exist mostly at the surface of cotton fibers after multicycle washing and using. Cotton textiles treated with a cellulase-containing detergent have a smooth surface. For these reasons cellulases are preferably applied in specialty detergents for delicate fabrics or in color heavy-duty detergents for colored textiles. Today many detergents contain blends of two, three, or even four enzymes [17].

4.3.4.2 Soil Antiredeposition Agents, Soil Repellent/Soil Release Agents

The main property expected of a detergent is to remove soil from textile fibers during the washing process. The soil removed is normally finely dispersed, but if the detergent formulation is sub-optimal or too little detergent is used, some may return to the fibers. The problem becomes especially apparent after multicycle washing as a distinct graying of the laundry, which then looks dull and dingy.

Redeposition of displaced soil can be largely prevented by carefully choosing the various detergent components (surfactants and builders), but the addition of special *antiredeposition* agents is also helpful. Such agents act through irreversible adsorption, i.e they cannot be removed from textile fibers or soil particles by water, which prevents the soil from adhering to the fibers [18–19]. This phenomenon is called soil repellent effect. Classical antiredeposition agents are carboxymethyl cellulose (CMC) derivatives bearing relatively few substituents. Analogous derivatives of carboxymethyl starch (CMS) have played a similar role. These substances are effective only with cellulose-containing fibers such as cotton and blends of cotton and synthetic fibers. With the increasing replacement of these natural fibers by synthetics, on which CMC has virtually no effect, the need has arisen to develop other effective antiredeposition agents and soil repellents. Anionic derivatives of terephthalic acid poly(ethylene glycol) polyesters have proved to be very effective soil repellents, particularly on polyester fibers and polyester–cotton blends. They impart hydrophilic properties to these fibers and thus strongly repel oily/greasy soil.

4.3.4.3
Foam Regulators

For soap detergents, which are very popular in regions with low per capita income worldwide, foam is understood as an important measure for washing performance. With detergents based on synthetic surfactants, however, foam has lost virtually all of its former significance. Nonetheless, most consumers – apart from those using horizontal axis drum-type washing machines – still expect their detergent to produce voluminous and dense foam. The reason for this seems to be largely psychological (i.e., foam provides evidence of detergent activity and it hides the soil). Consequently, detergents designed for use in vertical-axis washing machines (i.e., products mainly for the non-European market) are quite frequently given the desired foam characteristics by incorporating small amounts of foam boosters. Compound types suited to the purpose include

- Fatty acid amides [20]
- Fatty acid alkanolamides [21, 22]
- Betaines
- Sulfobetaines [23]
- Amine oxides [24]

In Europe, horizontal axis drum-type washing machines are very common; with such machines, only weakly or moderately foaming detergents are permissible. Thus, foam boosters have lost their former significance in the European market. Especially at high temperatures, heavy foaming may cause overfoaming in drum-type machines, often accompanied by considerable loss of active ingredients. For these reasons, foam regulators - often somewhat incorrectly described as "foam inhibitors" – are commonly added to minimize detergent foaming tendencies [25–28].

The detergents on the market that are based on LAS/alcohol ethoxylates can be effectively controlled by soaps with a broad chain length spectrum (C12–22) [29, 30].

Intensive investigations have shown that specific silica-silicone mixtures or paraffin oil systems are considerably more universal in their applicability and that their effectiveness is independent of both water hardness and the nature of the surfactant-builder system employed [31–33]. Therefore, most heavy-duty detergents in Europe have silicone oil and/or paraffins as foam depressors. Soap has almost lost its importance as a foam regulator. Silica-silicone systems, frequently called silicone antifoams, are usually commercially available as concentrated powders. The key silicone oils used for antifoams are dimethylpolysiloxanes.

4.3.4.4 Corrosion Inhibitors

Washing machines currently on the market are constructed almost exclusively with drums and laundry tubs of corrosion-resistant stainless steel or with an enameled finish that is inert to alkaline wash liquors. Nevertheless, various machine components are made of less detergent resistant metals or alloys. To prevent corrosion of these parts, modern detergents contain corrosion inhibitors in the form of sodium silicate. The colloidal silicate that is present, deposits as a thin, inert layer on metallic surfaces, thereby protecting them from attack by alkali.

4.3.4.5 Fluorescent Whitening Agents

Properly washed and bleached white laundry, even when clean, actually has a slight yellowish tinge. For this reason, as early as the middle of the 19th century, people began treating laundry with a trace of blue dye (blueing agents, e.g., ultramarine blue) so that the color was modified slightly and a more intense visual sensation of improved whiteness was produced. Modern detergents contain fluorescent whitening agents (FWA), called *optical brighteners* to accomplish the same purpose [34]. Fluorescent whitening agents are organic compounds that convert a portion of the invisible ultraviolet light into longer wavelength blue light. It is well-known that the yellowish cast of freshly washed and bleached laundry is a result of partial absorption of the blue radiation reaching it, so the reflected light is partially deficient in the blue region of its spectrum. The radiation emitted by whitening agents makes up for this deficiency, and the laundry becomes both brighter and whiter.

The major commercially available optical brighteners for laundry products are based on four basic structural frameworks: distyrylbiphenyl, stilbene, coumarin, and bis(benzoxazole). Most detergents currently on the market contain fluorescent whitening agents. However, their use on certain pastel fabrics can cause unwanted color shifts, which explains the high popularity of special detergents that lack FWA (e. g., color detergents and detergents for woolens).

4.3.4.6 Fragrances

Fragrances were first added to detergents in the 1950s. Their presence is more than simply a fad or a matter of fashion. Thus, apart from their role in providing detergents with a pleasant odor, an important function of fragrances is to mask certain odors arising from the wash liquor during washing. This becomes particularly important as more washing machines find their way into the living areas of homes. Fragrances are also intended to confer a fresh, pleasant odor on the laundry itself [35–37]. This is why long-lasting fragrances on dry laundry, resulting either from detergents or from fabric softeners have become a more and more important factor in the 1990s.

Detergent fragrances are generally present only in very low concentrations, usually $<1\%$. They are all complex mixtures of many individual ingredients.

4.3.4.7 Dyes

Until the 1950s, powder detergents were more or less white, consistent with the color of their components. Thereafter, products were commonly encountered in which colored granules were present along with the basically white powder: certain components had been deliberately dyed to make the finished detergent more distinctive, featuring speckles. Uniformly colored detergents have also appeared on the market, and the idea of introducing coloring agents has become quite common. The preferred colors for both powders and liquids are blue, green, and pink.

4.3.4.8 Fillers

The usual *fillers* for powder detergents are inorganic salts, especially sodium sulfate. Their purpose is to confer the following properties on a detergent:

- Flowability
- Good flushing properties
- High solubility
- No caking of the powder even under highly humid conditions
- No dusting

Compact detergents usually contain no fillers.

4.4 Physical Parameters in Detergency

4.4.1 Introduction

Washing and cleansing are processes that involve many interfacial effects, which is why a fundamental description of detergency has to be very complex. If we cluster the different processes involved, we can distinguish the following main steps in the cleaning process:

- wetting of the substrate to be cleaned or washed
- dissolution of the detergent formulation
- complexation or removal by ion exchange of the ions of the washing liquor
- interaction of the detergent or cleanser with the stains
- removal of the stains from fabric
- stabilization of the soil in the washing liquor
- modification of the substrate (e.g. by softener in the rinse cycle).

All these processes occur either consecutively or simultaneously and are influenced by a range of interfacial parameters [38].

4.4.2
Surface Tension and Wetting

What characterizes surfactants is their ability to adsorb onto surfaces and to modify the surface properties. At the gas/liquid interface this leads to a reduction in surface tension. Fig. 4.1 shows the dependence of surface tension on the concentration for different surfactant types [39]. It is obvious from this figure that the nonionic surfactants have a lower surface tension for the same alkyl chain length and concentration than the ionic surfactants. The second effect which can be seen from Fig. 4.1 is the discontinuity of the surface tension-concentration curves with a constant value for the surface tension above this point. The breakpoint of the curves can be correlated to the critical micelle concentration (cmc) above which the formation of micellar aggregates can be observed in the bulk phase. These micelles are characteristic for the ability of surfactants to solubilize hydrophobic substances in aqueous solution. So the concentration of surfactant in the washing liquor has at least to be right above the cmc.

The presence of electrolytes increases the adsorption of anionic surfactants at the gas/liquid interface. This leads to a reduction of the surface tension at an equal solution concentration [39] and to a strong decrease of the cmc. The effect can be in the magnitude of several decades. Mixtures of surfactants with the same hydrophilic group and different alkyl chain length or mixtures of anionic and nonionic surfactants have a similar effect. An aspect that has been underestimated for a long time regarding the mechanisms of washing and cleaning is the kinetics of surface effects. Especially at lower concentrations time might have a strong influence on the surface and interfacial tension. The reason for this dependence is the diffusion of surfactant molecules and micellar aggregates to the surface which influences the surface tension on newly generated surfaces. This dynamic effect of the surface tension can probably be attributed to the observation that optimal washing efficiency is usually achieved well above the critical micelle concentration.

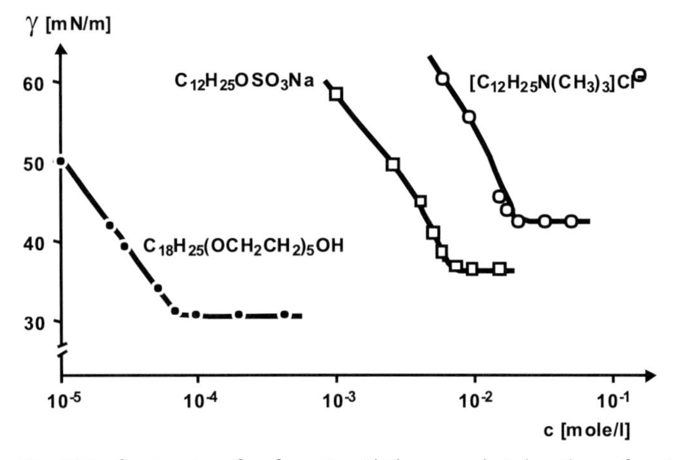

Fig. 4.1 Surface tension of surfactants with the same chain length as a function of concentration [38].

Tab. 4.2 Critical surface tension of polymer solids [40].

Polymer	γ_c *at 20°C, mN/m*
polytetrafluoro ethylene	18
polytrifluoro ethylene	22
poly(vinyl fluoride)	28
polyethylene	31
polystyrene	33
poly(vinyl alcohol)	37
poly(vinyl chloride)	39
poly(ethylene terephthalate)	43
poly(hexamethylene adipamide)	46

Connected with the parameter surface tension is the wetting process of the surface, e.g. fabrics or hard surfaces. The wetting process can be described by the Young equation:

$$\gamma_s = \gamma_{sl} + \gamma_l \cos\theta \qquad (4.1)$$

γ_s = interfacial tension solid/gas
γ_{sl} = interfacial tension solid/liquid
γ_l = surface tension liquid/gas
θ = contact angle

Complete wetting of a solid is only possible if a drop of the liquid spreads spontaneously at the surface, i.e. for $\theta = 0$ or cos $\theta = 1$. The limiting value cos $\theta = 1$ is a constant for a solid and is named critical surface tension of a solid γ_c. Therefore, only liquids with $\gamma_l \leq \gamma_c$ have the ability to spontaneously spread on surfaces and wet them completely. Tab. 4.2 gives an overview of critical surface tension values of different polymer surfaces [40]. From these data it can be concluded that polytetrafluoroethylene surfaces can only be wetted by specific surfactants with a very low surface tension, e.g. fluoro surfactants.

4.4.3
Adsorption at the Solid/Liquid Interface

The physical separation of the soil from the fabrics is based on the adsorption of surfactants and ions on the fabric and soil surfaces. For a pigment soil the separation is caused by an increased electrostatic charge due to the adsorption [41, 42]. In the aqueous washing liquor the fabric surface and the pigment soil are negatively charged due to the adsorption of OH^--ions and anionic surfactants. This leads to an electrostatic repulsion. The nonspecific adsorption of surfactants is based on the interaction of the hydrophilic headgroup and the hydrophobic alkyl chain with the pigment and substrates surfaces as well as the solvent.

Typical examples of adsorption isotherms of sodium dodecyl sulfate onto different surfaces are shown in Fig. 4.2 [41]. Fig. 4.2 also demonstrates the effect of the

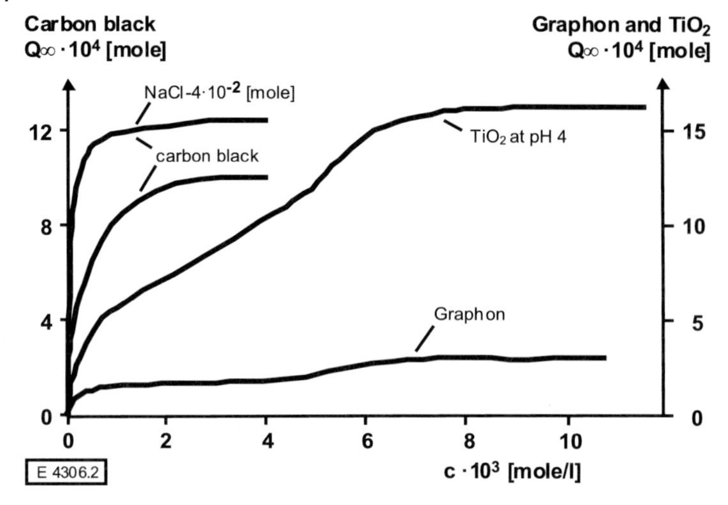

Fig. 4.2 Equilibrium adsorption of sodium n-dodecyl sulfate on carbon black, TiO_2, and Graphon at room temperature [41].

addition of electrolytes which are present in the washing process. The amounts of anionic surfactant adsorbed increase in the presence of ions. This is due to a decreased electrostatic repulsion of the negatively charged hydrophilic groups of the anionic surfactant in presence of electrolytes. Therefore the adsorption density in equilibrium can be significantly enhanced. A similar effect can be observed when comparing an anionic and a nonionic surfactant with the same alkyl chain length adsorbed onto a hydrophobic solid. The nonionic surfactant gives higher adsorbed amounts at the same concentration than the anionic surfactants. This is especially valid at low concentrations, whereas at very high concentrations both surfactants reach the same plateau value. For a hydrophilic solid surface this effect can be just the opposite due to a higher affinity of anionic surfactant to the surface via specific interactions.

The electrolyte effect for the adsorption of anionic surfactants which leads to an enhancement of soil removal is valid only for low water hardness, i.e. low concentrations of calcium ions. High concentrations of calcium ions can lead to a precipitation of calcium surfactant salts and reduce the concentration of active molecules. Therefore, for many anionic surfactants the washing performance decreases with lower temperatures in the presence of calcium ions. This effect can be compensated by the addition of complexing agents or ion exchangers.

4.4.4
Liquid/Liquid Interface

The phenomena at the liquid/liquid interface are of outstanding importance for the removal of oily soil from the surface. The interfacial tension is one of the decisive parameters in the rolling-up process. This parameter vary considerably, de-

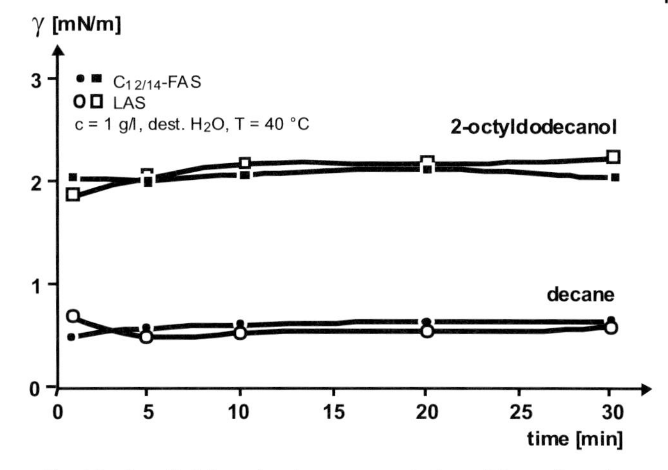

Fig. 4.3 Interfacial tension between a solution of $C_{12/14}$-fatty alcohol sulfate (FAS) and linear alkylbenzene sulfonate (LAS) and two different oils as a function of time [43].

pending on the surfactant structure and the type of the oily soil [43]. Fig. 4.3 shows this for two different oils and two anionic surfactants. For both surfactants the interfacial tension is the same with lower values for the nonpolar decane. As the interfacial tension should be minimized in detergency, there is a need to further reduce interfacial tension in formulations. A possible way is again to create mixed adsorption layers of suitable surfactants. For example, the interfacial tension of the system water/olive oil as a function of composition for a surfactant mixture containing the anionic surfactant sodium n-dodecyl sulfate with the nonionic surfactant nonylphenol octaethylene glycol ether shows a pronounced minimum at a certain concentration ratio for a constant total surfactant concentration. Even small additions of one surfactant to another can lead to a significant reduction of the interfacial tension. Thus, the interfacial tension can be used to optimize detergent formulations.

4.5 Phase Behavior of Surfactant Systems

The phase behavior of the surfactant systems is decisive both for the formulation of liquid and solid products and the mode of action of the surfactants in soil removal during the washing and cleaning process. Due to the different phases of surfactant systems at higher concentrations e.g. the flow properties can vary very strongly dependent on concentration and type of the surfactants. This is of crucial importance for the production and handling of liquid products. In addition to this the phase behavior influences the dissolution properties of solid detergents when water is added, forming or preventing high-viscous phases. The phase behavior can also have a significant impact on the detergency [44]. If there is no phase change for the surfactant water system, a linear dependence of the detergency on temperature is observed. The surfactant is in an isotropic micellar solution at all temperatures.

Tests with pure ethoxylated surfactants have revealed that a discontinuity is observed with respect to oil removal versus temperature in cases of the existence of dispersions of liquid crystals in the binary system water/surfactant. Fig. 4.4 shows that the detergency values for mineral oil and olive oil, i.e. two oils with significantly different polarities, are at different levels. It also demonstrates that in both cases a similar reflectance vs. temperature curve exists. In the region of the liquid crystal dispersion, i.e. between 20 °C and 40 °C, the oil removal increases significantly. For olive oil, a small decrease in detergent performance is observed. The macroscopic properties of the liquid crystal dispersion seem to be responsible for the strong temperature dependence. It can be assumed that fragments of liquid crystals are adsorbed onto fabric and oily soil in the $W + L_\alpha$ range during washing.

During the oil removal from fabrics or hard surfaces ternary systems occur where three phases coexist in equilibrium. These systems are also referred as three-phase microemulsions. The effects were studied in detail for alkyl polyglycol ethers [45]. Depending on temperature different phases exist, having a three-phase region between the temperature T_l and T_u. When these three phases are formed, extremely low interfacial tensions between two phases are observed. Because the interfacial tension is generally the restraining force, with respect to the removal of liquid soil in the washing and cleaning process, it should be as low as possible for optimal soil removal. Other parameters such as the wetting energy and the contact angle on polyester, as well as the emulsifying ability of e.g. olive oil, also show optima at values of the same mixing ratio at which the minimum interfacial tension is observed.

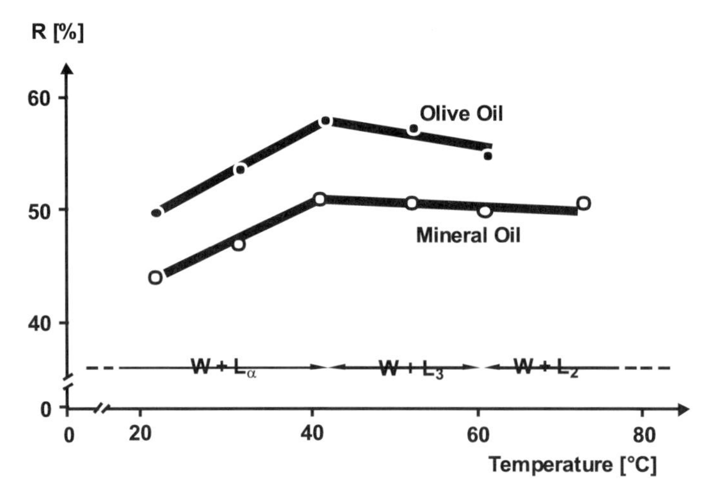

Fig. 4.4 Phase behavior of the polyoxyethylene alcohol $C_{12}E_3$ and detergency, 2 g/L surfactant [44].

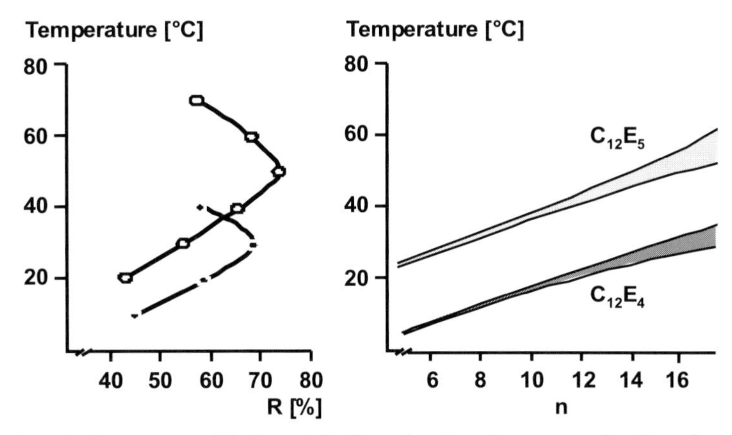

Fig. 4.5 Detergency of $C_{12}E_4$ and $C_{12}E_5$ against hexadecane as a function of temperature (left side) and the corresponding three-phase ranges for these surfactants as a function of the number n of carbon atoms of alkanes [46].

Fig. 4.5 (right) represents the three-phase temperature intervals for $C_{12}E_4$ and $C_{12}E_5$ vs. the number n of carbon atoms of n-alkanes. The left-hand graph of Fig. 4.5 shows the detergency of these surfactants for hexadecane. Comparison of both graphs in Fig. 4.5 indicate that the maximum oil removal occurs in the three-phase interval of the oil used (n-hexadecane) [46]. This means that not only the solubilization capacity of the concentrated surfactant phase, but probably also the minimum interfacial tension existing in the range of the three-phase body are responsible for the maximum oil removal.

4.6 Foaming

Foaming and the control of foam is an important factor in the application of detergents and cleansers. This regards high-foaming systems for e.g. manual dishwashing detergents as well as low-foaming systems for use in laundry or dishwashing machines. The foam properties of the products are mainly governed by the surfactant system and the use of anti-foams (4.3.4.3.). Besides this the chemical composition of the product or the washing liquor, e.g. electrolyte content and soil strongly influences the foam properties. Physical-parameters like temperature and pH-value or mechanical input in the system have additionally taken into account.

The basis for the foam properties is given by interfacial parameters. Although correlations have been shown between a single parameter and foam properties, there is still a lack in a general correlation between interfacial properties and the foam behavior of complex systems in detergency. The simplest approach to correlate interfacial parameters to foam properties is the comparison of the surface activity measured by the surface tension of a surfactant system and foam stability.

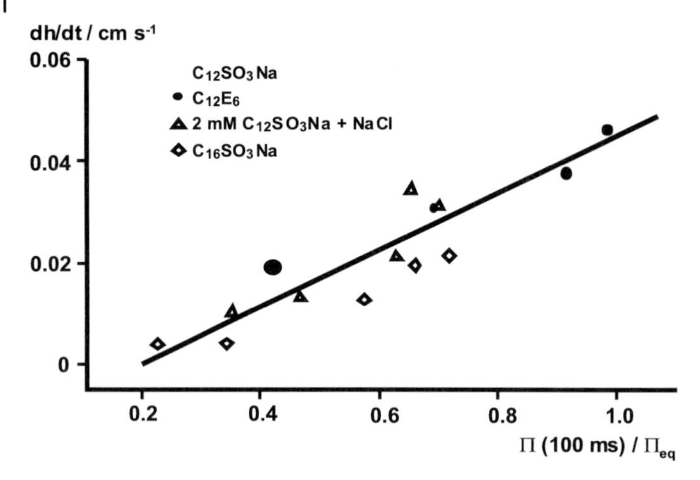

Fig. 4.6 Correlation of relative dynamic surface pressures with foam kinetics data *dh/dt* as a function of the type of surfactant, alkyl chain length and salt concentration [47].

This has been done for a series of pure surfactants. Within a specific class of surfactants the surface tension directly correlates to the foam stability of the surfactant-water system. As foam generation and also foam stability are dynamic processes generating and reducing surface area, in a surfactant-water system the diffusion of the surfactant to the surface and the change in surface coverage, at least locally during bubble generation and drainage of the film, is a more useful way of explaining foam properties. If one distinguishes between foam formation and foam stability, a good correlation has been found between the relative dynamic surface pressure derived from the time-dependent dynamic surface tension and the rate of foam formation (Fig. 4.6) [47]. The specific time for the relative dynamic surface pressure was chosen empirically. The correlation of the two parameters is valid for different surfactant types and the addition of electrolyte. The effect can be explained by the micellar kinetics of the surfactant solution and the diffusion of the molecules and micelles to the surface.

4.7 On-line Sensorics in Detergent Development

4.7.1 Product Relevant Physico-chemical Parameters and their Availability for On-line Determination

The knowledge of physico-chemical parameters like surface tension, conductivity, turbidity and the pH of the washing liquor is important for the improvement of existing washing and dishwashing detergent formulations further development of new alternatives. Today's advanced physico-chemical and analytical methods make it possible to gain more and more information on each specific step during the diverse washing and dishwashing processes. Relevant steps can be dissolution behavior and dissolving sequences, stability of ingredients, interactions, reactions, stain and soil removal, water softening, bleach activity and many more. Besides the already mentioned parameters, other important measurable information to be taken into account are water hardness, bleach and surfactant concentration and the enzyme activity. Tab. 4.3 gives an overview of the most relevant parameters and the available analytical methods for their on-line determination. Most of these parameters can be measured during the washing or dishwashing application using commercially available equipment like standard electrodes or photodiodes.

Fig. 4.7 shows the typical behavior of a solid detergent product during the first 20 minutes of the washing process as the result of on-line sensorics in a washing machine. It can be seen that during the dissolution process pH value, conductivity and peroxide content of the washing liquor increase, while the surface tension decreases simultaneously. The core information that can be derived from the on-line experiment is that the washing power product is available within a very short time. Detailed analysis of pH, bleach concentration and surface tension at each moment during the washing process also shows that the wetting, cleaning, solubilising, dispersing and stain removal properties of the detergent product can be exactly tailored for the desired application.

It becomes obvious that on-line experiments offer distinct advantages over the traditional off-line laboratory beaker methods. Firstly, they give information on the 'real life' situation in which a product has to fulfill its task. Secondly, the on-line measurement is the only correct way to ensure the practicability of theoretical product concepts. Nearly all products for washing and dishwashing are complex mixtures of different ingredients, which makes it hard to use simple beaker experiments with single substances as a means to differentiate surface effects, since the complex surroundings in the application play an important role. Additional effects due to adsorption, interfacial dynamics and unwanted ingredient interactions

Tab. 4.3 Parameters for on-line sensorics in detergency.

Parameter	***Method***
pH	pH-sensitive electrode
Ionic strength	conductivity electrode
Water hardness	ion-selective electrodes (Ca^{2+}, Mg^{2+})
Surfactant concentration	surfactant-selective electrode, potentiometry
Turbidity	light transmission/extinction, photodiode
Surface tension	bubble pressure tensiometer
Peroxide concentration	peroxide sensor, i.e. iodometry, electrodes
Temperature	thermo-electrode, Pt-100
Soil removal	fluorescence probe
Enzyme concentration	bio-sensor
Foam height	conductivity, light extinction

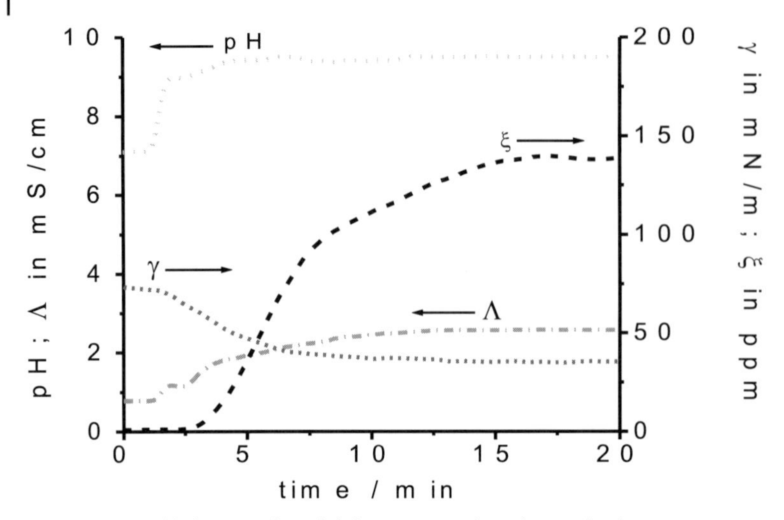

Fig. 4.7 Typical behavior of a solid detergent product during the first 20 minutes in a commercially available washing machine. Relevant parameters (pH value, conductivity Λ, surface tension γ, peroxide content ξ) were detected by on-line sensorics.

may also lead to unclear characterisations. An in-situ access to the relevant data is therefore of high interest. This can be achieved by using on-line sensors in the specific applications.

4.7.2
On-line Determination of Dynamic Surface Tension by the Bubble-pressure Method

Several interesting methods and ways to conduct the relevant on-line experiments have been published during the last years. The most important advancement can certainly be regarded in the development of reliable dynamic surface tensiometers [48–55]. The on-line determination of the surface tension of a washing or dishwashing liquor leads to highly interesting information on the behavior of surface-active substances during the cleaning application. In 1986, the first use of a dynamic bubble-pressure tensiometer for the sensorical analysis of active ingredients in the automatic dishwashing process was described [48]. It was later shown that the surface tension and the machine washing result can be correlated, so that the correct dosage of a detergent product can be calculated [49]. Using this concept of the 'optimum washing point', it was further shown that depending on the dosage, the relation between surfactant content of the washing liquor and bound/adsorbed surfactant could be optimized [49].

The sensoric principle of the dynamic bubble-pressure tensiometer is based on the differential LAPLACE pressure between two capillaries from which a controlled gas flow is released. At the lower end of a capillary which points into the liquid, a gas bubble is formed which increases its radius with increasing gas pressure (see

Fig. 4.8). At maximum pressure, the radius of the bubble is equal to the radius of the capillary r_K; on further increasing the pressure the difference decreases, and the gas bubble is released from the capillary. This maximum pressure p_{max} can be measured and correlates to the surface tension γ_{dyn} via the modified LAPLACE-equation:

$$\gamma_{dyn} = \frac{r_K}{2}(p_{max} - p_{stat}) = \frac{r_K}{2}(p_{max} - \rho g h_E) \quad (4.2)$$

The hydrostatic pressure p_{stat} corresponds to the density ρ of the liquid multiplied by the gravity constant g multiplied by the immersion depth of the capillary h_E but can be eliminated by using a parallel second capillary with much larger radius [49] or by relative measurements of p [50]. Usually, capillaries with radii of 0.5–2.5 mm and gas flow rates of 5–20 ml/min are used while the pressure difference is below 1000 Pa (100,000 Pa = 1 bar).

Recently, a very practical bubble pressure tensiometer was developed using elegant pressure transducer mechanics which only needs one capillary made from a high-tech polymer [51, 52]. The tensiometer is able to measure at different immersion depths but needs calibration in order to make the resulting data comparable to surface tension values from other sources. It was shown in a series of measure-

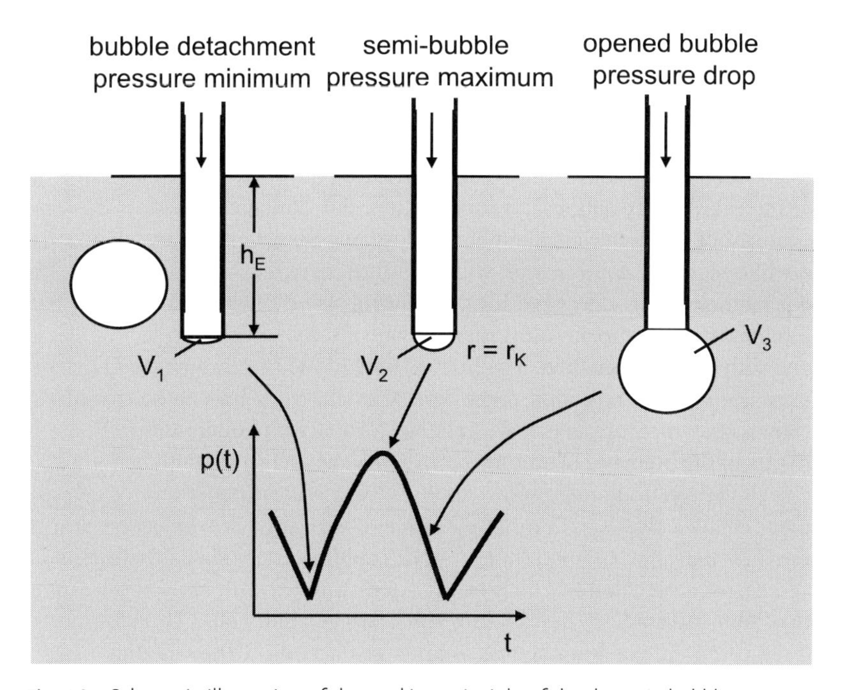

Fig. 4.8 Schematic illustration of the working principle of the dynamic bubble pressure method. If the bubble radius equals the capillary radius, maximum pressure is detected. The pressure minimum occurs on bubble detachment.

ments that the change of surface tension during a commercial dishwashing process could be followed using this setup [51].

An investigation using another commercially available dynamic bubble-pressure tensiometer (Lauda MPT 2) was performed in order to follow the dissolution behavior of detergent tablets in a household washing machine [53]. A flow-measuring cell with a continuous wash-liquor supply via a bypass was constructed. A constant flow rate through the bubble pressure tensiometer was used in the dissolving test in order to keep the surface age constant and to make the received dynamic surface tension data comparable. Dissolving experiments were performed with sodium salts of C_{12-18} alkyl sulfate and C_{10-13} alkylbenzene sulfonate. While the alkylbenzene surfactant was dissolved within 9 mins in a regular washing cycle, the alkyl sulfate did not dissolve at all at 30 °C due to its higher KRAFFT temperature and the ability to form gel phases on contact with water. Reproducibility of the measurements was very good.

The use of microcavity electrodes for generating gas bubbles and a washing process in which the dynamic bubble pressure method based on such electrodes is used for on-line control of a washing machine was covered in a patent application [54]. The bubbles can for example be generated by electrolysis reactions in which H_2 is produced. Bubble formation frequency is monitored by measuring the overpotential changes at the gas-generating electrode. The surface tension can then be obtained by measuring the bubble frequency change at constant electrolysis current.

4.7.3 Other Accessible On-line Parameters

The on-line determination of water hardness ions was investigated [55]. It is well established that free hardness ions can negatively affect the detergency by leading to increased redeposition and incrustation. For ecological reasons, the 'fast' ion exchange builder materials like phosphates have been replaced by 'slow' ion exchangers like zeolite in combination with co-builders such as polycarboxylates. Two different methods were compared for the determination of calcium ions in a basic washing liquor: an electrochemical method which uses a calcium-selective electrode [56], and a photometric method. The first method has to be examined very carefully in the presence of surfactant ions, since several interactions have to be considered. It was shown that measurements are reliable for calcium-builder-interactions. Measurements in the presence of anionic surfactants showed strong interactions of ionic species in the electrode membrane, so that this method doesn't seem to be suitable for more detailed analysis. Combining method two with a flow injection process, a very sensitive detection method for Ca^{2+}-ions could be established. The method takes advantage of a special dye which forms a red complex with specific absorption at 570 nm. After calibration, the calcium ion concentration could be continuously monitored, even in the presence of anionic surfactants. Other studies using calcium ion-selective electrodes in the presence of anionic surfactants indicated that electrodes made from PVC matrix membranes were superior to other membrane systems due to their higher resistance to the leaching-out of membrane components by the surfactants [57]. The interference of Ca^{2+} ion-selective electrodes by anionic surfactants was attributed to the different solubilities of calcium-surfactant complexes in the membrane materials. PVC-membrane based systems showed the expected values even in the presence of wash liquor. The use of electronic resistivity sensors for on-line hardness determination [58] or turbidity sensors for the support of the rinsing systems in automatic washing and dishwashing machines [59] have also been described (see also 4.8.).

One of the most interesting parameters in the washing process is the on-line determination of bleach-active components like peracetic acid or hydrogen peroxide. Dissolved peracetic acid is always in equilibrium with water, hydrogen peroxide and acetic acid. The comparison of different analytic methods led to the development of a photometric flow injection analysis system, which enables the detection of free peracetic acid in the presence of hydrogen peroxide [60]. Iodometric redox titration is the basis of the system, which can be followed photometrically if the washing solution is filtered through a dialysis membrane. Hydrogen peroxide was quantitatively removed by adding the enzyme katalase during the measurements. Pulsed amperometric measurements can also be used in order to ensure the right dosage of peracetic acid during washing and cleaning applications [61].

On-line sensorics of the chemical oxygen demand (COD) in rinsing solutions was investigated by another electrochemical method that produces OH-radicals with high oxidation potential by applying an electrical current [62]. The OH-radicals are able to oxidize substances in water rapidly. From measuring the oxidizing current, the amount of residual soiling can be derived. Such an application has a high potential in commercial dying and rinsing processes, where substantial amounts of water and processing time can be saved.

4.8 Sensors in Household Appliances

4.8.1 Introduction

Besides the already described determination of physico-chemical parameters with classical methods of chemical analysis and measurement technology there are various efforts in research and technology to implement new methods and technologies within household appliances. The devices, which are integrated in the different applications, have to fulfill some basic requirements that are common to all types of sensors [63]. First of all the particular parameter has to be determined sensitive within the given (concentration) range. Furthermore, the recognition must be completely reversible to ensure reliable consecutive measurements. Closely related to the requirement of fully reversible measurements is the stability and lifespan of the devices needed for a household appliance. Furthermore, the measurements must not be affected by changing boundary conditions, e.g. a sufficient cross-selectivity must be given.

Taking all these prerequisites into account, the use of chemical and physical sensors within household appliances is considerably restricted, and only a few applications are already on the market. In the field of bioanalytics, sensors are already used for bioprocess-monitoring and biomedical applications. In this area highly specific recognition processes can be used in sensors that only require a short lifespan, due to operating conditions etc. [64].

Generally the different sensors can be categorized into physical, biological and chemical sensors.

Applications for physical sensors are dominated by the automotive industry. Engines are monitored and controlled by various sensors, and most cars are equipped with systems to enhance safety for brakes, steering, chassis, and even crashes. Controlling airflow and temperature in the cabin enhances the comfort of drivers.

Chemical sensors are widely used to monitor hazardous and combustible gases [65]. Applications include safety control in industrial applications, surveillance of boilers and other devices which are operated with natural gas as well as more sophisticated areas like cooking control and odor determination [66].

Biosensors normally offer highly specific molecular recognition reactions like enzyme/substrate-, antigen/antibody-, DNA/DNA-, or protein-interactions [67]. Due to their specific sensing principles and set-up they are limited to special applications and boundary conditions. The limited stability and reproducibility of these devices requires higher standards of maintenance and recalibration.

4.8.2 Sensor Applications

Different sensor applications have already been described in the relevant literature. Most of the systems make use of physical sensors, while (bio)chemical sensors have been described to a lesser extent. The various applications can be classified by the parameters to be detected.

Applications focus on turbidity, conductivity, water hardness, mass, and concentration, which are measured in automatic washing and dishwashing machines.

4.8.2.1 Turbidity

The turbidity level or more exactly the optical transmittance and reflectance provides a range of information about the washing and rinsing water in automatic washing and dishwashing machines. The most important parameters are soil and detergent concentration, both of which have an impact on turbidity.

Measurement set-ups for turbidity can be realized quite easily. A light source and a detector are mounted within the washing drum or preferably in the connected piping. The reliability of the measurement can be enhanced by combining several light sources and detectors and by detecting passing light as well as reflected light. Most common emitters are light-emitting diodes; the radiation is measured with conventional photodiodes.

A possible application could be a washing machine that detects detergent-related and soil-related information from several light transmission devices that monitor the whole washing process from start to finish [68]. For applications in an automatic dishwasher the concentration of dirt and detergency in the rinsing water can also be monitored simultaneously [69]. To prevent inaccuracies which might occur due to air bubbles within the continuously pumped rinsing water, the sensors need to be carefully designed. With respect to the dependency between soil and detergent concentration their absence can be measured more easily with optical methods. A possible application is the monitoring of the rinsing process [70]. With detailed information about the contamination of the rinsing liquid the rinsing process can be optimized with respect to the necessary amount of water and rinsing cycles.

4.8.2.2 **Conductivity**

In washing machines and automatic dishwashers the conductivity of the washing liquor is affected by a range of parameters. Conventional washing agents contain several ionic compounds like water softeners, bleaching agents, fillers, and ionic surfactants the dissolved stains also consist of different ionic parts. Hence, the conductivity of the washing liquor can be used to monitor different parameters. The systems in use mainly vary in the technical layout of the conductivity sensor and the specific application.

Conductivity sensors are most commonly used for safety purposes in household appliances. Presence and absence of washing liquor, detergency, and water softener can be easily measured and proper operation ensured [71]. The various applications mainly differ by their design of electrode geometry and methods for electrical measurement. Due to the close relation between ionic conductivity and water hardness, the automatic water softener in an automatic dishwasher can be controlled by a conductivity sensor [72]. To isolate the transmission of the measured value from the process controller, the conductivity sensor could incorporate an opto-electronical coupling [73]. Thus, protective insulation of the electrodes in a washer-dryer could be ensured.

4.8.2.3 **Water Hardness**

The hardness of water used in washing machines and automatic dishwashers varies greatly, depending on sources and pretreatment. High water hardness leads to a reduced performance of detergency whereas extremely soft water causes corrosion of glassware in automatic dishwashers. The simplest method is to determine water hardness indirectly by the conductivity of the rinsing water as already described. However, this concept is not applicable in the presence of soil or detergency.

Calcium can be directly determined with ion selective electrodes [74]. Especially for cations there are different ion selective materials available for fast and selective determination.

The limitations of simple conductivity measurement can be overcome using a combined impedimetric and potentiometric technique [75].

4.8.2.4 Concentration

The required concentration of detergency is extremely important in the field of washing machines and detergency. Two kinds of monitoring methods can be distinguished – direct determination of active substances in the washing liquor and indirect methods, which rely on the measurement of masses and flows.

The amount of remnant detergency in a storage tank can be quite easily monitored using a level indicator [76] or by the total mass of the tank [77]. The technical set-up of level and mass detectors is quite complex and is therefore limited to high-volume and professional applications. The flow of liquid detergents can be monitored by conventional flow meters or pumping rates [78]. Simultaneous control of the amount of water fed into the washing tub would permit to calculate the concentration of detergency in the washing liquor.

This technology can also be applied for pre-dosed detergents like tablets [79], which offer the advantage of an easier determination of detergent already fed into the washing tub.

The concentration of detergents in the washing liquor can also be determined by measuring their conductivity [80]. Due to the influence of water hardness and soil the accuracy is limited.

Direct determination of surfactants in complex matrices can also be carried out using ion-selective electrodes. Depending on the membranes and additives used, the detergent electrodes are optimized for the detection of anionic surfactants [81], cationic surfactants [82], and even nonionic surfactants [83]. The devices are sensitive to the respective group of surfactants but normally do not exhibit sufficient stability and reproducibility for their use in household appliances. With further optimization of membrane materials, plasticizers and measurement technology, surfactant-selective electrodes offer high potential for future applications.

Agents for chemical bleaching rely on different types of peroxides. Potentiometric or amperometric biosensors that detect the highly specific and sensitive reaction of enzymes like katalases with their corresponding substrates can be used for on-line measurement [84]. The sensors can be manufactured with simple technologies at moderate cost, but their stability is not sufficient for integration in household appliances.

4.8.2.5 Miscellaneous

A quite important parameter for the operation of washing machines is their load. There are several ways to measure the wash load, e.g the force required to rotate the drum/tub/pulsator could be measured and the required amount of water worked out accordingly, or the machine load could be measured by a remote pressure-sensing device with piezoresistive transducers and wireless communication

[85]. Even in a rotating tub the actual load can be measured on-line, and the machine can be controlled on the basis of these data.

In some other washing machines that have recently entered the market, the wash load is detected by a weight sensor that measures the elongation of the spiral springs where the tub is suspended inside the washing machine. This is known as a "load size indicator". Based on information from the detergent manufacturer, it tells the customer how much detergent to use.

In automatic dishwashers, amount of dirt on crockery and cutlery can be measured with a radar sensor [87] or with a device that uses the reflectivity of an integrated surface [88]. Thus, the load of the washing liquor with soil and detergency can be measured.

4.8.3
Conclusion

Most of the relevant substances within household appliances cannot be measured directly with any kind of sensor. Several applications have been described that can indirectly determine one or several parameters. Sensors which rely on changes of weight, turbidity or conductivity have already implemented, while direct determination of substances with chemical or biochemical sensors is still problematic because of insufficient stability and reproducibility.

4.9
Detergent Products with Built in Sensor Functions

4.9.1
Introduction: "Intelligent" Detergent Products

In the previous sections a variety of sensors have been described which play a role in detergency or in the development of detergents. These sensors generally convert a physical environmental parameter into an electronic signal. In this section a completely different type of sensor will be introduced: examples of chemical sensors within detergent products will be discussed. These systems react to changes in environmental parameters, e.g. the environment in a dishwasher or a washing machine. In this way the release of active ingredients of detergent or cleanser products can be triggered at a certain moment during the cleaning process. Thus, these systems allow the development of a new generation of powerful multifunctional "intelligent" products that can notably increase convenience for the consumer.

The relevant parameters to trigger the release of actives can be the temperature or pH-value of the washing liquor, the salt concentration or even the presence of water. In some automatic dishwashing tablets that have been recently introduced, for instance, the difference in temperature of the main wash cycle and the rinse cycle [89], or the difference in the pH-values of these two stages is utilized to trigger the release of rinse aid. This component is normally added automatically by the dishwasher, but it has to be topped up after several cycles. The intelligent product, however, combines the function of the detergent and the rinse aid, saving the housewife the trouble of refilling.

Heavy-duty detergent tablets for textile laundry, as mentioned in chapter 4.2.4, use the presence of water to activate a sophisticated disintegration system that dispenses the tablets in a few seconds even though these products have to be stable and hard during storage and application [90, 91, 92]. These two systems are discussed in more detail in the following sections.

4.9.2
Automatic Dishwashing Detergent (ADD) Tablets

As shown in the upper part of Fig. 4.9, the whole washing program in an automatic dishwasher consists of several phases or cycles. At the very start there is the pre-wash cycle with pure, cold water in order to remove coarse residues of food. Then the main wash cycle follows. At this stage the temperature rises to typically 50 °C or 55 °C and due to the release of the alkaline detergent the pH-value increases. One or more intermediate rinse cycles remove residues of the washing liquor of the main wash cycle. Both temperature and pH-value decrease. In the last rinse cycle the temperature increases once again, this time even to a higher value of 65–70 °C. The pH value decreases again. The rinse aid is being dosed at this stage.

As mentioned above, these changes of parameters can be used by products with integrated sensor functions to release the rinse aid at the right moment. In the market two different systems are currently available. One system uses the temperature change, another the change in pH value. The principal mode of action of these two systems is the same: The active substance, in our case the rinse aid, is

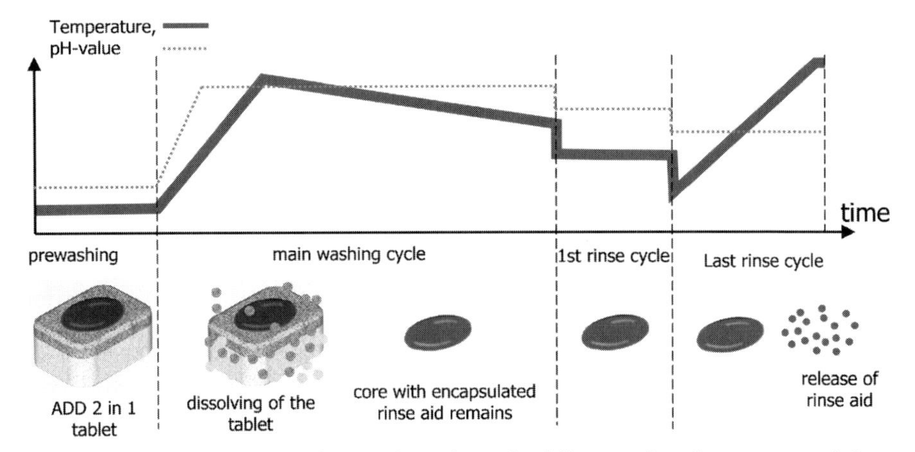

Fig. 4.9 Temperature and pH-value (qualitatively) in the different cycles of an automatic dishwashing program and the dissolution of a tablet with temperature triggered release of rinse aid.

encapsulated into a sensor material. This sensor material changes its properties on variation of the external parameters. It can, for instance, be solid and stable under one set of conditions and dissolve or melt on changing the conditions.

In the case of a temperature-sensitive product the rinse aid is embedded into a substance e.g. a wax with a melting point higher than the temperature of the main wash cycle (e.g. 50 °C or 55 °C) but lower than the temperature of the rinse cycle that is most likely above 65 °C. Thus, the encapsulation with a melting point between 60 °C and 65 °C resists the conditions in the main wash cycle but melts in the rinse cycle and releases the rinse aid. This process is shown in the lower part of Fig. 4.9.

Basically there are two different ways of embedding an active substance in encapsulating material: A solid core can be surrounded by a discrete encapsulating wall or otherwise a multitude of many tiny compartments of the active substance can be embedded into a sponge-like structure of encapsulating material. For a temperature-sensitive product the latter system is used. A fine emulsion of rinse aid, mainly a liquid surfactant in molten wax, is prepared and filled into the cavities of the detergent tablets where it immediately solidifies, enclosing the active substance until the temperature rises again above the melting point – during the application in the rinse cycle of the dishwashing machine.

Another product has basically a built-in pH sensor. In this case the rinse aid is encapsulated into a material that is solid and insoluble in water at an alkaline pH-value. Thus it is able to withstand the conditions in the alkaline main wash cycle. When the pH-value decreases, due to the dilution of the washing liquor with pure water in the rinse cycles, the sensor encapsulation material becomes soluble and releases its contents [93, 94].

4.9.3 Laundry Detergent Tablets

As explained above, detergent tablets can also be regarded as products with a built-in sensor function sensitive to the presence of water. The first detergent tablets were brought to the market as long ago as in the 1960s in the USA, but they were not very successful, possibly because of their poor disintegration properties due to the lack of such a sensor function. Since the mid 1980s a detergent tablet has been available in Spain, but only with a small market share. The breakthrough came in 1997 when an efficient system that disintegrates the tablets in just a few seconds became available for the first time. Since then, detergent tablets have been enjoying great popularity because of their many benefits compared to conventional detergents. They are easy to handle and simply to dose. The dosage instructions on the package are easier to understand, and there is no need for a dosing aid. Thus, under and over dosage can easily be avoided. Moreover, this very compact form of detergent reduces the volume of the package. Summing up, the above-mentioned advantages of detergent tablets increase the convenience of the product and lower its environmental impact.

These advantages can only be significant if the tablets are considerably stable and hard. They must be hard enough to withstand mechanical stress by packaging and transport. On the other hand, as already mentioned, one of the key requirements of laundry detergent tablets is their fast disintegration. Upon their first contact with water, be it in the washing machine dispenser of a European drum-type washer or in the basket of top-loading agitator (North America) or pulsator machines (East Asia), tablets have to disintegrate within seconds or at least a minute. But the disintegration of hard tablets poses special challenges. Detergent tablets contain a high amount of surfactants. These substances, however, form liquid crystals on contact with water which are gel-like pastes. These highly viscous pastes can hamper the disintegration of the tablet because they can prevent more water from penetrating. Thus, one important requirement for the disintegration system is that the disintegrating effect is faster than the gel formation. Otherwise the tablet would remain undissolved in the dispenser tray or – even worse – in the laundry.

Basically four different types of disintegration systems are known:

- Effervescence: these systems that consist for instance of a carbonate or bicarbonate salt and an organic acid like e.g. citric acid, develop carbon dioxide gas on contact with water and are well known in pharmaceutical/health food tablets. A few market products exist that make use of this system but for reasons explained above this system is not that suitable for detergent tablets with a high content of surfactants.
- Highly water-soluble substances like e.g. Na (K) acetate or Na (K) citrate that are mixed to the active substances of the tablet. They enable the water to penetrate immediately and weaken the binding forces inside the tablet. These systems are sometimes used in detergent tablets.
- Water-soluble coatings. Strictly speaking these systems are not genuine disintegrants but give the tablets high mechanical stability if their core has only been weakly compressed and would easily disintegrate anyway.
- Granular swelling agents. These systems typically consist of cross-linked hydrophilic polymers like e.g. cross-linked poly-(N-vinyl pyrrolidone) or hydrophilic but insoluble polymers, e.g. cellulose that swell to a considerable volume within a few seconds on contact with water. These systems are powerful disintegrants and are used particularly in quick-release detergent tablets put in a dispenser tray.

The development of these systems in the late 1990s enabled the detergent tablet to gain a notable market share and to facilitate the housewife's work, offering her a high-performing, convenient product.

4.10 References

1 Ullmann's Encyclopedia of Industrial Chemistry (6th edition)
2 Th. Müller-Kirschbaum, E. Smulders: "Facing future's Challanges – European Laundry Products on the Treshold of the 21st Century", in A. Cahn (Ed.): Proceedings of the 4th World Conference on Detergents, AOCS Press, Champaign IL, USA 1999, pp 93–106.
3 H. Andree, P. Krings, H. Verbeek, *Seifen Öle Fette Wachse* **107** (1981) 115–119.
4 D. Nickel, Proceedings of the World Surfactant Congress (CESIO), 2000, 1106–1111.
5 G. Jakobi in H. Stache (ed.): Tensid-Taschenbuch, 2nd ed., Hanser Verlag, München-Wien 1981, pp. 253–337.
6 M.F. Cox, T.P. Matson, J.L. Berna, A. Moren, S. Kawakami, M. Suzuki, J. Am. Oil Chem. Soc. 61 (1984) no. 2, 330.
7 W. Kling, Münch. Beitr. Abwasser Fisch. Flußbiol. **12** (1965) 38.
8 G.C. Schweiker, J. Am. Oil Chem. Soc. **58** (1981) 58–61.
9 G. Jakobi, M.J. Schwuger, Chem. Ztg. **99** (1975) 182–193.
10 Unilever, DE 1 291 317, 1969.
11 G. Reinhardt: "New Bleach Systems" in A. Cahn (ed.): Proceedings of the 4th World Conference on Detergents, AOCS Press, Champaign IL, USA 1999, 195–203.
12 Noury van der Lande, DE 1 162 967, 1960.
13 J. Mecheels, Seifen Öle Fette Wachse **108** (1982) 31–34.
14 Procter&Gamble, US 4 412 934, 1983.
15 H. Upadek, B. Kottwitz: "Application of Amylases in Detergents" in J.H. Ee, O. Misset, E.J. Baas (eds.): Enzymes in Detergency, **Vol. 69**, Surfactant Science Series, 1997, pp. 203–212.
16 H. Andree, W.R. Müller, R.D. Schmid, J. Appl. Biochem. **2** (1980) 218.
17 B. Kottwitz, H. Upadek, Seifen Öle Fette Wachse **120** (1994) 794–799.
18 J. Stawitz, P. Höpfner, Seifen Öle Fette Wachse **86** (1960) 51–52.
19 P.G. Evans, W.P. Evans, J. Appl. Chem. **17** (1967) 276–282.
20 BASF, DE 806 366, 1948.
21 Colgate Palmolive, CH 300 865, 1950.
22 Procter&Gamble, CH 314 599, 1951.
23 Henkel, DE-AS 1 135 606, 1961
24 H.Y. Lew, J. Am. Oil Chem. Soc. **41** (1964) 297–300.
25 E. Schmadel, Fette Seifen Anstrichm. **70** (1968) 491.
26 Henkel, DE 1 257 338, 1968.
27 J. Perner, G. Frey, K. Stork, Tenside Deterg. **14** (1977) 180–185.
28 H. Distler, D. Stoeckigt, Tenside Deterg. **12** (1975) 263–265.
29 Procter&Gamble, DE 1 056 316, 1966.
30 Procter&Gamble, DE-AS 1 080 250, 1960.
31 Procter&Gamble, DE-OS 2 338 464, 1974.
32 Dow Corning, DE-OS 2 402 955, 1974.
33 G, Rossmy, Fette Seifen Anstrichm. **71** (1969) 56.
34 H. Bloching, Chem. Ztg. **99** (1975) 194–201.
35 V. DiGiacomo, Soap. Chem. Spec. **1** (1967) Jan., 79–82, 119.
36 E. Obrocki, Fette Seifen Anstrichm. **73** (1971) 327–329.
37 W. Sturm, G. Mansfeld, Tenside Deterg, **15** (1978) 181–186.
38 W. von Rybinski in Handbook of Applied Surface and Colloid Chemistry, ed. K. Holmberg, Wiley 2001, Chap. 3
39 Berth, P. and M.J. Schwuger, Tenside Det. **16**, 3–12 (1979)
40 Shafrin, EG. and W.A. Zisman, J. Phys. Chem. **64**, 519 (1960)
41 Schwuger, M.J., Ber. Bunsenges. Phys. Chem. **83**, 1193 (1979)
42 Dobias, B., X. Qiu and W. von Rybinski, Solid-Liquid Dispersions, Marcel Dekker, New York, 1999
43 Nickel, D., H.D. Speckmann and W. von Rybinski, Tenside Surfactants Det. **32**, 470 (1995)
44 Schambil, F. and M.J. Schwuger, Colloid Polymer Sci. **265**, 1009 (1987)
45 Kahlweit, M., Tenside Surfactants Det. **30**, 83 (1993)

46 Benson, H.L., K.R. Cox and J.E. Zweig, Happi, 50 (1985) and Kahlweit, M. and R. Strey in Proceedings Vth Int. Conf. Surface Colloid Sci., Potsdam New York, 1985
47 Engels, Th., W. von Rybinski and P. Schmiedel, Progr. Colloid Polym. Sci. **111**, 117 (1998)
48 A.J. Rosenthal, S.N. Thorne, 'Surface-tension as a controlled variable in mechanical dishwashing' JAOCS 1986, **63**, 931–934.
49 T. Müller-Kirschbaum, E.J. Smulders, 'Das On-line Tensiometer', SÖFW-Journal 1992, **118**, 427–434.
50 K. Tauer, C. Dessy, S. Corkery, K.-D. Bures, 'On-line surface tension measurements inside stirred reactors', Colloid Polymer Sci. 1999, **277**, 805–811.
51 L. Schulze, K. Lohmann, 'Diagnosing the condition of washing and rinsing liquids with a new surface measurement technique', Tenside Surf. Det. 1999, **36** 384–386.
52 A. Boettger, J. Krolop, R. Muenzner, L. Schulze, German patent application DE 19529787 A1, 1997.
53 B. Villis, 'The dissolving behavior of surfactants in household washing machines', Tenside Surf. Det. 2000, **37**, 52–55.
54 A. Volanschi, P. Bergveld, W. Olthuis, R. van de Heijden, International patent application WO 9618877, 1996.
55 H. Krüssmann, J. Bohnen, G. Rohm, ‚Zur Analytik freier Härteionen in Gegenwart von Waschflotteninhaltsstoffen', SÖFW-Journal 1995, **121**, 173–176.
56 R.A. Llenado, 'Potentiometric response of the calcium selective membrane electrode in the presence of surfactants', Anal. Chem. 1975, **47**, 2243–2249.
57 A.J. Frend, G.J. Moody, J.D.R. Thomas, B.J. Birch, 'Studies of calcium ion-selective electrodes in the presence of anionic surfactants', Analyst 1983, **108**, 1072–1081.
58 A. Boettger, J. Krolop, R. Muenzner, L. Schulze, German patent application DE 19541719 A1, 1997.
59 I. Schulze, C. Engel, G. Czyzewski, German patent application DE 19846248 A1, 2000.
60 H. Krüssmann, J. Bohnen, 'On-line-Analytik für Peressigsäure', Tenside Surf. Det. 1994, **31**, 229–232.
61 L. Bütfering, A. Werner-Busse, F. Siepmann, German patent application DE 4412576 A1 (1995)
62 J. Rieker, T. Guschlbauer, ‚Wasch- und Spülprozesse: Automatische Steuerung mittels neuer Sensoren', Textilveredlung 1999, **34**, 4–12.
63 W. Göpel and K.-D. Schierbaum in Sensors – A Comprehensive Survey: Vol 2. Chemical and Biochemical Sensors (Eds. W. Göpel, J. Hesse, J.N. Zemel) VCH, Weinheim, 1991, p. 1–27.
64 J. Janata, M. Josowicz, P. Vanysek, D.M. DeVaney, Anal. Chem. 1998, 70, 179R–208R.
65 A.J. Ricco, R.M. Crooks, Acc. Chem. Res. 1998, **31**, 200-324.
66 J.W. Gardner, Microsensors: principles and applications, John Wiley & Sons Ltd., Chichester, 1994.
67 U.E. Spichiger-Keller, Chemical sensors and biosensors for medical and biological applications, Wiley-VCH, Weinheim, 1998.
68 S. Sato, Japanese patent application JP5096081, 1993.
69 R. Rieger, C. Wolf, A. Rizzo, German patent application DE19949801 (2001).
70 I. Schulze, C. Engel, G. Czyzewski, German patent application DE19846248 (2000).
71 Y. Nishioka, M. Tokunaga, S. Maruyama, Japanese patent application JP2000170663 (2000).
72 K. Zucholl, European patent application EP900765 (1998).
73 R. Anderer, M. Mayr-Willius, German patent application DE19808839 (1999).
74 R.P. Buck, Electrochemistry of Ion-Selectrive Electrodes. In: Comprehensive Treatise of Electrochemistry, Vol. 8 (Eds. R.E.W. White, J.O'M. Bockris, B.E. Conway, E. Yeager), Plenum Press, New York (1984).
75 R. Holze, Electroanalysis 1997, **9**, 265–266.

76 T. Kawano, Japanese patent application JP11128590 (1999).

77 K. Nishiyama, K. Ihsii, Y. Ikuno, K. Aono, Japanese patent application JP11123322 (1999).

78 T. Imai, Japanese patent application JP6086894 (1994).

79 Y. Yamaguchi, Japanese patent application JP4367695 (1992).

80 Y. Takeda, K. Miwano, N. Fujita, European patent application EP0506137 (1992).

81 D. Knittel, E.W. Schollmeyer, SÖFW-Journal 1998, **124**, 338–342.

82 S. Alegret, J. Alonso, J. Bartroli, J. Baro-Roma, J. Sanchez, M. de Valle, Analyst 1994, **119**, 2319–22.

83 T. Masadome, T. Imato, N. Ishibashi, Anal. Sci. 1990, **6**, 605–606.

84 I. Willner, E. Katz, Angew. Chem. 2000, **112**, 1230–1269.

85 M. Lazzaroni, E. Pezzotta, G. Menduni, D. Bocchiola, D. Ward, conference records of the 17th IEEE Instrumentation and Measurement Technology Conference 'Smart Connectivity: Integrating Measurement and Control, 2000, 478–482.

86 C. Beier, U. Hein, R. Herden, German patent application DE 10005991 A1

87 G. Biechele, S. Hillenbrand, K. Roth, H. Schrott, German patent application DE10034546 (2001).

88 H. Poisel, J. Weber, German patent application DE19904280 (2000).

89 Henkel KGaA "In View of Tomorrow" (2000), p. 36

90 F. Schambil, M. Böcker, G. Blasey, Chimica Oggi **18**(3/4) (2000), 34–36

91 F. Schambil, M. Böcker, Tenside Surf. Det. 36 (2000) 1, 48–50

92 M. Böcker, A. Machin, F. Schambil, Comun. Jorn. Com. Esp. Deterg. **30** (2000), 13–19

93 G. Wäschenbach, R. Wiedemann, E. Carbonell, E. Endlein, K.-L. Gibis, German patent application DE 198 34 172 A1 (2000)

94 G. Wäschenbach, R. Wiedemann, E. Carbonell, L. Hertling, N. Wolf, German patent application DE 198 34 180 A1 (2000)

5
Sensor Related Topics

5.1
Thin Film Temperature Sensors for the Intelligent Kitchen

M. Muziol

5.1.1
Introduction

In recent generations of household appliances, there has been a noticeable increase in the use of electronic circuits. The use of such control circuits has enhanced the functionality of appliances, helped economize on resources, and brought about effective and lasting savings in household costs. Moreover, these intelligent systems are helping to improve further the convenience and efficiency of domestic work. Exact temperature measurements, combined with carefully designed electronic controllers, extend the lifetime of both the appliance and its components. The use of this type of technology is spreading from traditional technological and engineering sectors to a growing number of areas in everyday life. To be able to measure temperatures precisely, the electronic control circuit requires a signal from a suitably accurate temperature sensor. The versatility of platinum temperature sensors based on thin-film technology suits ideally the wide variety of domestic applications.

5.1.2
Platinum Temperature Sensors

Today's platinum temperature sensors comprise structured and encapsulated platinum thin-film elements. These are created under clean room conditions using techniques from the semiconductor industry such as sputtering, vapour deposition, photolithographic structuring and laser trimming. At this stage, the reproducibility of the method is tested with a view to subsequent large-scale production. The platinum metal structure is deposited on a aluminium oxide ceramic substrate and then covered with a thin layer of glass. The bond pads for creating electrical connections to the meander shaped platinum structure are also sealed using a special glass-ceramic paste (Fig. 5.1).

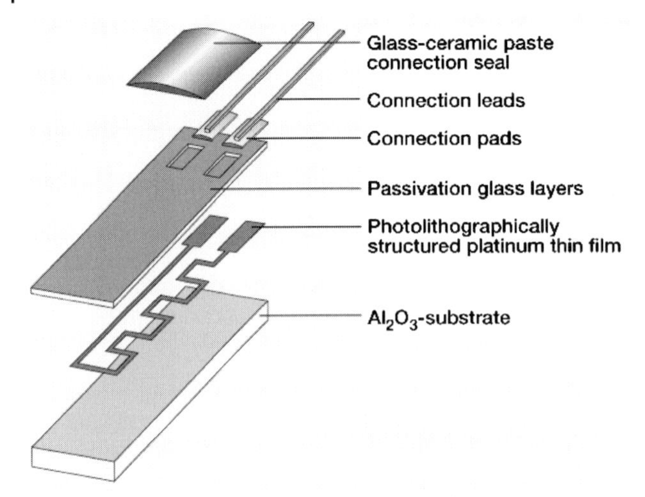

Fig. 5.1 Schematic diagram of the structure of a platinum temperature sensor.

5.1.2.1 The Variety of Forms for Platinum Temperature Sensors

Platinum thin-film sensors are available in a broad variety of forms. Depending on the proposed application, the sensor may be unhoused, encapsulated in ceramic, configured as a surface mounted device (SMD) or located within a plastic housing with rigid pins. An overview of the variety of different forms currently available is given below (Fig. 5.2).

5.1.2.2 Temperature Dependent Changes in Electrical Resistance

Platinum sensors exploit the fact the electrical resistance of platimum metal is temperature dependent. This behaviour can be described by a polynomial of the form

$$T_t = R_0(1 + at + bt^2) \,. \tag{1}$$

Here R_0 is the resistance at 0 °C, and a and b are coefficients whose values are specified in internationally agreed standards covering platinum temperature sensors, for example DIN EN 60751. The coefficient b is so small that for most applications one can assume a linear relation between R_t and the temperature t.

For each type of sensor, a classification framework sets precision tolerance ratings. Platinum temperature sensors have a positive temperature coefficient which is defined as:

$$\mathrm{TC} = (R_{100} - R_0)/(100 \cdot R_0) \tag{2}$$

The temperature coefficient TC is the gradient of the linear approximation of the polynomial I between 0 °C and 100 °C. In the DIN EN 60751 standard, the

Fig. 5.2 Platinum temperature sensors in a range of different shapes.

temperature coefficient for platinum temperature sensors is specified as 0.003850/°C.

5.1.2.3 Advantages of Platinum Temperature Sensors

An economic large-scale production of platinum temperature sensors is now possible due to the application of modern manufacturing techniques based on thin-film technology. This development has led to price levels which were previously the exclusive domain of NTC elements.

By modifying individual process stages, a broad range of variants differing in structure, resistance values and geometry can be realized. Apart from classical temperature sensing applications, platinum combination sensors also permit other physical quantities to be monitored. Such combination sensors are used, for example, as air mass meters in car engines. Furthermore, the high level of automation in the production process permits extremely precise miniature sensors to be manufactured which are capable of meeting the exacting demands required by physical, biochemical and medical applications. By fine-tuning the sensors properties to meet these requirements, even mass markets can be supplied with these advanced technological sensor devices.

5.1.3
High Tech in the Home

Given this situation, it is not surprising that, over the last few years, products based on platinum thin film technology have been finding their way into the home. With the growing use of electronic control systems in the new generation of domestic appliances, platinum temperature sensors have been more widely used in ovens where they have replaced electromechanical regulators such as capillary tubes, solid expansion thermometers and NTC thermistors. Typical sensor applications in the food preparation sector are shown in Fig. 5.3.

Even conventional hotplates can be fitted with a continuously variable temperature control thanks to platinum thin-film sensors. The encapsulated platinum sensor is positioned so that it is in contact with the cast iron plate from below and thus able to register the temperature of the hotplate. Normally the power supplied to the plate is regulated by a stage switch. The increased control sensitivity is made possible by incorporating electronic circuitry capable of interpreting and acting upon the sensor signals.

Similar technology is also being used in induction hotplates and radiant heating elements and in the silicon nitride hotplates currently under development and which are characterized by an excellent heat transfer between heat source, sensor and saucepan. Gas hotplates could also be provided with the latest in domestic temperature sensing technology by fitting them with electric temperature control circuitry.

Intelligent process control based on platinum temperature sensors and tailored electronics provides an effective method of improving both kitchen safety and the user friendliness of kitchen appliances, and also contributes to reducing energy consumption.

A major growth area for platinum sensor technology is to be found, above all, in ovens. The sensors are used to fulfil two main functions:

The sensors measure the temperature in the oven during the baking process. The precise temperature measurement achievable contributes effectively to reducing hunting oscillations. The electronic control systems now in use have also made it possible to store different heating profiles which the oven can use to react intelligently to the different types of food being baked. Thanks to this type of technology there is no longer a need for cooking by trial and error or having to adapt the baking process to the quirks and peculiarities of a particular oven.

The use of catalytic converters in modern domestic ovens means that odours in the kitchen arising from baking and roasting are now a thing of the past. The catalytic converters are fitted into the oven-venting shaft and convert unpleasant smelling substances into CO, CO_2 or other non-odorous gases. To provide the best possible control of the catalytic odour eliminator, platinum temperature sensors are employed to ensure that the catalyst is working in its optimum operating range, typically 500–550 °C. An analogous function is fulfilled by sensors that are used to control the pyrolytic cleaning process. This involves heating the entire oven to a temperature of around 550 °C, a temperature high enough to make oven

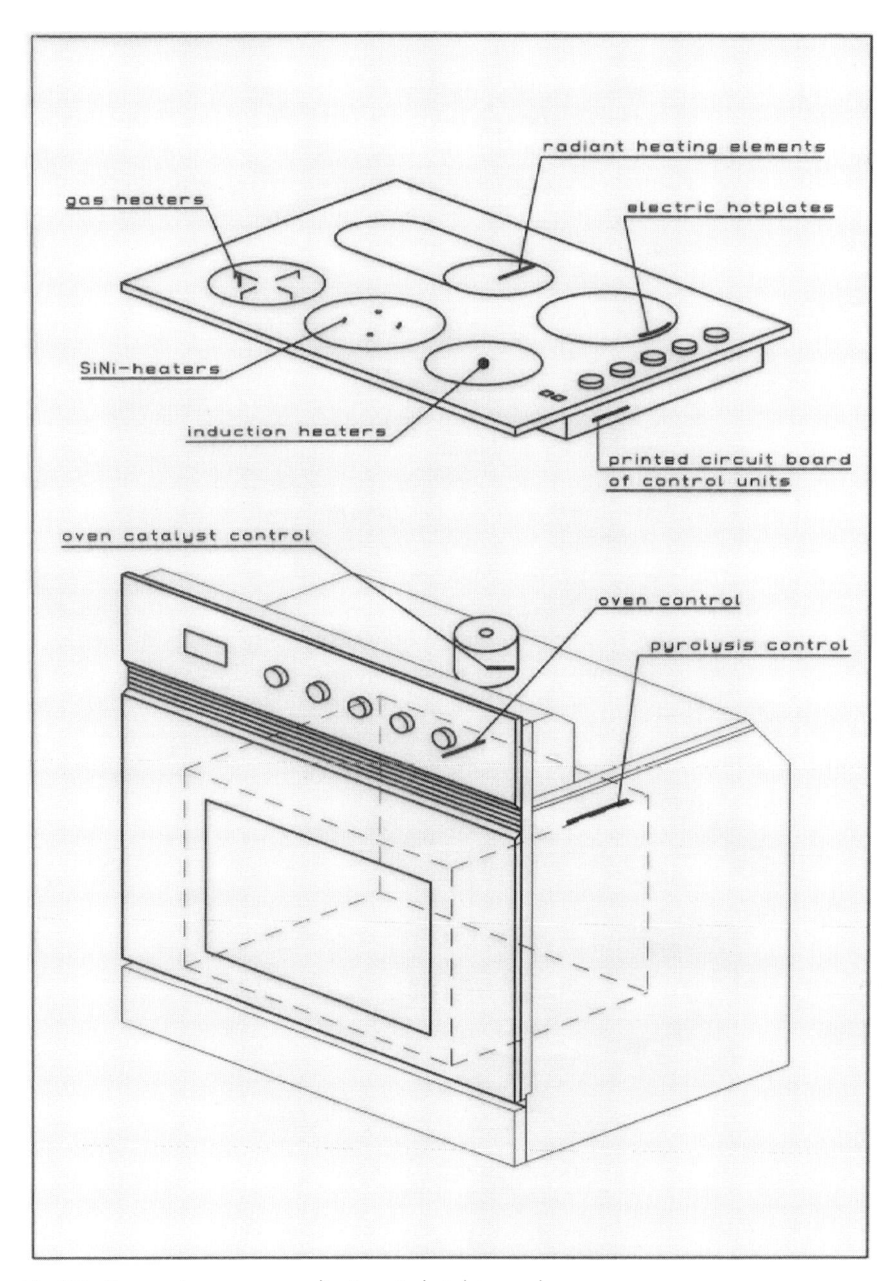

Fig. 5.3 Temperature sensor applications in hotplates and ovens.

cleaning with the usual chemical cleaners superfluous. The high temperature pyrolyzes any organic compounds present and the ash residues can then be easily removed from the oven with a brush. Platinum high temperature sensors provide a reliable means of monitoring and regulating the temperature and thus safeguard the safe and complete pyrolysis of food deposits in the oven. Economic considerations also make platinum temperature sensors attractive to users, providing affordable systems to measure high temperatures. In contrast to conventional thermoelements, platinum sensors do not require complicated analysis electronics and are insensitive to electromagnetic field influences.

Examples of Applications

Fig. 5.4 presents a number of platinum sensors currently being used in kitchen appliances. The probe with the flange and the long stainless steel housing was developed for use in ovens. Also available are models equipped with high-voltage insulation which can be built in if demanded by the control electronics. The shorter, tapered sensor is a customized high temperature probe which was built to order for a client. On the right of Fig. 5.4 is a sensor designed for measuring surface temperatures. The sensor, which is integrated into a plastic housing, is easily mounted via the borehole. Apart from these special probe geometries, there is also a universal sensor in a ceramic housing available. Situated in the centre of the picture is an SMD component on a long ceramic plate that is used for temperature measurements up to 750 °C. In addition to the designs shown, Heraeus Sensor-Nite is able to develop and manufacture sensor geometries and housings that are specially tailored to meet specific customer requirements.

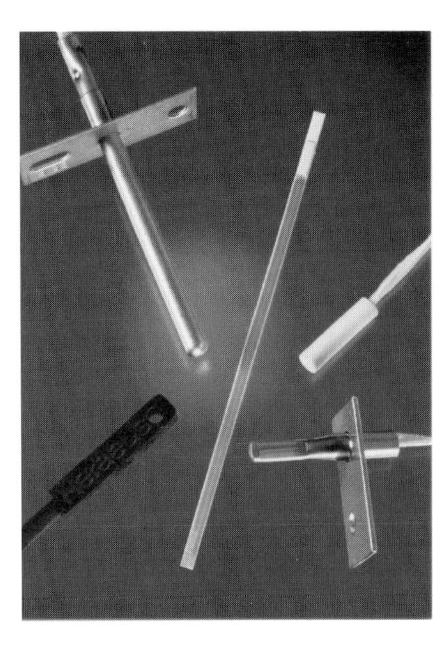

Fig. 5.4 A range of differently shaped sensors for use in domestic appliances.

5.1.4
Future Prospects/Outlook

In principle, platinum sensors can be used for applications in the temperature range –200 °C to +1000 °C. In the kitchen, temperatures typically ranging between 200 and 600 °C are measured. The ability to accurately measure lower temperatures, opens the possibility of using platinum sensors in fridges, freezers, washing machines and tumble dryers. Given these prospects, it can be said that the end of the road leading to intelligent kitchen and household appliances has not yet been reached.

Supported by the knowledge and experience at Heraeus, the precious metals and technology company, which is one of Germany‘s fifty largest companies, Heraeus Sensor-Nite continues to be very active in the field of sensor technology. For many years Heraeus Sensor-Nite has been successfully involved in the development, application and serial production of platinum temperature sensors. In all its activities, Heraeus Sensor-Nite combines the experience of the Heraeus company with today‘s cutting edge technology. This fusion of traditional know-how with modern methods of production and analysis has made Heraeus Sensor-Nite one of the leading high technology companies serving the environment:

In addition to developing products for the domestic appliance market, Heraeus Sensor-Nite is also active in the automotive sector, and in the fields of calorimetry, electronics and medicine. The goal that Sensor-Nite has set itself for both its present and future activities is to increase process safety whilst reducing depletion of resources.

5.2
Reed Switches as Sensors for Household Applications

U. Meier

5.2.1
Market Aspects

Reed sensors are used in a wide range of applications in household appliances. Increasing security regulations and comfort requirements generate continually new applications. In the past these have been particularly floating-level indicators, door detectors and position sensors. In addition to new applications, existing mechanical micro-switches are also replaced by reed sensors. By contrast reed relays are very rarely used in household appliances. Mostly the required switching load exceeds the maximum switching current of reed relays. Exceptions are reed relays for high-voltage-applications or for high-isolation-voltage over open contact.

5.2.1.1 Further Technical Development

The trend towards miniaturization continues. The original reed switch is getting continually smaller and therefore can be integrated at minimal space. Its magnetic sensitivity has been improved and stabilized. New materials allow the use of even smaller magnets without reducing the distance between reed switch and magnet.

5.2.1.2 Development of Reed Sensors in Household Applications

Market tendencies can be distinguished according to continents:

Europe has been the biggest market up to now. Future applications will be in environmental protection and security; e.g. water-meters and door detection devices. We expect an increase of appr. 15% p.a. The US market shows a delayed response. It might reach the present-day European level of appr. 25 million reed sensors p.a.in 10 years' time. Asia the market is growing at an above-average rate. We expect annual increases of 20%.

5.2.2 Reed Switch Characteristics

5.2.2.1 Introduction

The reed switch was first invented by Bell Labs in the late 1930s. Most of the manufacturers of reed switches today produce very high quality and very reliable switches. This has given rise to unprecedented growth.

As a technology, the reed switch is unique. Being hermetically sealed, it can exist or be used in almost any environment. Very simple in its structure, it is a hybrid of many manufacturing technologies. Decisive for its quality and reliability is its glass to metal hermetic seal, in which the glass and metal used must have exact linear thermal coefficients of expansion. Otherwise, cracking and poor seals will result. Whether sputtered or plated, the process of applying the contact material, usually rhodium or ruthenium, must be carried out with precision in ultra-clean environments similar to semiconductor technology. Like semiconductors, any foreign particles present in the manufacture will cause losses, quality and reliability problems. Over the years, the reed switch has shrunk in size from approximately 50 mm (2 inches) to 6 mm (0.24 inches). These smaller sizes have opened up many more applications, particularly in RF and fast-time domain requirements.

5.2.2.2 Reed Switch Features

1. Ability to switch up to 10,000 Volts
2. Ability to switch currents up to 5 Amps
3. Ability to switch or carry as low as 10 nanoVolts without signal loss
4. Ability to switch or carry as low as 1 femtoAmp without signal loss
5. Ability to switch or carry up to 6 GigaHertz with minimal signal loss
6. Isolation across the contacts up to 10^{15} Ohm

7. Contact resistance (on resistance) typical 50 MilliOhms
8. In its off state it requires no power or circuitry
9. Ability to offer a latching feature
10. Operating time in the 100 Microsecond to 300 Microsecond range
11. Ability to operate over extreme temperature ranges from –55 °C to 200 °C
12. Ability to operate in all types of environments including air, water, vacuum, oil, fuels, and dust laden atmospheres
13. Ability to withstand shocks up to 200 Gs
14. Ability to withstand vibration environments of 50 Hz to 2000 Hz at up to 30 Gs
15. Long life. With no wearing parts, load switching under 5 Volts at 10 mA, will operate well into the billions of operations

5.2.2.3 The Basic Reed Switch

A reed switch consists of two ferromagnetic blades (generally composed of iron and nickel) hermetically sealed in a glass capsule (Fig. 5.5). The blades overlap inside the glass capsule with a gap between them, only making contact with each other in the presence of a suitable magnetic field. The contact area on both blades is plated or sputtered with a very hard metal, usually rhodium or ruthenium. These very hard metals make very long lifespan possible if the contacts are not switched with heavy loads. The gas in the capsule usually consists of nitrogen or some equivalent inert gas. In order to increase their ability to switch and standoff high voltages, some reed switches have an internal vacuum. The reed blades act as magnetic flux conductors when exposed to an external magnetic field from either a permanent magnet or an electromagnetic coil. Poles of opposite polarity are created, and the contacts close when the magnetic force exceeds the spring force of the reed blades. As the external magnetic field is reduced so that the force between the reeds is less than the restoring force of the reed blades, the contacts open. The reed switch described above is a 1 Form A (normally open (N.O.) or Single Pole Single Throw (SPST) reed switch.

When a permanent magnet is brought into the proximity of a reed switch the individual reeds become magnetized with the polarity of the attracting magnet, as shown (Fig. 5.6). When the external magnetic field becomes strong enough the

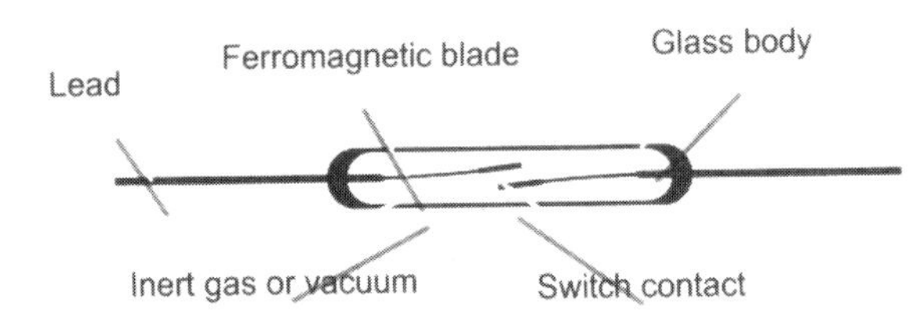

Fig. 5.5 The basic hermetically sealed form 1 A (normally open) reed switch and its component makeup.

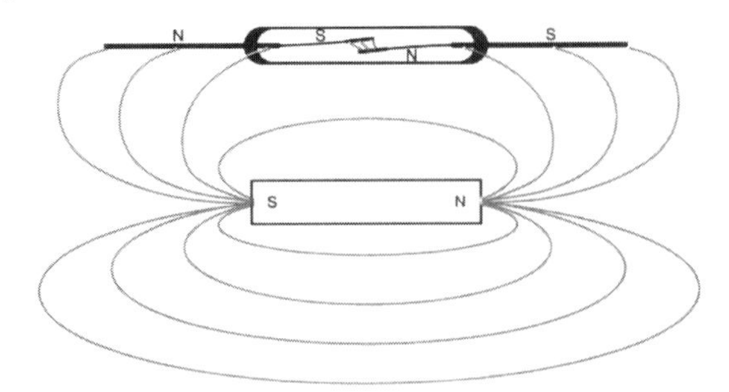

Fig. 5.6 The basic operation of a reed switch under the influence of the magnetic field of a permanent magnet. The polarization of the reed blades occurs in such a manner to offer an attractive force at the reed contacts.

magnetic force of attraction closes the blades. The reed blades are annealed and processed to remove any magnetic retentivity. When the magnetic field is withdrawn the magnetic field on the reed blades also dissipates. If any residual magnetism existed on the reed blades, the reed switch characteristics would be altered. Proper processing and proper annealing clearly are important steps in the manufacturing process.

5.2.3 Magnetic Response in a Variety of Arrangements

The switching behavior depends on the way the magnet is set up:

5.2.3.1 Direct Operation

A perpendicular approach with the magnet parallel to the reed switch gives one operation at max. operation distance (Fig. 5.7). This method of operation offers medium sensitivity to stray magnetic fields, large switching courses and hysteresis and high switching position tolerance.

The magnet moving parallel to the reed switch allows a maximum of three switching operations while traveling the length of the switch (Fig. 5.8). One operation is achieved by a small magnet movement. This arrangement is extremely sensitive to stray magnetic fields and needs very accurate adjustment if a single operation is required.

A parallel motion with the magnet perpendicular to the reed switch allows two switching operations (Fig. 5.9). The distance over which the switch opens is very

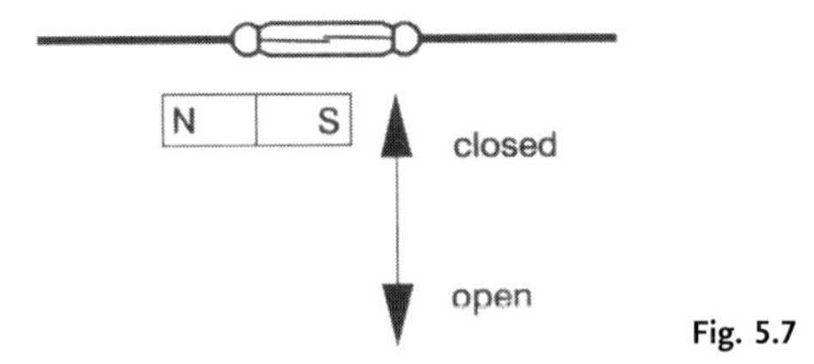

Fig. 5.7

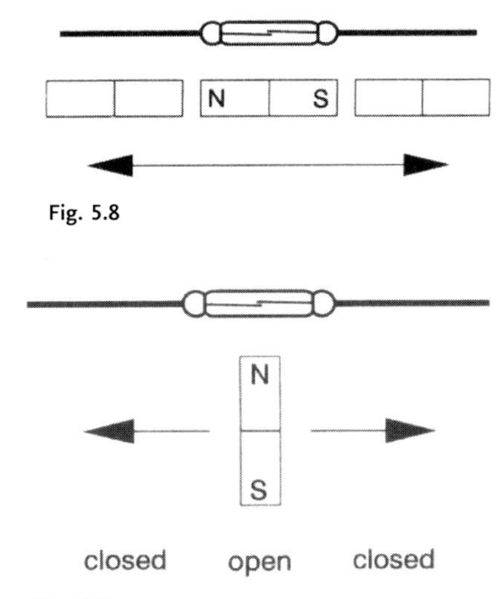

Fig. 5.8

Fig. 5.9

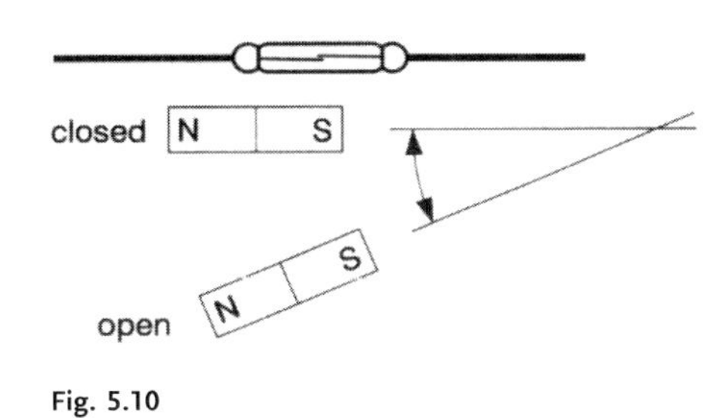

Fig. 5.10

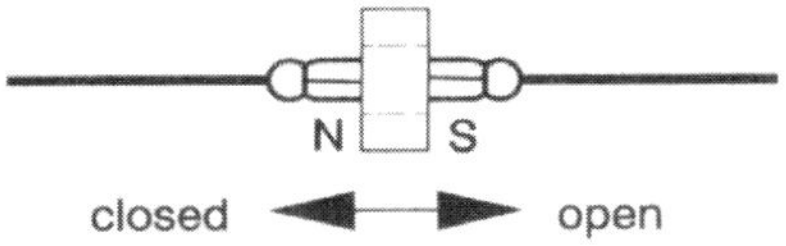

Fig. 5.11

short, thus requiring only small movement to operate the switch. There is very low sensitivity to stray magnetic fields when a strong magnet is used. No adjustment is necessary but requirements on mechanical tolerances are higher.

An angular motion which requires a large angle of movement to operate the switch (Fig. 5.10). This offers a medium sensitivity to stray magnetic fields and shorter switching distances than method A.

Using a ring magnet with an axial motion, three closures can be obtained (Fig. 5.11). This method offers short switching distances and low sensitivity to stray magnetic fields. If a single operation is required, a very accurate adjustment is necessary.

5.2.3.2 Indirect Operation

With both the magnet and the switch fixed, the switching operation is achieved by moving ferromagnetic material between the switch and the magnet (Fig. 5.12). Strong magnets are required.

This method operates the switch with a rotating magnet. The magnet can either be parallel or perpendicular to the switch (Fig. 5.13). These methods offer very low sensitivity to stray magnetic fields and easy adjustment if a precise switching is not crucial to the operation.

5.2.4 Handling of Reed Switches in Various Sensor Applications

5.2.4.1 Cutting and Bending of a Reed Switch

Many users of reed switches for sensor applications try to make their own sensors. Often, however, they do not observe some basic precautions and preventive measures to ensure reliable operation of the switch. Below we try to cover the key areas that users and manufacturers must be aware of.

Modifications in reed switch can severely damage them if not properly carried out. Primarily, this is because the reed lead is large by comparison to the glass seal. A balance has to be achieved between reed switch sensitivity and mechanical strength. If the lead of the reed switch were much smaller than the glass, seal stress and glass breakage would not be an issue. However, to achieve the sensitivity and power requirements in the reed switch, a larger lead blade is necessary.

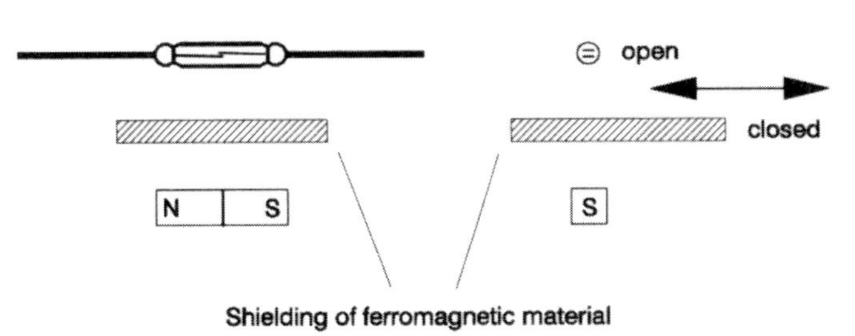

Fig. 5.12

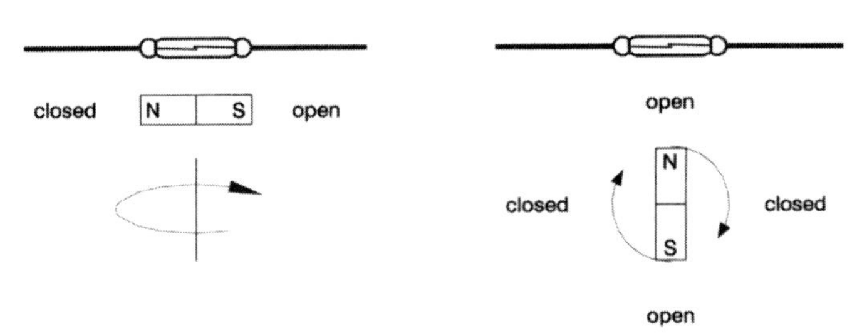

Fig. 5.13

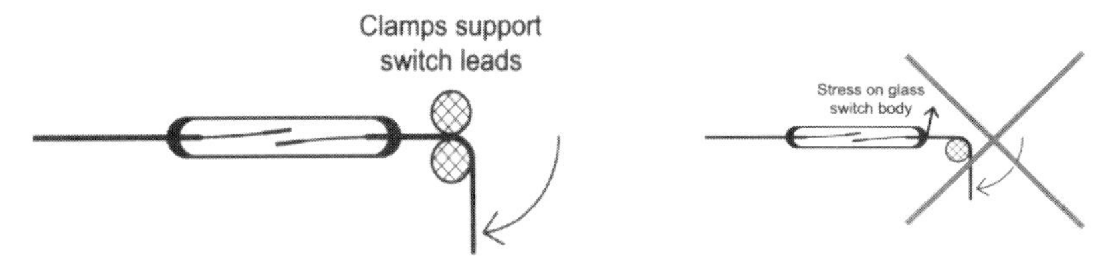

Fig. 5.14 Presentation of the proper and improper way of bending a reed switch. (Supporting the switch lead while bending is a must.)

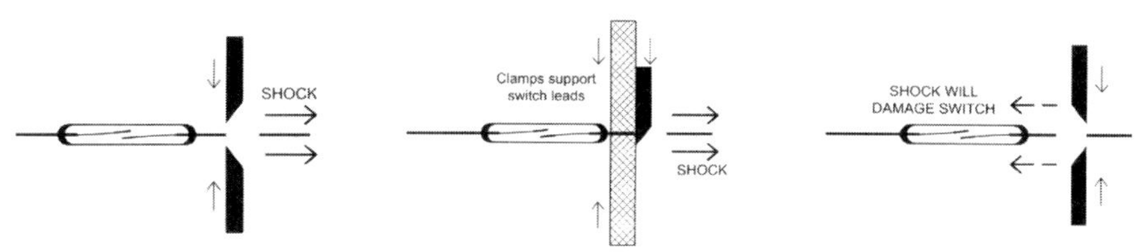

Fig. 5.15 Properly supported the switch lead while cutting is required, otherwise damage can occur to the reed switch.

With that in mind, it cannot be emphasized enough, any forming or cutting of the reed switch leads must be done with extreme caution. Any cracking or chipping of the glass indicates damage. Internal damage can occur with no visible signs on the seal. In these instances, seal stress has occurred, leaving a torsional, lateral, or translational stress in the seal. This produces a net force on the contact area that can affect the operational characteristics (pull-in and drop-out), contact resistance and durability.

Most reed switch suppliers can perform value-added cutting and shaping of the leads in a stress-free environment, using proper tools and equipment. This is often the most economical approach for users, although it may not seem so at the time.

Often, users will choose to make their own modifications, and only when experiencing manufacturing and quality problems with the product will they go back to the reed switch manufacturer and let him carry out the desired modifications. In Figs 5.14 and 5.15, the proper approach for cutting and/or bending the reed switch is shown. The effect on the pull-in and drop-out characteristics of cutting and bending the reed switch will be explained in more detail later.

5.2.4.2 **Soldering and Welding**

Often, a reed switch has to be soldered or welded. Reed switches are usually plated with suitable solderable plating. Welding is also quite easy on the nickel/iron leads of the reed switch. However, both, if not carried out properly, can cause stress, cracking, chipping or a rupture in the reed switch. When soldering or welding, it is best to keep as far away from the glass seal as possible. This may,

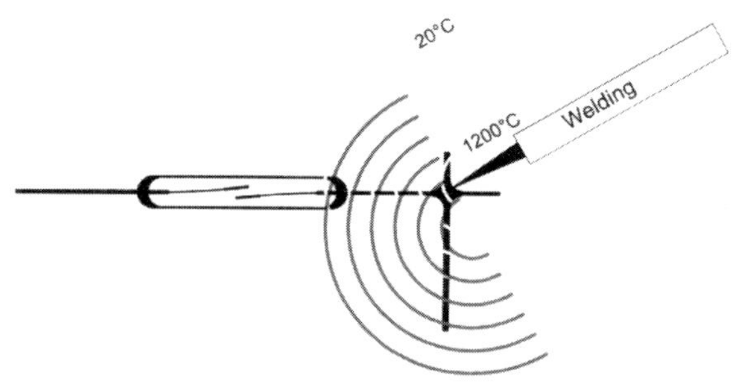

Fig. 5.16 Soldering and welding can generate a heat front to the glass to metal seal of the reed switch causing potential damage.

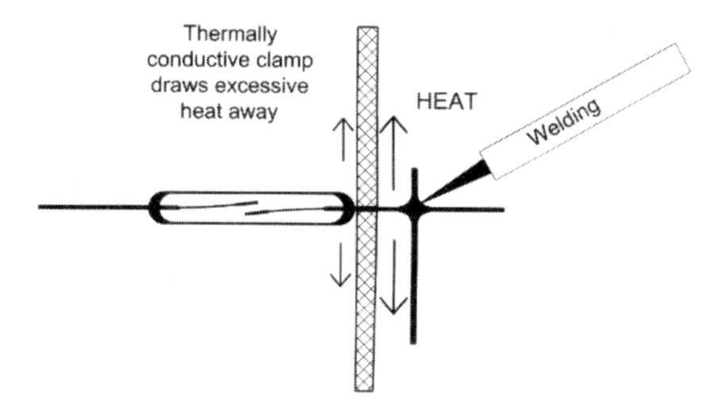

Fig. 5.17 Use of heat sinking or preheating reed switches for soldering or welding can prevent heat stress damage.

however, not always be possible. Welding in particular can be dangerous if it has to be done very close to the seal. Temperatures of up to 1000 °C can reach the seal. While the heat is arriving first at one end of the seal, the temperature at other end of the seal may well be just 20 °C. Thus, a dramatic thermal gradient stretches across the seal and can disrupt it in many ways that can all result in a faulty reed switch operation (see Fig. 5.16). In a similar manner, soldering close to the seal can have the same effect, but to a lesser extent because of the lower solder temperatures involved (200 °C to 300 °C). Two ways to increase the likelihood of success are by heat-sinking the lead of the reed switch (see Fig. 5.17) or by preheating the reed switch and/or assembly.

Most commercial wave soldering machines have a preheating section before the PCB or assembly is immersed into the solder wave. the thermal shock is reduced As the ambient temperature has already been raised before the solder wave, the thermal gradient in the reed switch seal is reduced.

5.2.4.3 Printed Circuit Board (PCB) Mounting

Reed products mounted on PCBs can sometimes be a problem. If the PCBs have a flex attached to them after wave soldering, it may be necessary to remove it when mounting the board in a fixed position. When the flex is removed, the hole distance, where a reed switch for instance may be mounted, can change by a small amount. If no allowance is made for this small movement when the PCB is mounted, the reed switch seal will end up absorbing the movement, which leads to seal stress, glass chipping or cracking. Care should be taken in this area, particularly when very thin PCBs are used and flexing or board distortion is common.

5.2.4.4 Using Ultrasonics

Another approach to making a connection to a reed switch is ultrasonic welding. Reed switch sensors and reed relays may also be sealed in plastic housings by ultrasonic welding. In addition, cleaning stations use ultrasonics. In all these areas the reed switch can be damaged by the ultrasonic frequency. Ultrasonic frequencies range from 10 kHz to 250 kHz, and in some cases even higher. Not only the resonant frequency of the reed switch and its harmonics, but also the resonant frequency of the assembly in which the reed switch resides has to be taken into account. At certain frequencies and under certain conditions, the contacts can be severely damaged. If using ultrasonics in any of the above conditions, be very cautious and perform exhaustive testing to make sure there is no interaction or reaction with the reed switch.

5.2.4.5 Dropping Reed Switch Products

Dropping the reed switch or a reed sensor on a hard object, typically on the floor of a manufacturing facility, can induce a damaging shock to the reed switch. Shocks above 200 Gs should be avoided at all cost (see Fig. 5.18). Dropping any of the above on a hard floor from 20 cm or more (more than one foot) can and will

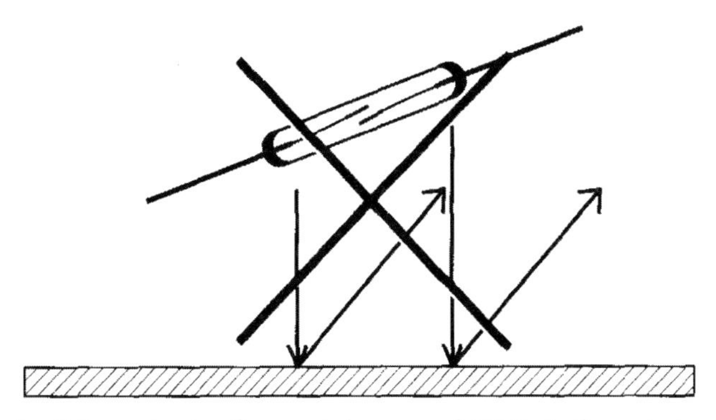

Fig. 5.18 Dropping the reed switch on a hard surface can induce several 100 Gs to the contacts many times altering the switch characteristics.

often destroy a reed switch where G forces greater than 1000 Gs are not uncommon. Not only can the glass seal crack under these circumstances, but the reed blades may be dramatically altered. The gaps may have been drastically increased or the gaps may be closed, due to these high G forces, Simple precautions, such as placing rubber mats at assembly stations can eliminate these problems. Also, operators should be instructed that if a reed product has been dropped it cannot be used until has been re-tested.

5.2.4.6 Encapsulating Reed Switch Products

Further damage can occur to a reed switch during attempts to package the reed switch by sealing, potting, or encapsulating. Whether this is done by a one or two part epoxy, thermoplastic encapsulation, thermoset encapsulation, or other approaches, damage to the glass seal can occur. Without any buffer, the encapsulants crack, chip or stress the glass seal. Using a buffer compound between the reed switch and the encapsulant that absorbs any induced stress is a one way to eliminate this problem. Another approach would be to match the linear coefficient of thermal expansion with that of the reed switch, thereby reducing stress as the temperature fluctuates. However, keep in mind, this approach does not take into consideration the shrinkage that occurs in most epoxies and encapsulants during the curing stage. Sometimes a combination of both approaches may be the best way to seal a product with a reed switch.

5.2.4.7 Temperature Effects and Mechanical Shock

Temperature cycling and temperature shock that occur naturally in a reed switch application must be taken into consideration. Again, temperature changes creating movement with various materials due to their linear coefficients of thermal expansion will induce stress to the reed switch if not properly dealt with.

5.2.5 Applications

5.2.5.1 Measuring the Quantity of a Liquid, Gas and Electricity

Water or fluid flow can be easily measured by mounting a propeller just outside the water pipe and connecting it underneath the plastic casing of the meter. The water flows trough the pipe and spins the propeller. A magnet is mounted on the propeller and a reed sensor is mounted outside of the plastic casing. In this case, each propeller rotation is counted as the magnet rotates past the reed sensor. The rotations are taillied and electronic circuitry converts the rotations to volumes of water (or other liquid) flowing through the pipes. In a similar manner, the flow of gas and electricity can also be measured.

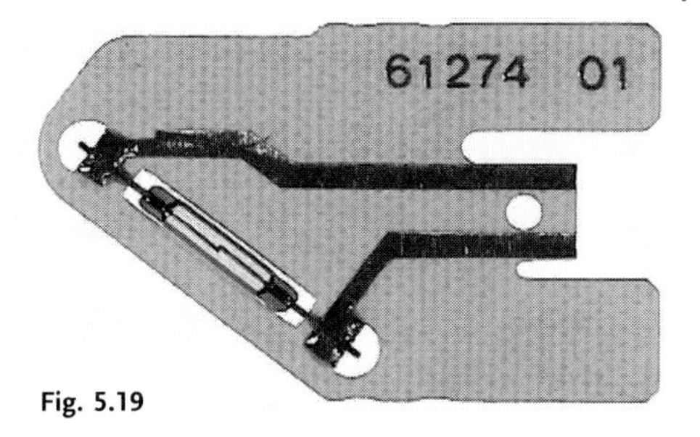

Fig. 5.19

5.2.5.2 Detecting Water Flow

In this application, the sensor recognizes the movement of water. The reed switch, in going from an open to a closed state, produces a fast response to the initiation of water flow; in turn, an action sequence is initiated.

Applications such as electric water heaters, air conditioners, etc. represent some examples. A baffle plate with a magnet mounted to it is used in the water flow line. When water begins to flow, the baffle plate moves parallel to the water flow. A reed switch is strategically positioned to pick up this movement, and once the magnetic field has been sensed, the reed switch closes. In the case of a water heater, it instantly detects the water flow, and in turn triggers the heating element to be turned on. It takes much longer to detect a fall in temperature when cold water is added to the tank. Using a reed switch can save valuable heating time particularly when high water usage is involved.

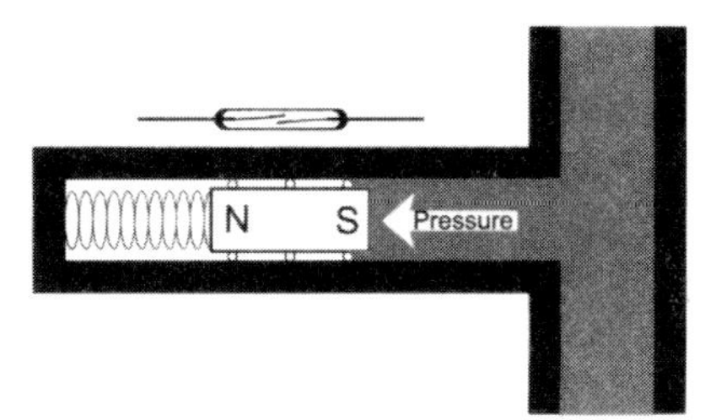

Fig. 5.20 Pressure sensor, can be calibrated for precise pressure points

5.2.5.3 Water Level Sensors

In many of the newer coffee machines for commercial or domestic applications that are not connected to the water mains, a warning light indicates a low water level. In this case a permanent magnet is built into a float in the removable plastic water tank. In the machine itself, a reed switch is mounted behind a plastic or aluminium cover. The positioning of the reed switch and the magnetic strength of the magnetic float guarantee the switching of the contacts when the water level drops to the level of the reed switch sensor's position. By design, the switch sen-

Fig. 5.21

sor remains closed even after the water level continues to drop well below the minimum water level. The construction of the water tank is usually designed to ensure that at least 2 cups of water remain once the minimum water level has been reached. The reed switch will only open (and thereby turn off the low water light) when the water level has been sufficiently refilled and the magnetic float is above the minimum water level mark.

In the case of coffee machines with a permanent water connection, the same sensor approach applies except for the fact that there is no indicator light, but a water-switching valve is added instead. Generally speaking, this approach works well for all such fluid level applications.

5.2.5.4 Monitoring the Level of Water Softener and Clarifier in a Dishwasher

The containers for water softener and clarifiers are situated in the door of the dishwasher. The reed sensor is mounted next to the plastic casing that contains a magnet in a protective foam housing, floating on the liquid. When the level drops to a certain point, the floater activates the reed sensor, which in turn directly operates an indicator light on the front panel. The operator of the dishwasher is then alerted that softeners and/or clarifiers need to be refilled.

Sensing the water softener (alkaline) works in the same way. In this case the container is completely flooded with water at all times. The specific weight of the

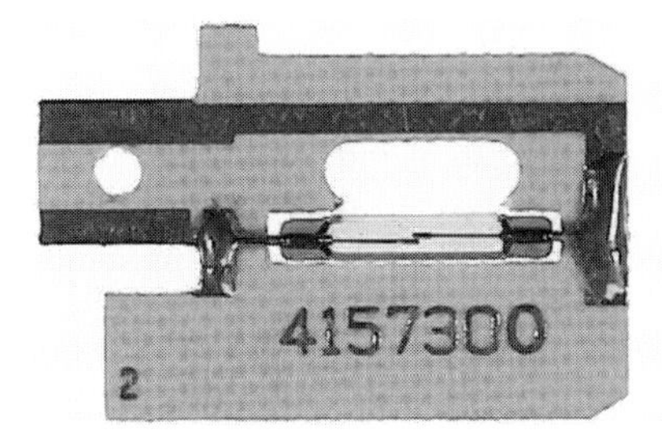

Fig. 5.22

floater is adjusted to the required alkalinity of the water. When the level drops, the reed sensor outside of the container will be activated to indicate the need to refill the softener.

5.2.5.5 **Protecting Dishwasher Spray-arms**

Another application for reed sensors in modern dishwashers is within the spray-arm control device. This application is very important in protecting the dishes, and even the dishwasher itself, from damage.

A reed sensor is mounted either above or underneath the spray-arm, outside of the washing area, while a magnet is placed on the spray-arm. The reed sensor counts the revolutions of the spray-arm. If for any reason the spray-arm is not working (Dishes inside the dishwasher could block the movement, or the spray-arm is faulty, etc.), the sensor will indicate on a small outside light or display that service is needed. An additional electronic device could also be activated with this reed sensor to stop the machine, so that the dishes or the machine itself will not be damaged.

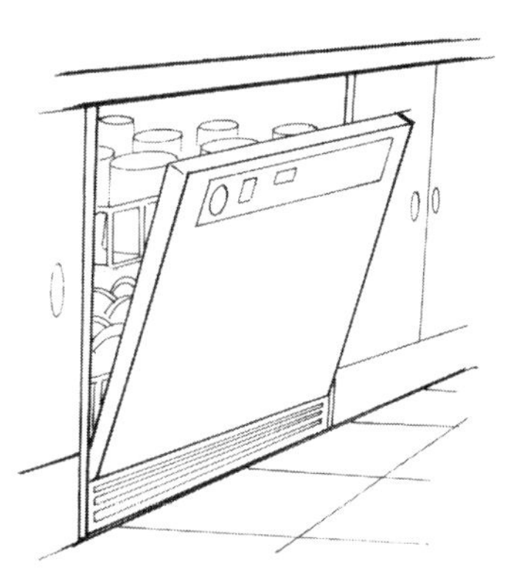

Fig. 5.23

5.2.5.6 **Preventing Condensed Water Overflow**

Commercial laundry dryers can not be easily vented to the outside in the same way that home laundry dryers are. Because of this fact, the moist air is passed through a condenser where the water will accumulate. If there is no direct water drain, the water is directed to a holding tank. The water level is monitored using a magnetic float. When the water rises to the full level, the reed sensor is activated and shuts down the dryer. Once the water is removed, the reed sensor is deactivated, enabling the dryer to operate.

If there is a direct outlet to drain the water, the collected water will be pumped out as soon as the level detection sensor has been activated. Here, the sensor serves the same purpose in detecting the full level, which in turn activates the pump and drains the water.

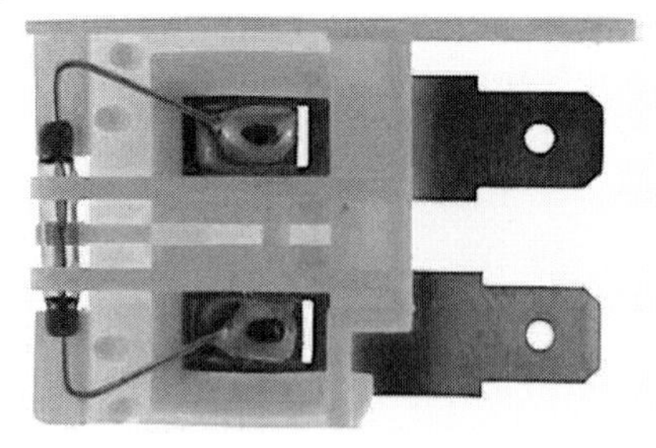

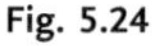

Fig. 5.24

5.2.5.7 **Detecting the Drum Position in Washing Machines**

Some top-loading washing machines have spinning drums that rotate perpendicular to the top surface of the washing machine. The drums these machines have doors for loading and unloading clothes that need to be aligned with the top flap of the machine. Since the drum rotates a random number of times, it will not necessarily stop at the right position for the doors to open, and the drum would then have to be rotated manually. Using a reed sensor approach has greatly simplified this and permits automatic positioning of the door to line up with the top surface door of the machine when the drum stops rotating.

This is accomplished by mounting the reed sensor on top of the washing machine chassis above the drum lid. The magnet is strategically mounted and positioned on the drum lid itself. An electronic counter counts the impulses for every rotation of the drum when the magnet passes the sensor. In this way, the logic in the electronic circuitry knows when the cycle is finished. After the washing procedure the drum turns slowly, and the electronic logic detects this motion as a final rotation. It then stops the drum with the drum lid coming to rest at the top of the washer for easy unloading reloading of the clothes.

With this type of washing machine, it is also important to make sure that the washing machine cover is in place while the washing machine is running. Opening the door while the machine is running could be a safety hazard. To prevent this happening, again a reed sensor is used. It is mounted underneath the door of the washing machine, and the magnet is placed right above the sensor in the

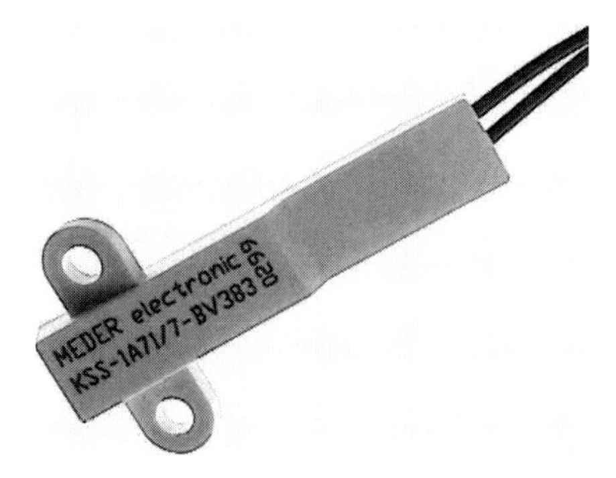

Fig. 5.25

door. If the door is opened while the machine is running, the magnet moves away from the sensor, deactivating it. This in turn triggers an interruption to the wash cycle and stops the drum. Once the door is closed the reed sensor reactivates and the wash cycle resumes.

5.2.5.8 Safety Control in Appliance Doors

The white industry of refrigerators, freezers, microwave ovens, stoves, etc. requires safety elements that detect the status (open/closed) of the appliance doors.

These door sensors are designed in many sizes and shapes, depending upon the specific application. Many are custom-built. Generally, both a reed switch and a magnet are used, and in many cases, added circuitry is built into a PCB for smart sensoring. If the sensor is not activated after a specified period of time, an alarm will sound, alerting that the door is ajar. In the case of a freezer, several hundred dollars worth of frozen meat and other food can be saved from spoilage if an open door is detected.

The reed sensor is usually mounted in the chassis of the appliance and the permanent magnet is placed in the doorframe. Thus, when the door is closed, the magnet's position is above or parallel to the sensor. When someone opens the door, the circuit is broken.

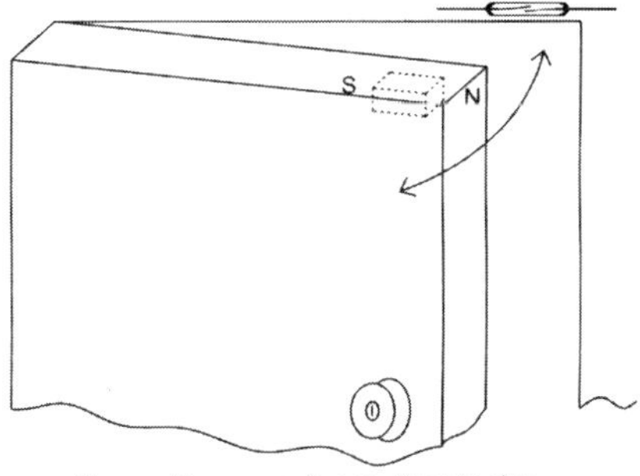

Fig. 5.26 Door position sensor for security and safety

5.2.5.9 Reed Sensors in Stove Applications

Knob control on stoves can now become a thing of the past. First seen on flameless ceran stoves, reed sensors are used to control the various power settings for each burner. The reed sensor eliminates the need for knobs, electronic gadgetry and hall effect sensors; the latter two of witch require added power and additional circuitry. In addition, this particular reed sensors design allows for child proofing, in that burners will not be indiscriminately turned on. Here is how it works: On the surface of the oven and near one of the ceran elements, a sliding device with an integrated magnet is installed. reed switches are mounted just below the surface. As the sliding device (magnet) is moved, it will trigger the various reed

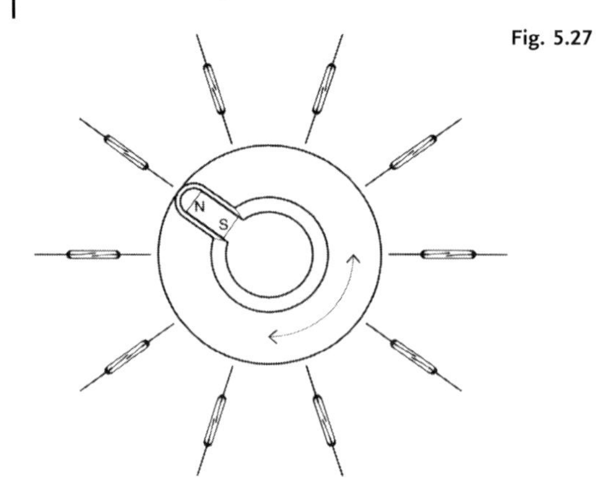

Fig. 5.27

switches, which in turn will activate different power settings. When the stove is not in use, the sliding magnet may simply be removed and stored in a safe place. So, if children are playing in the area, they will not inadvertently turn on power to any of the stove elements.

5.2.5.10 **Electric Toothbrushes**

Electric toothbrushes are commonly used in tougher environments than many other consumer household items. In the bathroom, they may be dropped several feet, they are exposed to human saliva (acidic) and various tooth pastes (generally alkaline), they have to withstand hot and cold water and are expected to survive for many years.

Most of these issues do not pose a problem as long as the electronics and switching circuitry are not directly exposed to any of the above conditions. The key potential area for exposure is the on-off switch of the device, where only a slide switch is used, for it may become contaminated very quickly and ultimately corrode. This will result in faulty switching unless the slide switch is hermetically sealed. Hermetically sealed switches can be very expensive.

Another approach, which has met with wide approval, has been the use of a reed switch/magnet combination. The magnet is moulded into a plastic sliding device that can move up and down. a reed switch that senses the position of the magnet is mounted on a pc board inside the electric toothbrush. In the down po-

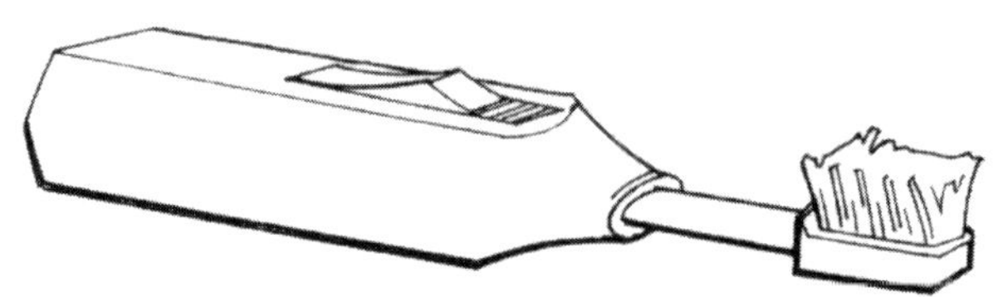

Fig. 5.28

sition, the switch remains open; in the up position, the magnet activates the reed switch and thereby turns on the motor.

Using this latter approach, the electric toothbrush is able to achieve many years of operation without failure because it is designed to withstand the rigors of its normal environment.

Lately, we have noted that SMD reed sensors are preferred for this type of application.

5.2.5.11 **Level Control of Jet-Dry Cleaning Liquid**

Many modern dishwashers have a container for Jet-Dry cleaning liquid. The use of Jet-Dry eliminates unsightly water stains from clear glassware. This liquid or its equivalent is typically filled once a month or whenever the storage container is empty. The traditional approach to monitor the level of Jet-Dry liquid was to use a little window that showed the liquid level. This, however, was not a very precise indicator of the true level of the liquid left in the container.

To replace the old-fashioned "eye-control" method of detecting the level of the liquid, appliance manufacturers now use reed sensors to show when the container for Jet-Dry needs refilling. The reed sensor is mounted underneath, or next to, the liquid container. A floater with an internal permanent magnet is placed in the container and floats on the liquid. If its level falls to a minimum, the floater activates the reed sensor; which, in turn, activates a light on the front control panel of the dishwasher, signaling the need to refill the Jet-Dry container.

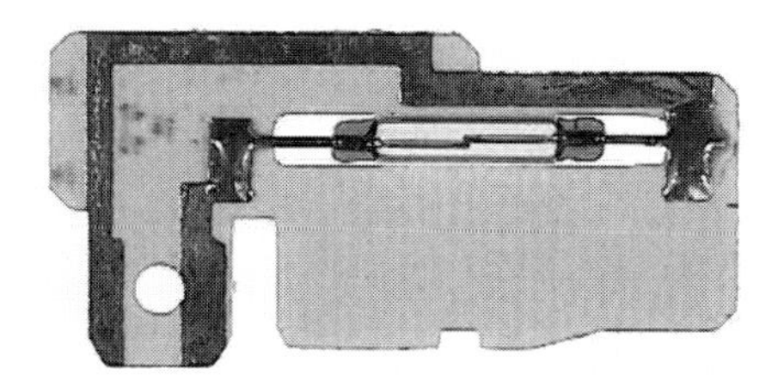

Fig. 5.29

5.2.5.12 **Reed Sensors in Carpet Cleaners**

When changing from normal vacuuming to a wet mode, the rotors of the vacuum cleaner must generally be changed. This makes the cleaning process more effective. Also, the required RPM level of the rotor is reduced for a better 'scrubbing action' of the rotor. These upright vacuum cleaners generally use a magnet/reed sensor approach because of the simplicity of its design. This is accomplished by using a reed switch to sense the magnetic field of a permanent magnet. Since the reed switch is hermetically sealed, all switching takes place in a perfectly clean atmosphere, undisturbed by the dusty, dirty environment of a carpet cleaner.

Generally, the permanent magnet is mounted on only one of the rotor assemblies. The rotor is chosen on the basis of what RPM speed is considered the de-

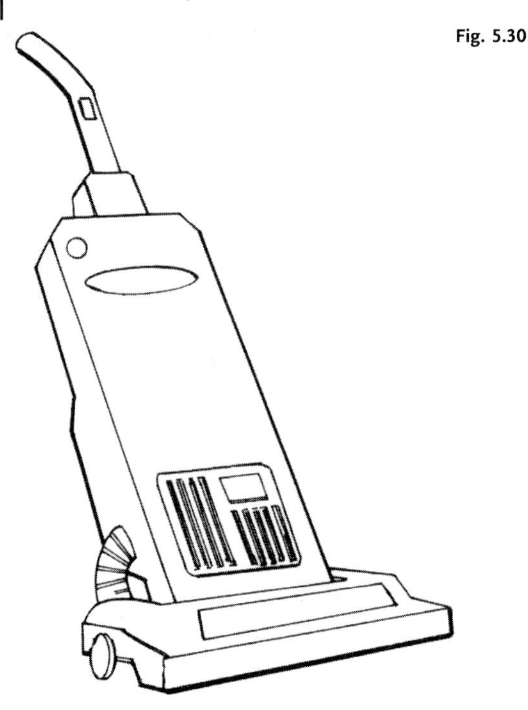

Fig. 5.30

fault level. The reed sensor is mounted on the carpet cleaner itself, in such a position to detect the magnetic field of the permanent magnet when its assembly is inserted in the carpet cleaner. The reed sensor then detects the permanent magnet, which closes the reed sensor contact. This closure then activates the appropriate circuitry leading up to the motor, which in turn adjusts the RPM level. Proper rotor operation is adjusted accordingly to suit the rug cleaning process.

5.2.5.13 **Detecting Movement or End Positions**

Massage chairs, special lifts for bathtubs, hospital beds, etc. all have electronic adjustments that can be controlled by reed sensors. In applications like these, no special voltage or current is necessary. Therefore, there is no risk of dangerous electric shock.

Reed sensors are often used as end position sensors. The reed sensor is mounted at the end position for a given movement. A magnet is strategically placed or mounted on the moving part. When the magnet approaches the reed sensor, it activates and triggers the appropriate circuitry, which in turn stops the movement.

Instead of sensing only the end position, it is also possible to monitor different positions for movement. For this purpose, several reed sensors are mounted at the various positions requiring sensing. When the magnet passes these positions, they are registered by the sensors. Several magnets could also be used with one sensor as an alternative. If the positions and motions are complex, a combination of both is also possible.

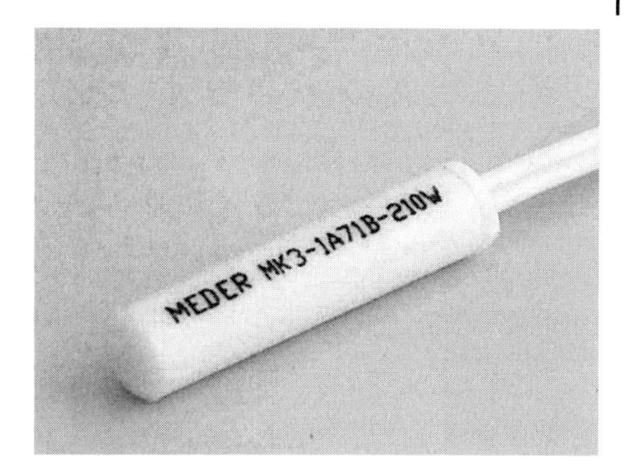

Fig. 5.31

5.3 Gas Sensors in the Domestic Environment

U. Hoefer and M. Meggle

5.3.1 Introduction

As a consequence of the growing automation in nearly every area of our environment there is an increasing demand for various kinds of sensors that make our life easier healthier and safer. Fire alarm sensors for example are well known and are nowadays installed in nearly every public building. However, especially in homes and other buildings, there is still a great demand for a large number of gas sensors and this number is going to increase, due to new technologies and the construction of energy conserving houses. The need is not only for gas sensors to detect harmful and/or explosive gases like carbon monoxide and natural gas. People spend on average more than 90% of his life in buildings and are very sensitive to bad indoor air quality. It is likely that uncomfortable room climate is a factor contributing to what is known as sick building syndrome, which is an unspecific feeling of being unwell and uncomfortable. The complaints can be linked to time spent in a building but no specific illness or cause can be identified. Apart from temperature, the comfort of inhabitants is mainly affected by humidity and the concentration of carbon dioxide in room air, which can be directly measured with suitable gas sensors. But there are about 10,000 additional air components which can affect health and which cannot be measured continuously with reasonable effort and costs [1]. Most of these air components are released by building materials as well as fungi and biochemicals, which makes maintenance complicated. In this chapter we intend to give an overview of applications for gas sensors in households, including systems detecting the main indoor pollutant CO_2, gas sensors detecting CO (garage, combustion control, pyrolysis in kitchen ovens), oxygen (combustion control) and natural gas which consists mainly of CH_4 (indoor alarm systems).

5.3.2 Sensor Principles

Depending on the gas type and application, different gas detection and sensor principles may be suitable. Beside the price, there are other parameters like accuracy, power consumption, long-term stability, lifetime, selectivity and sensitivity which have to be taken into account. However, whether a sensor is developed into a product depends on the relation between additional functionality and additional cost. In the subsequent chapters we intend to give an overview of common sensor principles and their potential use in household appliances. In order to give some information about existing (commercialized) sensors and sensor systems each subchapter is completed by an amendment listing a selection of relevant gas sensor products.

5.3.2.1 Metal Oxide Semiconductor Gas Sensors

Conductance (metal oxide) gas sensors are based on a reversible change in resistance when in contact with the gas type to be detected. This resistance change is due to a reaction of the ambient gas with surface-adsorbed and/or crystal lattice oxygen [2, 3] of the metal oxide. The most widely used sensor materials are semiconducting metal oxides like SnO_2, In_2O_3, Ga_2O_3, TiO_2, WO_3 and ZnO. Basically, the selectivity of metal oxide sensors may not be ideal but can be improved by using various methods:

- changing the operation temperature during continuous operation or using temperature pulses ("power activation mode", see also chapter 5.3.4),
- deposition of catalysts and doping [4],
- deposition of molecular filters,
- using various contact electrode geometries [5, 6],
- various electronic readout techniques (ac-/dc-measurements) [7].

Because of its high sensitivity to a wide range of gases, SnO_2 is often used as sensor material in applications where low concentrations of one or more gas components have to be measured. Due to its poor selectivity, the signal to be evaluated is mostly comprehensive. By using different sensors in an array, combined with mathematical pattern recognition methods, it is possible to increase the accuracy of measurement or to achieve some classification. In the food industry for example it is possible to differentiate between different smells after performing a calibration procedure. Also fresh and spoilt food can be distinguished. Due to a certain long-term instability of the single sensor signals, the calibration of the array has to be repeated periodically. Unfortunately SnO_2 sensors are also very sensitive to changes in ambient humidity and are not stable when used directly in flue gas where the oxygen concentration is low. Additionally the working temperature of SnO_2 is about 100 °C to 400 °C, which makes the sensors suffer from pollution effects when exposed to a soot-charged atmosphere.

Some of these disadvantages can be avoided by using Ga_2O_3-based gas sensors. Ga_2O_3 is an n-type semi-conducting metal oxide which is operated at the compar-

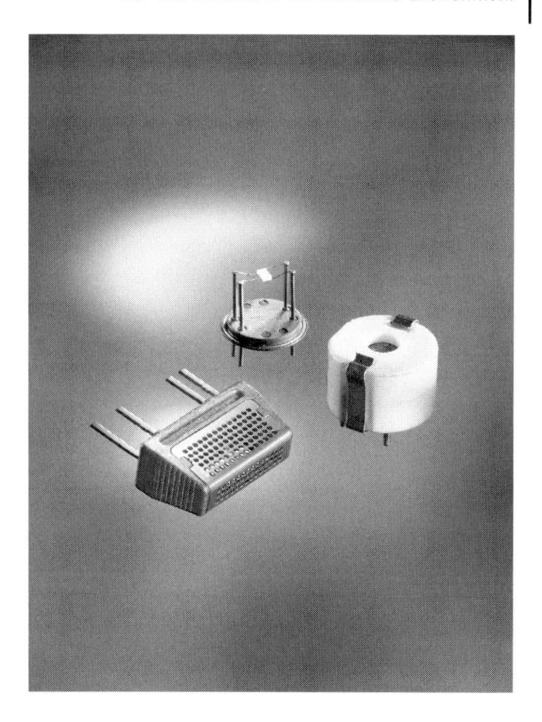

Fig. 5.32 Ga_2O_3-semiconductor gas sensors (STEINEL Solutions AG).

ably high temperatures of about 500 °C to 850 °C. Organic residues such as soot are burnt off during operation. This self cleaning effect, the stability also in atmospheres with low oxygen concentrations (<1 vol. %) and the general stability in harsh and corrosive environments are essential prerequisites for the direct use of gas sensors in hot flue gas of domestic fuel burners or self-cleaning ovens. Due to the high operation temperature, the cross-sensitivity of Ga_2O_3-sensors to humidity is comparably small and can be neglected in many applications.

The lifetime of metal oxide gas sensors in general is several years. Nevertheless poisoning effects can occur when exposed to silicones. An example of commercial Ga_2O_3-semiconductor gas sensors is shown in Fig. 5.32.

5.3.2.2 **Pellistors**

Pellistors are used to detect flammable gases like CO, NH_3, CH_4 or natural gas. Some flammable gases, their upper and lower explosion limits and the corresponding self-ignition temperatures are listed in Tab. 5.1. This kind of gas sensor uses the exothermicity of gas combustion on a catalytic surface. As the combustion process is activated at higher temperatures, a pellistor is equipped with a heater coil which heats up the active catalytic surface to an operative temperature of about 500 °C. Usually a Platinum coil is used as heater, embedded in an inert support structure which itself is covered by the active catalyst (see Fig. 5.33). The most frequently used catalysts are platinum, palladium, iridium and rhodium.

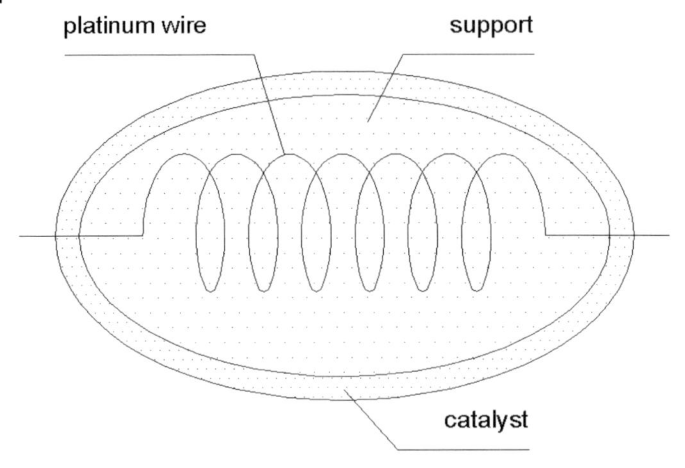

Fig. 5.33 Principal pellistor setup.

During the reaction of the hot catalyst surface with a flammable gas the temperature of the device increases. The Platinum coil itself serves at the same time as a resistance thermometer. The resistance increase of the coil then is a direct measure for the amount of combusted gas. Usually the amount of heat that develops during combustion is small and amounts to 800 kJ/mol for methane, for example [8]. Therefore the sensor is connected in a bridge circuit to a second resistor which shows the same setup as the pellistor but is catalytically inactive. The bridge voltage is then controlled by the temperature difference of the two sensors (see Fig. 5.34).

This type of gas sensor is usually not able to differentiate between gases because each flammable component contributes to the sensor signal. The lifetime in general is not limited, but aging effects can occur when the sensors are exposed to an atmosphere containing gases like silicone vapors, phosphate esters, sulfur-, halogen- and alkyl lead compounds. The most important poisoning components in households are silicones, which are, for example, present in fire retardants. This restriction also applies to metal oxide sensors.

Tab. 5.1 Explosion limit and self-ignition temperature of some flammable gases [9].

	H_2	CH_4	C_3H_8	CH_3OH	C_7H_{16}	NH_3	CO	$C_{10}H_{22}$
Lower explosion limit (LEL)/vol. %	4.0	4.4	1.7	6.0	1.1	15.0	7.5	0.7
Upper explosion limit (UEL)/vol. %	77.0	17.0	10.9	50.0	6.7	28.0	12.5	5.4
Self-ignition temperature/K	833	868	743	728	488	903	878	478

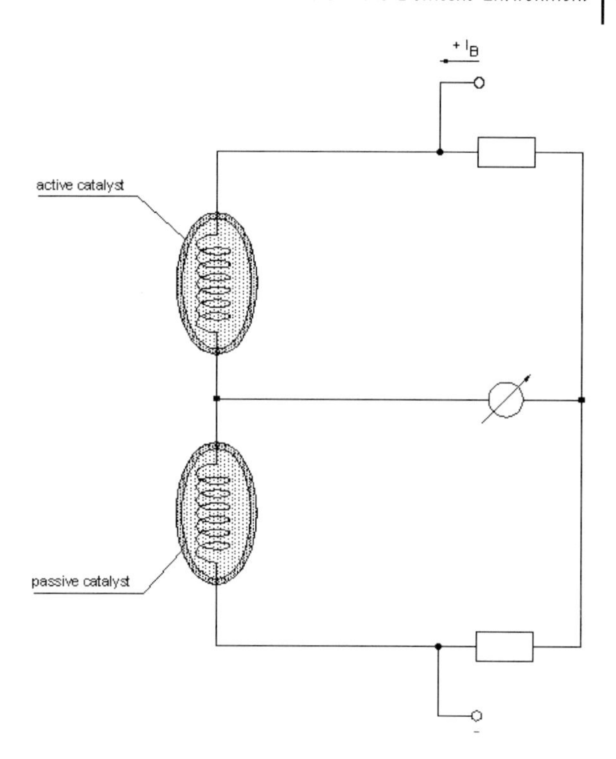

Fig. 5.34 Typical pellistor measurement circuit.

5.3.2.3 Liquid-state Electrochemical Gas Sensors

Most liquid-state electrochemical cells for gas detection purposes are based on either an amperometric or a potentiometric working principle. In all cases the gas to be detected diffuses across a gas-permeable membrane into a thin layer of a liquid electrolyte, where the gas is dissolved. The electrolyte layer is located between the measuring electrode and the membrane. Depending on the electrolyte type, acidic and/or alkaline gases like CO, CO_2, NO, NO_2, SO_2, and NH_3 can be detected. The chemical process and hence the sensor signal are temperature dependent, which is why the sensors must be equipped with a temperature compensation. In general the application of liquid-state electrochemical cells is restricted to atmospheres where the temperatures are in the comparatively moderate range of –40 °C to +50 °C. As a consequence it is not possible to use this kind of sensor for example directly in hot flue gases. Nevertheless they can be integrated in gas analysers using bypasses equipped with gas preparation units which are filtering and cooling the flue gas.

5.3.2.3.1 Amperometric Sensors

The principal setup of most amperometric electrochemical cells is based on three electrodes: a measuring electrode, the counter electrode and the reference electrode. After applying a voltage the dissolved gas will be electrochemically transformed at the measuring electrode and a current will be generated. If the voltage is high enough and the gas diffusion rate is limited, the current I_D is proportional to the gas concentration and is given by:

$$I_D = \frac{F \cdot A \cdot D \cdot c \cdot n}{d} \quad (1)$$

where F is the Faraday constant, A is the area of the measuring electrode, D is the diffusion coefficient, c is the concentration of the species, n is the number of electrons participating in the electrode reaction and d is the thickness of the diffusion layer.

By applying a specific voltage on the reference electrode it is possible to compensate non-linearities during exposure to higher gas concentrations and/or to increase the sensitivity and selectivity to different gas species. The lifetime of this kind of electrochemical cell is limited by the consumption of electrode material.

5.3.2.3.2 **Potentiometic Sensors**

In general, potentiometric electrochemical cells are based on the pH-measurement of the electrolyte. The pH is defined as the negative log-normal of the concentration of hydrogen ions. Fig. 5.35 shows the inner part of a commercial CO_2-sensor equipped with a pH-electrode and a reference electrode. Often glass electrodes are used in pH-metering systems. During gas exposure the gas diffuses through holes and the membrane into the electrolyte where it is dissolved. As a consequence the concentration of hydrogen ions is changed.

As an example the electrochemical reaction in the electrolyte of a CO_2-sensor can be described as:

$$CO_2 + H_2O \longleftrightarrow H^+ + HCO_3^- \quad (2)$$

$$HCO_3^- \longleftrightarrow H^+ + CO_3^{2-} \quad (3)$$

The potential between the pH-electrode and the reference electrode follows the Nernst equation – or more specifically

Fig. 5.35 Commercial CO_2-sensor. Left: electrolyte reservoir with gas diffusion holes. Right: tube like pH-electrode and reference electrode (STEINEL Solutions AG).

Tab. 5.2 Selected cross-sensitivities of the potentiometric CO_2-sensor shown in Fig. 5.35.

Gas type	1.25% CH_4	1500 ppm CO	500 ppm H_2	4 ppm HCl	50 ppm H_2S	120 ppm NH_3	25 ppm NO_2	15 ppm SO_2
Signal change	<1 ppm	<1 ppm	<1 ppm	<1 ppm	<1 ppm	<1 ppm	<1 ppm	<2 ppm

$$E = E_0 + 2.303 \frac{RT}{F} \log p_{gas} \text{ (Henry-Dalton law)[10]} . \quad (4)$$

In this case the sensor signal is proportional to the logarithm of the gas concentration.

The low power consumption of electrochemical cells is an advantage that has to be emphasized. The power consumption of the sensor which is shown in Fig. 5.35 for example is extremely small (in comparison to optical CO_2 measuring systems – refer to chapter 5.3.2.5) and amounts to less than 0.5 mW. Therefore they can also be integrated into portable gas analyzers or portable and/or battery-powered gas alarm systems. The lifetime of commercial electrochemical cells lies normally between 1 and 3 years; in special cases lifetimes of up to 8 years have been reported [11]. Typical t_{90}-response times (time until 90% of the sensor end signal is reached) vary from 30 s to about 120 s and are also dependent on the gas concentration. Cross-sensitivities to gases other than the target gas are low and can be neglected for many applications. As an example, Tab. 5.2 lists specific cross-sensitivities of the sensor shown in Fig. 5.35 to selected gases.

5.3.2.4 **Solid-state Electrochemical Sensors**

Solid-state electrolyte gas sensors are based on their predominant ionic conductivity. A well-known example is the ZrO_2 oxygen sensor. Nowadays most solid-state electrolyte sensors are used to measure the oxygen concentration in exhaust gases and metal- or glass melting. Because of their importance for household appliances (domestic fuel burner) this chapter is dedicated to O_2-ion conducting ZrO_2 cells, for example for measurement of the oxygen partial pressure in flue gases.

5.3.2.4.1 **Potentiometric Measurement**

In general ZrO_2 oxygen sensors consist of a tube-like solid-state ZrO_2 electrolyte where the electronic conductivity is based on oxygen ion charge carrier transport. The inner and outer surface of the yttrium-doped and stabilized zirconia tube is covered by porous platinum electrodes.

If the outer surface of the sensor is brought into contact with an exhaust gas containing a lower oxygen partial pressure than the surrounding atmosphere (which is in contact with the inner electrode), negatively charged oxygen ions will move through the electrolyte and cause the rise of a potential between the inner and the outer electrode. Usually the inner electrode is in contact with the ambient air which is used as a reference oxygen pressure reservoir (P_R). The voltage U_{12} between the electrodes is related to the measured oxygen partial pressure P, in the analysed gas, by the Nernst equation:

$$U_{12} = \frac{RT}{4F} \ln\left(\frac{P}{P_R}\right). \tag{5}$$

Where R is the gas constant, T is the temperature, and F is the Faraday constant. Caused by the logarithmic correlation between the gas concentration and the voltage signal, the potentiometric measurement is best suited for measurements of small amounts of oxygen. A well-known application of this principle has been realized in the so called lambda-probe for automotive applications where they are used to control the lambda value within a small interval around $\lambda = 1$. The lambda-value is defined by the relation between the existing air/fuel ratio and the theoretical air/fuel ratio for a stoichiometric mixture composition:

$$\lambda = \frac{(m_{air}/M_{fuel})_{mixture}}{(m_{air}/m_{fuel})_{stoichiometric}} \tag{6}$$

Potentiometric lambda-probes are not suitable for measuring the oxygen concentration at $\lambda \gg 1$.

5.3.2.4.2 Amperometric Measurement

Amperometric cells or so-called electrochemical pumping cells are based on the current correspondent flux of oxygen (charge carriers) through the device. Applying a voltage at the electrodes of the device shown in Fig. 5.36 will result in transport of oxygen into the cell and out of it respectively. By limiting the inwards diffusion of oxygen using a diffusion hole, this flux of charge carriers is related to an electronic current which is then proportional to the oxygen partial pressure

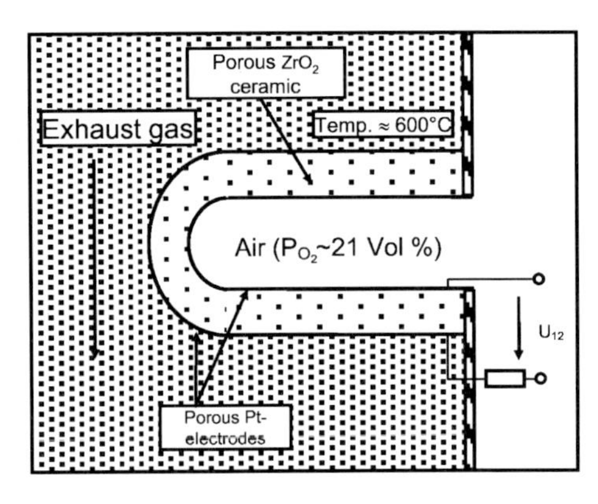

Fig. 5.36 Potentiometric ZrO_2 oxygen sensor.

which is to be analyzed. The main advantage of these devices is their higher sensitivity to small variations of oxygen partial pressure, and they can also be operated at lambda values $\lambda \gg 1$. Their sensitivity is almost constant over a wider range of oxygen partial pressure.

5.3.2.5 **Optical Gas Detectors**

Optical gas sensors are based on the absorption of light by molecules passing through an analyte. Electronic and vibrational energy levels of molecules can be excited when exposed to light with a specific frequency. Extinction of light in proportion to the number of gas molecules can be a measure for gas concentration at a given temperature and pressure. These levels are very specific and gases can be detected via absorption of light with a defined wavelength. Commonly NDIR (non-dispersive infrared) systems are used.

CO_2-IR detection systems for example contain a light source emitting infrared light ideally at wavelengths around 4.2 μm (often a simple light bulb), an absorption path (the so called cuvette), a spectral filter and a detector (thermopile). The filter is normally integrated into the detector housing.

Some distortion factors such as temperature, humidity, aging and pollution can be (partially) compensated by using a second "reference" channel. This second channel is placed inside the detector housing and uses a similar measuring element, the same light source and nearly the same light channel, but a second filter. The reference channel then works under the same physical conditions as the measuring channel but is not exposed to CO_2. This ensures that the extinction of light in the measuring channel only is influenced by the amount of ambient CO_2

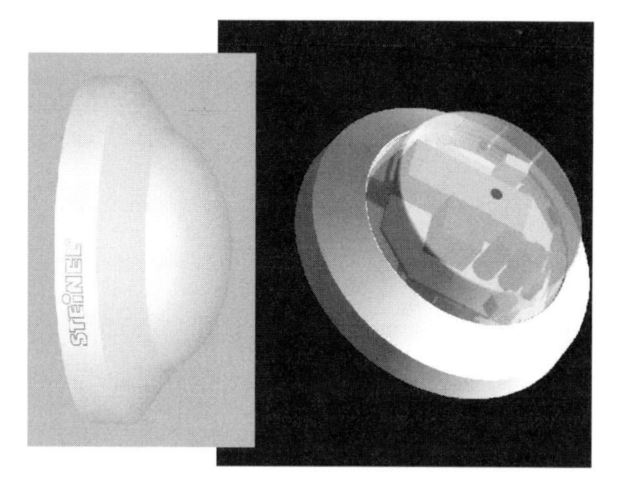

Fig. 5.37 Commercial two-channel infrared CO_2-detector. Measurement range 300 to 2000 ppm, accuracy ±100 ppm (STEINEL Solutions AG).

and not by aging of light bulb and other components. This is a common way for detecting and eliminating measuring errors.

Unfortunately, the sensor detects the number of CO_2 molecules and not the actual concentration. This means the signal will also depend on the air pressure. Our measurements and calculations showed that weather-dependent pressure changes (about ±20 ppm per year) will only change the CO_2 signal by less than 5%. On the other hand, the assembly should be adjusted for altitude once to avoid a systematical altitude-dependent pressure error (>10% for 600 m altitude difference).

Air quality measuring systems that detect for example the leading substance CO_2 (refer to chapter 5.3.3.3) do not require the high accuracy of expensive measuring systems. Therefore, not much effort needs to be put in the development and construction of CO_2 detectors. In general the accuracy of such systems amounts to ±10%, which is achievable for the cost of an electrochemical cell. Fast measurements aren't needed but nevertheless averaging to increase accuracy is possible and recommended.

5.3.3 Appliances

In this sub-chapter we intend to give an overview of household appliances using the different gas sensor principles mentioned in the previous chapters. The appliances can be divided into two categories: gas sensors for safety and for comfort. Natural gas and CO-alarm systems for example are safety-relevant whereas air quality measurement, control of self-cleaning of ovens etc. are more or less a matter of comfort or energy-saving.

5.3.3.1 Domestic Burner Control (Fuel Burners, Gas Condensing Boilers)

An oil burner or gas condensing boiler which does not operate at its optimum setting will not only decrease in efficiency, but the concentration of CO, HC and the content of soot in the exhaust gas will also increase. A measure for the quality of combustion is the well-known (in the automotive industry) lambda- (λ-) value (see also chapter 5.3.2.4).

A λ-value smaller than 1 means that there is an excess of fuel in the mixture. In this case the air/fuel mixture is called rich. If more air is in the mixture than needed for a complete fuel combustion ($\lambda > 1$) the term lean mixture is used. Ideally the combustion is complete at $\lambda = 1$. Real fuel cannot be combusted without an increase in CO and soot at λ-values smaller than 1.05. Due to changing operation conditions, for example a soiled burner, wear of the nozzle or leaky flaps, change of gas quality or changes of temperature and air pressure in the ambient atmosphere, the air/fuel ratio and thus flue gas composition can change over time. In order to minimize the risk of intoxication (see also chapter 5.3.3.3), explosion and pollution real (uncontrolled) fuel burners are adjusted to operate far beyond this limit in the excess (lean mixture) region. However, unfortunately effi-

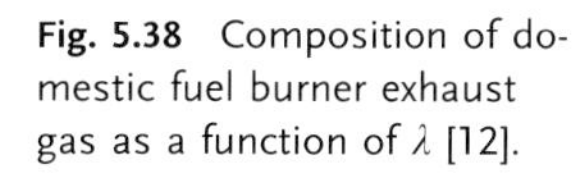
Fig. 5.38 Composition of domestic fuel burner exhaust gas as a function of λ [12].

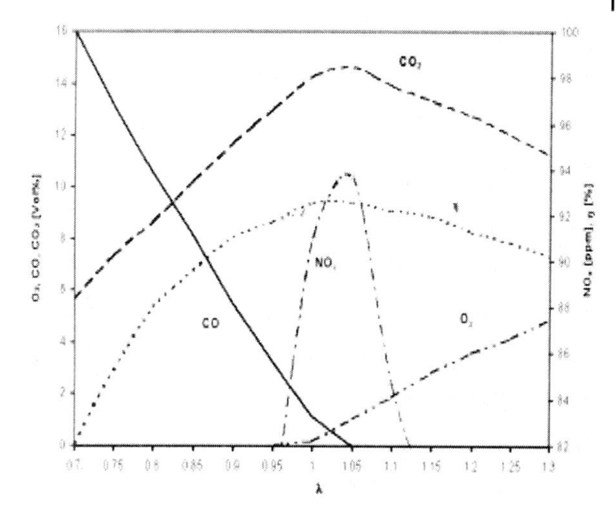

ciency η decreases with the increasing air supply. In order to fulfill the requirement to conserve fuel as well as minimize polluting emissions, fuel burners should be operated as close as possible to $\lambda = 1.05$. Some burner types, for example gas condensing boilers operate even at λ-values around 1.6. Consequently a direct measure of the combustion quality is the oxygen partial pressure or related components like carbon monoxide or carbon dioxide in the flue gas.

The aim of this chapter is to give an overview of commercially available gas sensing techniques with a focus put on sensors in the lower price range which are suitable for combustion control.

5.3.3.2 Sensor Requirements for Combustion Control

Most commercial flue gas analyzers use electrochemical cells and optical spectrometry. Because of the high temperature and dusty and harsh environment these

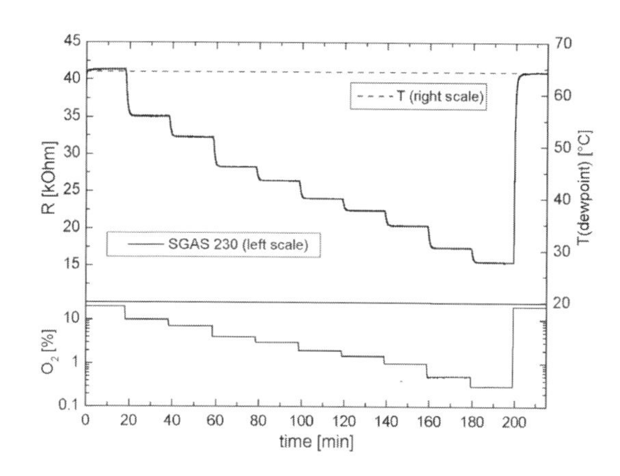

Fig. 5.39 Resistivity-change based (Ga_2O_3-) oxygen sensor in humidity saturated test gas.

kind of gas sensors cannot be placed directly in the exhaust pipe. In order to protect the sensors from damage, the flue gas has to be filtered from particles, cooled down and dried in special gas conditioner units. As a consequence, gas analyzers are very expensive and thus are not suitable for continuous measurements and integration into continuously operated burner combustion control systems. In order to be able to integrate gas sensors directly into combustion control units sensors are required which are long-term stable in high temperature, soot-charged and dusty environments sometimes containing corrosive components. Additionally, gas sensors in gas condensing boilers have to withstand high absolute humidity concentrations and condensation.

The most direct way to measure the combustion quality is to measure the oxygen concentration in the flue gas. Nowadays most oxygen sensors for use in com-

Tab. 5.3 Commercial oxygen sensors.

Company	*Name*	*Sensor type*	*Measurement range*	*Power consumption*	*Temperature range*
Dräger	Dräger Sensor O_2 6809720	Electrochemical cell	5 vol.% to 100 vol.% O_2	*	T=–20 to 40 °C
Gasmodul (Honeywell)	KGZ 20/21/22	ZrO_2 pumping cell	2 mbar to 1 bar	~ 10 W	Max. 400 °C
GfG	MWG ZD21	ZrO_2	0 to 25 vol.% O_2	3 W	T=–25 to 50 °C
Eichstädt-Elektronik	Typ A-01	Electrochemical cell	0 to 100 vol.% O_2	*	T=0 to 50 °C
Electrovac	Typ XXX-960	ZrO_2	1,0 to 96 vol.% O_2	*	Max. 350 °C
Figaro	KE 50	Electrochemical cell	0 to 20 vol.% O_2	*	T=5 to 40 °C
NTK	Oxygen Sensor	ZrO_2	0,1 to 95% O_2	2,0 W	T=–10 to 50 °C
Pewatron	DFCX-UW-CH	ZrO_2	50 ppm to 95 vol.% O_2	*	*
PSW-Elektronik	PSW-O2	ZrO_2	10 ppm to 96 vol.% O_2	1,6 to 2,0 W	Max. 350 °C
Steinel Solutions AG	SGAS 230	Ga_2O_3	0.1 to 21 vol.% O_2	1 W	Max. 650 °C
Sumatec	M-25	Electrochemical cell	0 to 100 vol.% O_2	*	T= 0 to 45 °C
Winter GmbH	TOXW	Electrochemical cell	0 to 21 vol.% O_2	0,5 W	T=–20 to 45 °C

* Information not available on website.

bustion flue gas are based on ZrO_2 (solid-state ionic conductor, conventional λ-probe and pumping cells), electrochemical cells (liquid) and resistance-based sensors (metal oxides). Fig. 5.39 shows the typical signal of a Ga_2O_3-semiconductor oxygen sensor in humidity-saturated test gas (simulation of the flue gas of a gas condensing boiler). The dewpoint stays around 65 °C.

Optical NDIR-methods are not suitable to detect oxygen due to the lack of a dipole moment. Tab. 5.3 gives an overview of commercial oxygen sensors.

Unfortunately the sensitivity of conventional Nernst type λ-probes is highest within a small interval at λ-values around 1 and therefore they cannot be used in domestic ($\lambda \approx 1.06 \ldots 1.6$) combustion controls (see chapter 5.3.2.4). Oxygen sensors which also show sufficient sensitivity in the lean region are based on ZrO_2 pumping cells but are rather expensive. Liquid electrochemical cells are very sensitive to higher temperatures and to corrosive gases and are therefore not suitable for direct use in the flue gas. Recently developed low-price alternatives work on the principle of selective changes in resistivity, using Ga_2O_3-metal oxide oxygen sensors which remain stable at high temperatures and in atmospheres with a low oxygen partial pressure (see Fig. 5.39).

Another possibility to analyze the combustion quality is to measure components in the flue gas which are directly related to the O_2-content like CO or CO_2. There are two main methods of measuring CO_2. One detection method uses electrochemical cells (see chapter 5.3.2.3), but both liquid and solid-state electrochemical CO_2-cells are not long-term stable if directly exposed to flue gas. A more promising approach is the measurement of the CO content by using modified high-temperature stable

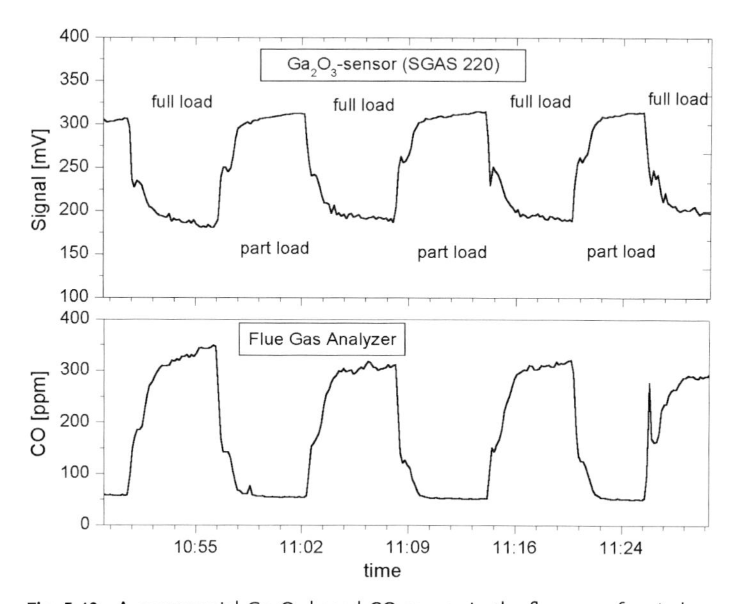

Fig. 5.40 A commercial Ga_2O_3-based CO-sensor in the flue gas of an independent vehicle heater (STEINEL Solutions AG SGAS 220)

semiconducting Ga_2O_3-sensors. This kind of sensor has already been tested for use in the flue gas of independent vehicle heaters (see Fig. 5.40).
Another low-cost alternative method is to measure parameters which can indirectly provide information about the combustion quality, such as the ionization current in the burner flame ("SCOT" method).

5.3.3.3 **Air Quality**

Sometimes people enter a room and notice at once that the air inside is "stuffy", while most people inside the room have been getting tired, but were unaware of the poor air quality. This is because our biological air quality sensor, the nose, gets used to slow changes of surrounding smells.

One simple possibility to have always fresh air inside a room is continuous ventilation, but this leads to additional power consumption for heating/cooling, draught and noise from outside – especially if the ventilation control is related to the maximum possible air requirement (i.e. volume). It is evident that an efficient, energy-saving method of regulating room ventilation is ventilation on demand, controlled by suitable air quality sensors. But which component is the leading substance indicating poor air quality? There are various ingredients of air which could be used as a parameter for air quality, some of which also depend on the room type. First of all there is the oxygen used up by humans or animals, so the amount of oxygen in the air of a room could be used as a direct measure for air quality. Unfortunately the consumption of oxygen is low and the detection of small amounts below the natural ambient oxygen concentration requires a very sophisticated gas sensor.

The second possibility to measure poor air quality is the detection of volatile organic compounds (VOC), but the number of different VOCs in room air is high, and often their impact on human health or comfort is not known. Additionally, the inconvenience caused by VOC's and "smell" is difficult to measure because it depends on human rating.

A straightforward way to measure air quality is to measure the carbon dioxide concentration, which is a natural biological metabolite and increases especially in rooms filled with people. An increase in CO_2 is mainly responsible for sleepiness and could therefore serve as a direct measure of poor air quality. The base concentration of CO_2 in the ambient air is around 400 ppm. In non-ventilated rooms the CO_2-concentration can amount to more than 1000 ppm. Some carbon dioxide concentrations and limits and their impact on human comfort are listed in Tab. 5.4.

CO_2 sensors are suitable for living rooms, offices and bedrooms, whereas for bathrooms (main parameter: humidity) and kitchens (main parameter: "smell") other non-specific sensors may be used. Therefore combinations of different sensors like humidity sensors, "smell"-sensors (unspecific VOC-sensors) and presence sensors could be the best solution. A simple movement detector with an additional manual override (ventilation on/off for a predetermined time) could be a low-cost alternative.

Tab. 5.4 Range of CO_2 concentrations and limits.

	CO_2 concentration
Clean air	350 ppm
Rural region	400 ppm–700 ppm
Comfort level (optimum conditions)	<800 ppm
Comfort level (acceptable conditions)	<1500 ppm
Threshold limit value	5000 ppm
Toxic warning level	15000 ppm

Fig. 5.41 CO_2-alarm system based on potentiometric electrochemical cells (STEINEL Solutions AG).

There are several approaches to measuring CO_2: NASICON based solid-state electrolyte sensors [13], ultrasonic methods (acoustical), photo-acoustic cells [14] (Siemens CO_2 Controller), optical IR-detection and liquid-state electrochemical cells.

There is no single sensor principle suitable for all applications. The advantages and disadvantages of some of the various detection principles are listed in Tab. 5.10. One main advantage of infrared optical sensors is their long time stability. The zero level (best possible air quality, ≈ 400 ppm CO_2) could be adjusted automatically via a controller with an eeprom. Using a lower voltage infrared source (bulb) increases lifetime dramatically. Thus, continuous measurements over many years are possible. However, electrochemical cells have also some advantages. They also have a relative long lifespan (Steinel Solutions AG guarantee a two-year-lifetime) and especially potentiometric electrochemical sensors an extremely low power consumption.

This could be used for battery-powered long time measurements. Tab. 5.5 gives an overview of available CO_2 sensors.

Tab. 5.5 Commercially available CO_2 sensors.

Company	*Name*	*Sensor type*	*Measurement range*	*Power consumption*	*Temperature range*
Automation Components Inc.	ACI/CO2	IR optical	0–2000 ppm	*	15 °C– 2 °C
General Monitors Inc.	IR7000	IR optical	0–5000 ppm	*	–40 °C–50 °C
GfG	POLYTEC-TOR G750	IR optical	*	*	*
Sierra Monitor Corp.	4102-89	IR optical	0–2000 ppm		5 °C–40 °C
Steinel Solutions AG	SGAS500	Electrochemical cell	0–100% 0–5%	<0.5 mW	2 °C–45 °C
Steinel GMBH	SGAS600	IR optical	0–5000 ppm	<1 W	15 °C–32 °C
Telaire	Ventostat 8001	IR optical	0–2000 ppm		15 °C–32 °C
TESTO	TE-535	IR optical	0–10000 ppm	6 h batt. lifetime	0 °C–50 °C
Texas Instruments	4GS-1	IR optical	0–2000 ppm	*	0 °C–50 °C
Thermo Gas Tech	67-0002	Electrochemical cell	0–5000 ppm	*	–20 °C–45 °C
Topac	MYCO2	IR optical	0–2000 ppm	*	0 °C–45 °C

* Information not available on website.

5.3.3.4 Indoor Detection of CO

Carbon monoxide (CO) is generated in incomplete combustion processes. In households the main sources are all kinds of fuel burners (fuel oil, wood, natural gas, coal etc.) and automotive exhaust gas. Carbon monoxide is an odorless and invisible gas, and, due to its affinity to hemoglobin, which is higher than that of oxygen, it reduces the blood's capacity to carry oxygen. Hence it is toxic, especially for unborn and small children as well as for the elderly or people with heart problems or anemia. Even small amounts of CO can be harmful. Tab. 5.6 gives an overview of the relation between CO concentration and the corresponding symptoms of intoxication.

Tab. 5.6 Symptoms of CO-intoxication [15].

Carbon monoxide concentration	*Symptoms*
30 ppm	German Threshold Limit Value (TLV)
200 ppm	Slight headache after 2–3 hours
400 ppm	Headache in the forehead after 1–2 hours, later throughout the head
800 ppm	Dizziness, nausea, convulsions in 45 min., unconciousness within 2 hours
1600 ppm	Dizziness, nausea and headache after 20 min., death within 2 hours
3200 ppm	Nausea, dizziness and headache after 5 to 10 min., death within 30 min.
6400 ppm	Nausea and headache after 1 to 2 min., death within 10 to 15 min.
12800 ppm	Death within 1 to 3 min.

5.3.3.4.1 Possible CO-sources

Nowadays houses are more air-tight due to energy-conserving measures. As a consequence, there is less fresh air coming into a home and not as many pathways for stale or polluted air to leave it. Especially in small rooms which are equipped with natural gas combustion equipment this means a considerable risk. In Japanese households, for example, food is usually prepared using natural gas cookers which are often situated in very small kitchens measuring less than 2 m^2. It is therefore possible that the oxygen content in the room decreases to less than 18%. Due to the lack of oxygen resulting from natural gas combustion, the CO-content can reach deadly levels. In Germany, for example, kitchens are bigger, and the danger is not as high as in Japan. Yet there are several cases of death each year in bathrooms where gas boilers have been installed [16].

Although in well-adjusted modern fuel burners the carbon monoxide concentration is comparably low (10 ppm to 30 ppm) missing, delayed or improper maintenance can be the reason for the generation of considerable amounts of more than 1000 ppm CO in the flue gas. Additionally, a possible reason for excessive carbon monoxide generation is when furnaces and boilers are starved of the oxygen needed to burn fuels completely. Especially newer houses can be so airtight that powered extraction fans in the kitchen or bathroom can produce a draft strong enough to overcome the draft in the furnace chimney and thus draw the toxic gases into the living space.

Further problems could include damaged or deteriorating flue liners, soot build-up, debris clogging the passageway, and animals or nesting birds could also obstruct the chimney. Oil flues need to be cleaned and inspected annually because deposits of soot may build up on the interior walls of the chimney. Tab. 5.7 gives an overview of possible harmful failures of indoor combustion systems.

Tab. 5.7 Combustion appliances and potential errors causing excessive CO generation [17].

Appliance	*Fuel*	*Typical errors*
Central Furnaces Room Heaters Fireplaces	Natural or Liquefied Petroleum Gas	Cracked heat exchanger; Not enough air to burn fuel properly; Defective/blocked flue; Maladjusted burner
Central Furnaces	Oil	Cracked heat exchanger; Not enough air to burn fuel properly; Defective/blocked flue; Maladjusted burner
Central Heaters Room Heaters	Wood	Cracked heat exchanger; Not enough air to burn fuel properly; Defective/blocked flue; Green or treated wood
Central Furnaces Stoves	Coal	Cracked heat exchanger; Not enough air to burn fuel properly; Defective grate
Room Heaters Central Heaters	Kerosene	Improper adjustment; Wrong fuel (not-K-1); Wrong wick or wick height; Not enough air to burn fuel properly
Water Heaters	Natural or Liquefied Petroleum Gas	Not enough air to burn fuel properly; Defective/blocked flue; Maladjusted burner
Ranges; Ovens	Natural or Liquefied Petroleum Gas	Not enough air to burn fuel properly; Maladjusted burner; Misuse as a room heater
Stoves; Fireplaces	Wood Coal	Not enough air to burn fuel properly; Defective/blocked flue; Green or treated wood; Cracked heat exchanger or firebox

The German guild of chimney cleaners states that in 1998 chimneys of more than 12 million gas boilers have been checked concerning the CO concentration in the flue. In 616,000 boiler systems the CO concentration was between 500 to 1000 ppm (maintenance recommended). In 375,000 cases the CO content amounted to more than 1000 ppm (complaint). It is estimated that in the USA 1,600 deaths occur yearly due to CO-intoxication [18].

In addition, around 10,000 cases of carbon monoxide-related injuries are diagnosed each year. Because the symptoms of prolonged, low-level carbon monoxide poisoning mimic the symptoms of a common flu (headaches, nausea, dizziness, fatigue), many cases are not detected until permanent damage to the brain, heart and other organs has occurred.

5.3.3.4.2 **Sensors for CO-indoor Monitoring**

CO-sensors and sensor systems for indoor use should meet several specifications regarding their sensitivity, cross-sensitivity and stability. A reasonable approach for the development of CO indoor alarm or monitoring systems is to comply with BS

Tab. 5.8 Alarm system requirements according to EN 50291.

CO-concentration	*Alarm not earlier than*	*Alarm not later than*
300 ppm	–	6 min.
100 ppm	10 min.	40 min.
50 ppm	60 min.	90 min.
30 ppm	120 min.	–

7860 or EN 50291 or equivalent standards. The hazardousness of CO is a function of exposure time and concentration. Therefore alarm systems should give a delayed optical and acoustical alarm scaled according to the indoor CO-concentration. Details are listed in Tab. 5.8.

EN 50291 also determines the limits of cross-sensitivities to H_2, CO_2, NO, SO_2, ethanol and hexamethyldisiloxane. The latter is known to cause poisoning of pellistors and metal oxide sensors.

Suitable methods for CO indoor monitoring are based on using metal oxide sensors (chapter 5.3.2.1), electrochemical cells (chapters 5.3.2.3 and 5.3.2.4), pellistors (chapter 5.3.2.2) and optical methods (chapter 5.3.2.5). An overview of CO-sensors and systems is given in Tab. 5.9.

Often there is no single sensor principle to suit all applications. The advantages and disadvantages of some different detection principles are listed in Tab. 5.10 and have to be balanced against each other according to the requirements of the specific environment. Metal oxide sensors for example are very robust even in harsh environments (like garages), they have a long lifetime but are relatively unspecific. Depending on the desired accuracy, all sensors normally have to be re-calibrated. An easy way to re-calibrate at least the zero signal is to measure the signal in moments when one can be sure that the ambient atmosphere contains no CO (for example after ventilation). Electrochemical cells are suitable in environments where no mains is available (battery operation) or low power consumption is required.

5.3.3.5 Natural Gas Detection and Alarm Systems

Another possible source of danger in households arises from natural gas installations. In general fixed natural gas fittings and installations in Europe are tight and do not release appreciable amounts of gas. Nevertheless there are several accidents every year where people are injured and buildings are damaged due to the explosion of natural gas/air mixtures. In most cases explosions are caused by improper handling, accidental damage of gas installations or intended discharge of natural gas into rooms. However, to date no installation of gas natural gas alarm systems is required in private houses in Germany.

Sometimes natural gas is discharged due to damage of gas installations resulting from natural disasters such as earthquakes. In Japan for example, where

Tab. 5.9 Gas sensors and sensor system for detection of carbon monoxide.

Company	*Name*	*Type*	*Measurement range*	*Power consumption*
AAA	S700	Electrochemical	0–100 ppm	*
Ados	592 CO	Electrochemical	0–300 ppm	*
Capteur	CAP07	Semiconductor	0–400 ppm	0.35 W
City Technology Ltd.	A3CO	Electrochemical	0–500 ppm	*
Dräger	DrägerSensor CO LS 6809620	Electrochemical	200–5000 ppm	*
Figaro	TGS 203	Semiconductor	0–300 ppm	0.3 W
Fis	SB-50	Semiconductor	10–1000 ppm	0.118W
Microsens SA	MSGS-3001	Semiconductor	5–1000 ppm	0.065 W
Pewatron	AF20	Semiconductor	*	max. 0.73W
Sensoric	CO 2E 300	Electrochemical	0–300 ppm	*
Steinel Solutions AG	SGAS 220	Semiconductor	10–10,000 ppm	~0.7 W
Winter GmbH	TCOW 400	Electrochemical	0–4000 ppm	0.5W

* Information not available on website.

Tab. 5.10 Advantages and disadvantages of three gas detection principles.

CO-sensing principle	*Advantage*	*Disadvantage*
Electrochemical	– High accuracy – Low power consumption – Mid price	– Limited lifetime (but at least several years) – H_2 cross-sensitivity – Re-calibration necessary
Optical	– High accuracy	– High price – Limited lifetime (bulb) – Possibility of soiling – Sensitive to changes in air-pressure (compensation) – High power consumption – Re-calibration necessary
Metal oxide	– Low price – High lifetime – Mid-range power consumption	– Cross-sensitivity to other gases – Low accuracy
Photoacoustic	– Mid-range power consumption – Multiple gas sources possible	– Only non continuous measurement possible – Mechanic components – Sensitive to changes in air-pressure (compensation)

earthquakes happen quite frequently, a law was introduced in 1986 that requires natural gas distributors to install gas detectors in the house of the final consumer [19].

5.3.3.5.1 Natural Gas Detectors

The chemical products from complete combustion of a hydrocarbon fuel are mainly CO_2 and H_2O (vapor). Combustion of gaseous fuel in air can occur in two different modes – one where fuel and oxygen is mixed during the combustion process, and the other where fuel and air are premixed (gas condensing boilers) and the fuel concentration must be within the flammability limits. In general the premixed situation allows the fuel to burn faster, i.e. more fuel is consumed per unit time.

Natural gas detectors should reliably detect the presence of gas concentrations far below the lower explosion limit of about 5% vol. In Germany many different cheap gas alarm systems can be purchased in DIY-market, but most of these devices are not based on any quality standard. Although high priced alarm and monitoring systems are available there is a lack of low cost reliable gas alarm systems for use in households which are based on appropriate standards.

The European norm EN 50194 contains a proposal concerning the construction of alarm systems for household gas combustion appliances, including norms for threshold limits, cross-sensitivities and long-term stability. One aspect of cross-sensitivity is the possibility of raising false alarms when the system is exposed to 2000 ppm ethanol. As ethanol is an ingredient of alcoholic beverages and cleaning chemicals, a sensor may well be exposed to it. Semiconductor metal oxide sensors are promising candidates to realize low-cost standardized natural gas alarm systems. In general these gas sensors are poorly selective and often show a considerable cross-sensitivity to ethanol. Apart from the measures mentioned in chapter 5.3.2.1, selectivity can also be improved by analyzing the sensor signal in a temperature-activated signal evaluation mode which is called "power activation mode" by the authors [20]. This method is suitable to differentiate between ethanol and methane by evaluating the signal slope when the operation temperature is raised. Fig. 5.42 shows the relevant logical operation diagram. The power activation mode prevents false alarms caused by ethanol at natural gas detection levels in the range from 2 to 20% LEL.

5.3.3.6 Other Appliances

Many appliances for gas sensors in households are still under development. For example, several attempts have been made in the past to include gas sensors in kitchen ovens to control cooking and frying processes as well as to control the self-cleaning process (pyrolysis).

Tab. 5.11 Selection of methane-gas sensors and sensor systems.

Company	*Name*	*Sensor type*	*Gas type*	*Measurement range*	*Power consumption*
AAA	IR 2100	Optical	Methane, propane	0–100% LEL	24 VDC 0,25 A
Ansynco	GA 90	Optical	Methane (CO_2, O_2)	0–100 vol. %	Battery ~ 6 h operation
Capteur	WL 02	Semi-conductor	Methane	500–10000 ppm	0.9 W
Electrim Tech	ESM 779	Semi-conductor	Methane (combustible gases)	10–25% UEG	12 V/24V/ 230 V 2–3 W
Figaro	TGS 842	Semi-conductor (SnO_2)	Methane (combustible gases)	500–10000 ppm	0.9 W
FIS	SP 12A	Semi-conductor	Methane	300–10000 ppm	0.38 W
GFG	MWG WT0238	Pellistor	Combustible gases	0–100% LEL	9 W
GMI	Gas-Test 3	Pellistor	Methane (combustible gases)	0–100% LEL	4 Monobat-teries 15 h
Indexa	GA 80	Semi-conductor	Methane (combustible gases)	0,8 vol.%	12 V/230 V <6 W
M.A.S. Consulting GmbH	EIT 4688 IR	Optical	Methane	*	*
Microsens	MSGS-3002	Semi-conductor (SnO_2)	Methane	100–10000 ppm	56 mW
Nemoto	NAP 55A/ 50A	Pellistor	Methane (combustible gases)	0–0,6 vol. %	0.45 W
Pewatron	AF 50	Semi-conductor	Methane, LP-Gas	500–10000 ppm	0.73 W
Sensoric	CTL 4 Series	Electro-chemical cell	Methane	0–5 vol. %	0.23 W
Steinel Solutions AG	SGAS 250	Semi-conductor (Ga_2O_3)	Methane, natural gas	100–10000 ppm	1 W
UST	Typreihe 1000	Semi-conductor	Methane	*	*
Winter	TBG	Semi-conductor	Methane (combustible gases)	0–100% UEG	2 W

* Information not available on website.

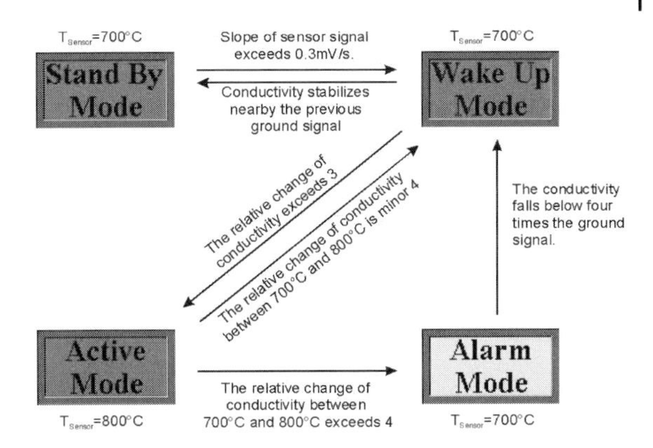

Fig. 5.42 "Power Activation Mode"-operation sequence for Ga_2O_3-natural gas sensors to avoid false alarm in the presence of ethanol and other interfering air components [20]. (With kind permission of D. Kohl, D. Skiera and M. Lämmer, Institute of Applied Physics, University of Giessen, Germany)

5.3.3.6.1 Cooking and Frying Control

Mainly two different approaches for the development of cooking controls have to be mentioned. The first, more empirical approach is based on the investigation of different non-selective sensors and/or sensor arrays and their signals during the cooking of different goods. In this approach it is assumed that the cooking good emits specific volatile compounds which indicate when the good is done. The objective is to find a specific signal or signal pattern when the "point of interest" is reached. Because of their good overall stability and robustness also in hot and harsh environments, mostly semiconductor metal oxide sensors and sensor arrays are investigated [21]. One major problem which has to be solved is the considerable amount of humidity which is generated during the cooking process. Unfortunately metal oxide sensors are known to show strong cross-sensitivity to humidity. An invention by Panasonic for example even evaluates the humidity signal of the sensor in a bread baker which increases strongly when the crust breaks.

The second, more theoretical approach is based on the chemical analysis of the volatile compounds which are emitted. The scope is to find and to identify one or more volatile compounds which are specific for the current status of the cooking process. A possible disadvantage of this approach is that after the identification of volatile components it cannot be assured that suitable gas sensors are available or can be developed.

5.3.3.6.2 Self-cleaning of Ovens

The control of the self-cleaning procedure (pyrolysis) of especially equipped kitchen ovens is another focus of development. The underlying idea is to burn the organic residues at elevated temperatures (around 400 °C) and to detect the emerging volatile compounds. In order to minimize energy consumption, the process time should be kept as short as possible. During this process considerable amounts of CO and CO_2 are released. A decrease in concentration of these compounds can thus be taken is an indicator of the end of process. The most direct method would be the detection of CO_2 in the flue gas. The most common CO_2-

sensors are based on electrochemical cells or optical NDIR-systems. Unfortunately, electrochemical cells are not stable at higher temperatures so that they cannot be placed directly in the flue gas. A possible solution could be the use of a gas-cooling bypass. Optical solutions are rather expensive and could be hampered by soiling. Other approaches use low-cost self-cleaning Ga_2O_3-metal oxide sensors which are stable at high temperatures, to detect CO prior its conversion into CO_2. The ultimate ambition of developers could lie in the combination of cooking and pyrolysis control in one system.

5.3.4
References

1 Endres, H. E.: Air Quality Measurement and Management. In Sensors in Intelligent Buildings, Sensors Applications Volume 2, edited by O. Gassmann, H. Meixner, Wiley-VCH Weinheim (2000), p. 85–103.

2 Yamazoe, N.; Miura, N.: Some Basic Aspects of Semiconductor Gas Sensors. Chemical Sensor Technology, Vol. 4, Ed. Yamauchi, S., Kodansha Ltd. Tokyo, 1992.

3 Madou, M. J.; Morrison, S. R.: Chemical Sensing with Solid-state Devices. Academic Press Boston, 1989.

4 Göpel, W.; Schierbaum, K. D.: Electronic Conductance and Capacitance Sensors. In: Sensors A comprehensive Survey. Vol. 2, VCH-Weinheim (1991) pp. 430.

5 Hoefer, U.; Steiner, K.; Wagner, E.: Contact and sheet resistances of SnO_2 thin films from transmission line model measurements. Sensors and Actuators B (1995), p. 59–63.

6 Hoefer, U.; Böttner, H.; Wagner, E.; Kohl, C.-D.: Highly Sensitive NO_2 Sensor Device Featuring a JFET-like Transducer Mechanism. Sensors and Actuators B 47 (1-3) (1998), p. 212–216.

7 Weimar, U.: Oxidgassensoren und Multikomponentenanalyse. Dissertation am Institut für theoretische Chemie, Eberhard-Karls-Universität Tübingen, 1993.

8 Walsh, P. T.; Jones, T. A.: Calorimetric Chemical Sensors. In: Sensors A comprehensive Survey. Vol.2, VCH-Weinheim (1991) pp. 531.

9 Homepage of the „Deutscher Wasserstoff-Verband": http://www.dwv-info.de/wiss_vgl.htm

10 Oehme F.: Liquid Electrolyte Sensors: Potentiometry, Amperometry, and Conductometry. In: Sensors A comprehensive Survey. Vol.2, VCH-Weinheim (1991) pp. 240.

11 Tränkler, H.-R.; Obermeier Ernst (Hrsg.): Sensortechnik Handbuch für Praxis und Wissenschaft. Springer Berlin u.a. (1998), p. 1129.

12 Pfannstiel, D.: Modellbildung, Simulation und digitale Regelung eines ölbefeuerten Heizkessels mit kleiner Leistung. VDI-Verlag Reihe 19 Nr. 58 (1992).

13 Kaneyasu, J.; Otsuka, K.; Setoguchi, Y.; Sonoda, S.; Nakahara, T.; Aso, I.; Nakagaichi, N.: A carbon dioxide gas sensor based on solid electrolyte for air quality control. Sensors and Actuators B, 66 (2000) 1–3, p. 56-58.

14 Staab, Joachim: Industrielle Gasanalyse, Oldenbourg Verlag (1994)

15 Landesinnungsverband der Schornsteinfeger in Schleswig Holstein: Toxische Wirkung von Kohlenmonoxid, http://schornsteinfeger-sh.de/.

16 Freiling, A.; Fromm, R.; Kohl, D.; Spahn, C.: Sicherstellung der Schutzfunktion von Gassensoren, Verbundprojekt KombiSens, Hrsg. TMS Innovation Consult, Mühlheim am Main, Beuth Berlin et al, TMS-Schrift 97-1 (1997).

17 United States Environmental Protection Agency: What You Should Know About Combustion Appliances and Indoor Air Pollution, http://www.epa.gov/iaq/pubs/combust.html.

18 JAMA, Vol. 261, No. 8 (1989)

19 Chiba, A.: Development of the TGS gas sensor. In: Chemical Sensor Technology,

Editor S. Yamauchi, Vol. 4, Kodansha, Tokio, 1994.
20 Kohl, D.; Lämmer, M.; Skiera D.: Safety Applications of Gas sensors, EU-Project 28114, semi-annual project report 07/01.
21 Ehrmann, S.; Jüngst, J.; Goschnick, J.: Automated cooking and frying control using a gas sensor microarray. Sensors and Actuators B 66 (2000) 43–45.

5.4 UV Sensors – Problems and Domestic Applications

O. Hilt and T. Weiss

Application areas for UV sensors in the household environment are introduced and the technological requirements and challenges of UV-sensing discussed. Different detection technologies with their strengths and weaknesses are explained. Finally, reasons that limit the use of UV sensors in household appliances are discussed and way outs are lined out.

5.4.1 UV Radiation – A General Introduction

UV-radiation is the part of the electromagnetic spectrum that lies between visible light and x-rays at a wavelength below 400 nm. Wavelengths shorter than 200 nm (VUV radiation) do not figure in this chapter because they are absorbed by air. The relevant spectrum is divided into UVA (320 nm–400 nm), UVB (280 nm–320 nm) and UVC (200 nm–280 nm) radiation. Compared to visible light, UV radiation generally has a greater chemical and biological impact because the photon energy of UV radiation is higher than that of visible light.

5.4.2 UV Radiation in Household Environments

5.4.2.1 Natural UV Radiation

The most common UV-source is sunlight. The UV part of the sun spectrum is restricted mainly to UVA and its intensity strongly decreases below $\lambda = 320$ nm, see Fig. 5.43. The UV intensity ranges up to 6 mW/cm^2, and this is only a few percent of the sun's intensity in the visible with up to 100 mW/cm^2. UV radiation has an impact on the human skin such as tanning or burning (erythema). Other sun UV-related effects are the degradation of organic materials and colors. The intensity and spectral distribution of UV changes with the azimuth of the sun, but also depends on the atmospheric weather condition. Especially the short-wavelength part below 330 nm strongly depends on the condition of the atmospheric ozone layer.

Most light sources for artificial lighting have a UV component in their emission spectrum. Usually, the glass encapsulation absorbs all radiation below about

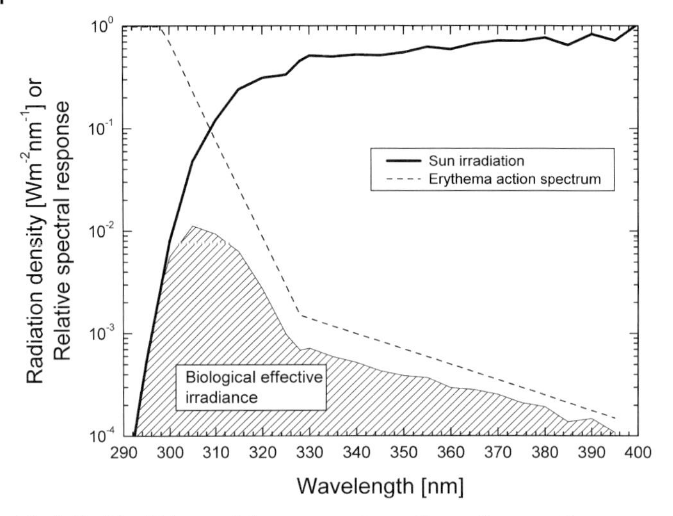

Fig. 5.43 The UV part of the sun spectrum, the erythema action spectrum and the biological effective solar irradiance. The biological effective irradiance has been obtained by folding the radiation density of the sun with the erythema action spectrum.

330 nm. Light bulbs work as a blackbody radiator at a temperature of approx. 2800 K. Temperatures up to 3300 K can be reached by filling a bulb with halogenated gas. The spectral radiation distribution has a tail below $\lambda = 400$ nm that strongly decreases at shorter wavelengths. Its magnitude rises with the temperature of the source of radiation.

Fluorescent lamps generate light through a low-pressure mercury vapor discharge that has strong emission lines in the UV, namely at $\lambda = 254$ nm and around 366 nm. The fluorescent layer is excited by the UV radiation and emits in the visible part of the spectrum. While remains of the 254 nm line are efficiently rejected by the glass tube, some fraction of the 366 nm radiation can be measured in the emission spectrum of the lamp.

Flames and related burning and combustion processes are an important source of UV radiation. Intermediate molecular species like OH, CO and CH groups are generated in an excited molecular state during the oxidation process of the fuel. Their lifetime is very short, usually in the range of nanoseconds, and they may emit UV-photons as they decay to their ground state. The intensity of combustion-related UV radiation lies in the nW/cm^2-range or below. Therefore, the UV component is often ignored when thinking of a flame emission spectrum. However, the occurrence of combustion-originated UV may be used for flame detection or fire alarms. The emission spectrum lies significantly below 330 nm, see Fig. 5.44, and can therefore be well distinguished from ambient UV radiation coming from the sun or artificial lighting.

5.4.2.2 Man-made UV Radiation

Sunbeds with fluorescent lamps that emit in the UVA and UVB are used for indoor tanning. They are supposed to simulate the solar UV spectrum and have therefore similar effects on the human skin. However, the intensity of radiation is often not monitored, and excessive exposure may cause serious dermatological health problems. A more detailed analysis reveals that the long-wave UV (UVA) is mainly responsible for the tanning while UVB radiation tends to be more dangerous.

UVC radiation can not only be used to sterilize tap water, but also for the treatment of air and sewage. Radiation between 250 nm and 265 nm is passes through water and is strongly absorbed by nucleic acids, i.e. any living creature present. This kind of radiation therefore efficiently kills all microorganisms in the water. It is a lucky coincidence that at 254 nm, the main emission line of mercury lamps lies within the range of effective UV sterilization.

5.4.3 Principles of UV Detection

Apart from a few applications, such as UV disinfection and lacquer hardening, the intensity of UV radiation is well below that of visible light in ambient daylight or indoor lighting. A UV sensor must therefore be insensitive to visible light, otherwise the detection signal would easily be drowned out by the visible fraction of the radiation spectrum. Sensors that fulfill this requirement have a selective spectral sensitivity in the UV range. There are two important selectivities, known as visible-blindness and solar-blindness.

A visible-blind UV sensor detects radiation only below $\lambda = 400$ nm and thus is sensitive to the UV radiation of sunlight. A solar-blind sensor does not react to sunlight and usually detects radiation below $\lambda = 300$ nm. An outside fire alarm sensor imposes one of the most stringent requirements for solar-blindness. It must be sensitive to 100 pW/cm^2 or less between 220 nm and 300 nm but should not react to direct sunlight that gives 100 mW/cm^2 between 320 nm and 720 nm.

5.4.3.1 UV-Enhanced Si Photodiode

The most common photodetector is a photodiode with a *p-n* or *p-i-n* junction made from crystalline silicon (Si). Photons are absorbed in the semiconductor, and an electron-hole pair is generated if the photon energy is above the band gap. There is a built-in electric field close to the *p-n* junction and the electron-hole pairs created there may escape from recombination and generate a net current. The band gap of silicon is E_g=1.1 eV, and Si photodetectors are thus sensitive to wavelengths below $\lambda = 1100$ nm. Quantum efficiencies above 80% can be reached, but below 400 nm sensitivity strongly decreases towards lower wavelengths. There, the absorption of Si increases and most of the photons are absorbed before they can reach the charge-density zone.

Special UV-enhanced Si photodiodes can be made by positioning the *p-n* junction close to the surface. Then, quantum efficiencies of 50% can be achieved for λ

between 200 nm and 400 nm. However, the photodiode remains sensitive in the visible. An inherently visible-blind photodiode cannot be made from silicon.

Visible- or solar-blind UV sensors can be made from a Si photodiode by additionally using an optical filter that transmits UV radiation only, see below. A more detailed explanation of the physics of UV photodiodes (made from Si as well as from other semiconductor materials) can be found in Ref. [1].

The most convincing argument for using Si photodiodes in UV detection is the availability of strong expertise in electronic Si devices. Processing and performance of opto-electronic Si devices have been optimized for decades, and the UV-enhanced photodiode is a high-performance niche product that can be produced at a reasonable price, thanks to these efforts. Its probably most serious drawback is the necessity of using filters for visible-blind applications, which considerably increases the cost of sensors and reduces their otherwise optimum sensitivity.

5.4.3.2 Crystalline Wide Band-Gap Semiconductors

Semiconductor photodiodes can be made visible-blind by using a semiconductor with a sufficiently high band gap. Promising materials are SiC (E_g=3.1 eV), GaN (E_g=3.3 eV) and its related compound AlGaN (E_g=3.3–5 eV, depending on the Al/Ga ratio) and diamond (E_g=5.5 eV) [1]. Sometimes, GaP (E_g=2.3 eV) is also used, but due to the low E_g GaP remains sensitive in the blue and green range.

Similar to silicon, crystalline silicon carbide is grown as an ingot and cut into wafers. The market share of SiC is still low but a strong increase can be expected due to its superior properties including charge-carrier mobility, heat-conductivity and maximum usable temperature. This is of particular interest for high-power electronics and highly integrated circuits. Visible-blind UV photodiodes based on SiC with performances similar to Si can be found on the market. However, their price is considerably higher than for Si photodiodes, and especially detectors with large sensing surfaces are very expensive.

GaN as a semi-conducting material for electronics is about to be launched on the market, especially for the use in blue- and UV-emitting LEDs and laser diodes [2]. The material is deposited on crystalline substrates like sapphire using thin-film epitactical techniques. Often, metal-organic chemical vapor deposition (MOCVD) is used. The necessity for such technologies limits the production rate and pushes up costs.

Very promising indeed is the ternary compound AlGaN. By shifting the Al/Ga ratio its spectral sensitivity can be tailored. The cut-off wavelength can be shifted between 380 nm and 310 nm [3]. Quantum efficiencies up to 50% have been obtained for SiC as well as for GaN, which is similar to the UV sensitivity of UV-enhanced Si photodiodes.

5.4.3.3 Polycrystalline Wide Band-gap Semiconductors

Although visible-blind UV photodiodes with good performances can be made from crystalline SiC and GaN, their introduction to the market is hampered by their high production costs. In many market segments, a photodiode made with polycrystalline thin films as wide band-gap semiconductor material could be an alternative. Production costs could be lowered considerably, at the price of lower sensitivity and a longer response time.

Polycrystalline GaN UV detectors have been realized with 15% quantum efficiency [4]. This is about 1/4 of the quantum efficiency obtained by crystalline devices. Available at a fixed price, however, their increased detection range may well compensate their lack in sensitivity. Furthermore, new semiconductor materials with a matching band gap appear as promising candidates for UV detection if the presumption of the crystallinity is given up. Titanium dioxide, zinc sulfide and zinc oxide have to be mentioned. The opto-electronic properties and also low-cost production processes for these compound semiconductors have already been investigated to some extent for solar cell applications [5].

5.4.3.4 **Fluorescence Converters**

Perfect semiconductor photodiodes with a supposed quantum efficiency 1 would generate one electron-hole pair per absorbed photon, regardless of the photon energy. Since photon energy increases with lower wavelengths, conversion efficiency, in terms of photocurrent per incident radiation power, decreases with lower wavelengths. As an example, a photodiode with quantum efficiency 1 would generate a photocurrent of 480 mA/W at $\lambda = 600$ nm but only 240 mA/W at $\lambda = 300$ nm. The sensitivity (photo current per light power) of a photodiode may therefore get increased if the radiation is efficiently converted to a higher wavelength. This concept sounds particularly interesting if UV radiation is converted to visible light, which can be measured by a standard Si photodiode. Standard materials from fluorescent lamp production, such as rare-earth aluminates and yttrium vanadate compounds can be used as fluorescence converters. However, to obtain a visible-blind sensor a UV-transmitting filter is still indispensable.

5.4.3.5 **Discharge Tubes**

In contrast to semiconductor photodiodes, discharge tubes are photo-emissive devices. Photons hit the metallic surface of a cathode, and if the photon energy is above the ionization energy an electron is emitted. An electric field between the cathode and an anode accelerates the electron in a low-pressure gas atmosphere, and the number of electrons are multiplied by the avalanche effect. The resulting current pulse leads to a breakdown of the acceleration voltage. The breakdown rate is then used for measuring the radiation intensity. Suitable cathode materials for UV sensors are molybdenum, tungsten and nickel with a cut-off wavelength of 300 nm, 275 nm and 250 nm, respectively.

Discharge tubes have an excellent detectivity and are solar-blind. These technological advantages have to be balanced against the need for acceleration voltage of some 100 V and complicated readout electronics. These requirements make sensing systems based on discharge tubes much more expensive than those based on photodiodes. The major field of application is therefore fire alarm monitoring in industrial environments. Another problem is the poor reliability of discharge tubes.

5.4.3.6 **Filters**

Filters that block visible light (and sometimes IR radiation) are necessary to make a broadband detector like a silicon photodiode visible-blind. UV-transmitting filters are made either of colored glass or as interference filters. Colored-glass filters like UG5 or UG11 of Schott have a maximum transmittance of typically between 320 nm and 370 nm. They block in the visible and radiation below 260 nm. A problem in using them together with a Si-photodiode is their re-appearing transmittance below 660 nm.

Interference filters consist of several evaporated dielectric layers on a glass or quartz substrate. Their transmittance can be tailored by choosing appropriate layers. A problem is their limited bandwidth of transmission which is usually above $\Delta\lambda = 30$ nm. Also, a substantial loss in sensitivity has to be accepted since the maximum transmission is limited to less than 40%.

5.4.3.7 **The Entrance Window**

The entrance window is a relevant part of the sensor. Common glasses as well as most transparent plastics block UV radiation. Standard UV-transmitting materials like quartz and sapphire are expensive. For UVA and to some extent the UVB special glasses but also plastics like some PMMA-derivatives or some silicon gels are available with a reasonable transmittance down to about 330 nm. For UVC applications with the important 254 nm line of Hg, quartz remains the standard solution.

5.4.3.8 **Concentrator Lens**

For low-intensity applications, a concentrator lens can be used to focus the radiation onto the detector surface. The detector area can be kept small, which is advantageous for the use of expensive wide-band gap UV-photodiodes. The lens must be made of quartz or another UV-transmitting glass and is often used as the entrance window of a standard TO housing. A lens limits the angle of acceptance to usually less than 10 degrees.

5.4.3.9 **Packaging**

Most UV photodiodes have a sealed metal TO package with an entrance window made of glass or special UV-transmitting glass. A TO housing is extremely reliable but it often counts for a non-negligible part of the sensor costs. Cheaper housings that should be favored for consumer-product applications like in household appliances are hard to find. Full-plastic encapsulation requires a UV-transparent plastic that also has good processing parameters. The solution to this problem is still elusive, probably because the market share does not seem big enough.

5.4.4 Household Applications

5.4.4.1 Personal UV Exposure Dosimetry

The impact of UV radiation on the human skin not only enhances pigment production (tanning), but also causes erythema (skin injury, sun burn). There is a huge market for a small portable UV dosimeter that gives an indication when sunburn has to be anticipated. Whereas such a consumer product would have to be cheap and straightforward, serious UV-dosimetry remains quite a challenge.

UV intensity leading to erythema is wavelength-dependent, as shown in the erythema action spectrum [6], see Fig. 5.43. Biological skin damage increases strongly with shorter wavelengths. On the other hand, solar radiation intensity decreases strongly below 340 nm. Combining these two dependencies gives a maximum erythemal effect of the sun between 300 nm and 320 nm, see Fig. 5.43.

A medically reliable UV sensor should have a spectral responsivity that closely follows the erythemal curve between 390 nm and 290 nm. So far, a photodiode with this specific sensitivity has not been available. Modeling the erythema spectrum with the help of filters also delivered only poor results. In fact, most available sunburn detectors vary in their spectral responsivity and may therefore only be used as an indicator for the actual UV charge.

5.4.4.2 Surveillance of Sunbeds

Monitoring the radiation of sunbeds serves two major purposes – finding out the current radiation intensity and establishing the (accumulated and biologically effective) dose the user has been exposed to. The intensity of the fluorescent lamps that are usually used as UV source decreases in time. As a consequence, a longer tanning session is needed for a particular dose.

In modern systems, the applied electric power is increased over time to compensate for the decreasing intensity. However, this adjustment uses the average degradation curve of a lamp as a reference, not the actual values of the lamps mounted in a particular sunbed. A radiation sensor could be used to control an active circuit that regulates the intensity of radiation.

Such a sensor should detect between $\lambda = 320 - 380$ nm and be exposed to an intensity in the range of 1–10 mW/cm^2 [7]. Visible-blindness is required since sunbeds are usually open to ambient light. Measuring the biologically effective radiation intensity or the radiation dose, integrated over time, is a more sophisticated procedure. As in outdoor UV-detectors, the (physical) spectral radiation-intensity distribution has to be weighted against a biological action spectrum to obtain the biologically active radiation intensity. There are different action spectra for skin cancer, erythema and tanning [8].

5.4.4.3 **Flame Scanning in Gas and Oil Burners**

In household heating systems running on gas or oil the presence of the flame has to be surveyed. If the flame extinguishes, a burner control system has to re-ignite the flame or the fuel supply has to be stopped. Several flame-sensing mechanisms are used in the control systems.

A flame can be sensed by its radiation, either in the infrared, the visible or the UV. Sensing on basis of visible or infrared radiation is advantageous with respect to the radiation intensity and the sensor-system costs. However, there are some drawbacks concerning the safety aspect of the sensing method. The sensing system has to make absolutely sure that the detected radiation has its origin in the flame. At elevated temperatures, the combustion chamber itself emits infrared radiation which may not easily be distinguished from the infrared radiation of the flame. Visible light may hit the sensor, coming not from the flame but from the outer environment if the cover of the burner is damaged or opened. Flicker-sensing electronics have been added to the radiation detectors to distinguish the varying radiation of the flame from other background radiation. However, the low-cost advantage is eventually lost. The UV radiation of a flame is of low intensity but unique. It can merely be affected by other radiation sources and since its occurrence is directly related to the combustion process it also allows an immediate response.

The yellow flame of traditional oil burners is often surveyed by flame guards that react on the visible emission. More modern oil heaters use blue-burning flames. There, the carbon black of the combustion process is redirected to the combustion area for a more complete combustion, thus giving higher efficiency. With the reduced carbon black their yellow emission in the visible also vanishes, leaving an almost invisible flame, see Fig. 5.44. For these blue flames surveillance based on UV emission is preferable

As can be seen in Fig. 5.44 as well, flames in gas heaters have a similar emission spectrum. Besides the UV surveillance ionization electrodes are often used in gas burners. The method is cheap and secure but it disturbs the combustion process since the electrode has to be placed close to the flames. New developments in gas heaters focus on catalytic combustion on a metal mesh. There, an ionization electrode would fail due to the lack of a flame. However, the characteristic UV emission is still present.

UV flame sensors are usually mounted at the rear of the burner and sense the back of the flame through an about 10 cm long narrow channel. This sensor position is chosen to avoid sensor temperatures above 60 °C, but UV-radiation intensity often does not exceed some 100 pW/cm^2. Accordingly, the requirements for the visible-blindness have to be stringent. The sensor sensitivity must be $10^4\times$ higher in the range of $\lambda = 280 - 360$ nm than above $\lambda = 400$ nm [9, 10]. The low radiation intensity and noisy environment at the burner require a sensitive detector with good electrical shielding. Often, the photodiode and a pre-amplifier are combined in one TO-housing.

A popular UV sensor designed for flame monitoring is the SFH 530 sensor by Infinion. It is a UV-enhanced Si photodiode with optical filter, concentrator lens and operational amplifier in a TO-39 package. Flame-sensing in gas ovens is another potential application of the described technique.

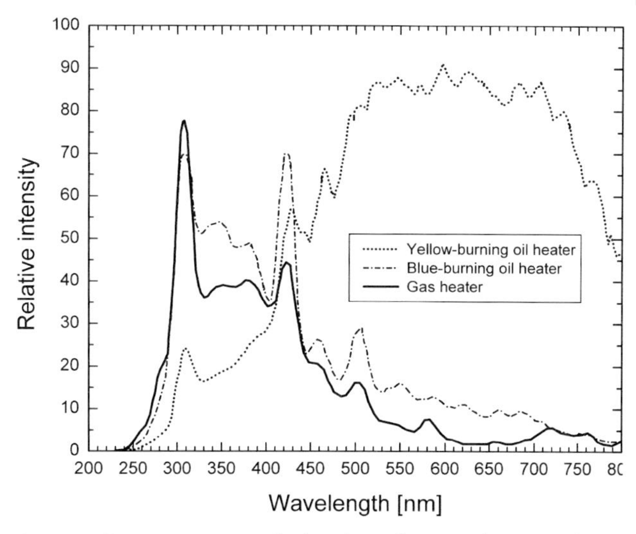

Fig. 5.44 Emission spectra of oil and gas flames in burners. The characteristic peak at $\lambda = 310$ nm determines the signal of a UV flame-monitor.

5.4.4.4 **Fire Alarm Monitors**

The excellent detection ability for flames makes UV sensing a good method for remote fire alarm-monitoring. UV radiation after the outbreak of a fire reaches a sensor much faster than heat or smoke. Also, the distance between sensor and fire is less critical. Requirements for the sensor are high sensitivity and excellent selectivity. Radiation intensities at the sensor position may be even lower and the ambient light conditions less restricted than for combustion controlling. When used outside, solar-blindness is a must. These stringent requirements make UV fire alarm monitors expensive, and they are used in industrial environments such as production floors or warehouses rather than in private homes.

5.4.4.5 **Water Sterilization**

Water sterilization systems based on UV radiation are available ranging from household sizes to systems capable of treating communal water supplies. In smaller systems, the water flows through a cylindrical reactor housing and is illuminated by a low-pressure Hg lamp situated on the cylinder axis inside. The required dose is at least 40 mJ/cm^2 [11], and typical radiation intensities are in the range of 10 mW/cm^2. The lifetime of the Hg-lamp usually is above 10,000 h but its intensity decreases with time. Also, the transparency of the glass housing of the lamp may be reduced due to algae or scale. To ensure complete sterilization, monitoring the radiation intensity seems appropriate. The UV sensors needed have to detect the intensity at 254 nm. German rules [11] require a selective responsivity for $\lambda = 240 - 290$ nm where the sensitivity above $\lambda = 290$ nm must not exceed 10% of the sensitivity in the detection range. A solar-blind system is required, but the demands on their selectivity are less stringent than for flame or fire-alarm monitors, as the intensity to be detected is much higher.

Another problem is degradation of the sensor due to the high UV dose. The radiation resistance of most photodiodes decreases with wavelengths. UV-enhanced Si photodiodes show a loss of 10% in sensitivity already after an accumulated dose of some hundred J/cm^2 at $\lambda = 254$ nm. This is the dose a sensor will have received over the lifetime of an Hg lamp. Special silicon nitride-protected photodiodes are stable up to 10^5 J/cm^2. A filter combined with an attenuator may help to achieve the required selectivity and reduce the exposure of the detector. However, the radiation stability of the filter has to be guaranteed.

5.4.5 The Market Potential of UV Sensors in Household Appliances

So far, UV sensors have been rarely seen in households although a small proportion of the applications described can be found. Sometimes other sensing methods are used (as in flame controlling) although they are less suitable from the technological point of view or they are simply omitted (as monitoring of sunbeds). Often, the major reason is the high price of the sensor.

Sufficiently cheap UV photodiodes are available but they are not visible-blind. Filters have to be used, but they raise the costs. Sufficiently selective photodiodes are also available but they are too expensive, mainly due to their only recently established technology. The sensor costs have been a limiting factor in two application fields of UV sensors, namely water disinfection and combustion monitoring, on the industrial as well as on the household scale.

In communal-scale water disinfection plants, UV monitoring systems are generally included. However, for household-scale systems they are not available or only as an optional accessory.

In power stations or other combustion units above a capacity of, say, 1 MW, UV sensors for flame monitoring are not unusual. Here, even combustion parameters like the air or fuel supply are controlled by sensing the UV emission spectrum of the flames.

To make a breakthrough in household appliances and other consumer product markets UV sensors have to become significantly cheaper while spectral selectivity as a major key feature must be guaranteed. Most of today's UV photodiodes are made from crystalline semiconductor materials. The cheaper materials (Si) lack spectral selectivity, and the wide band gap materials are very expensive. What they all have in common their top performance regarding sensitivity and speed. Crystalline photodiodes have risetimes of often below 1 s. However, the described processes to be sensed here are not faster than some milliseconds or even much slower. In order to obtain a reasonably-priced SiC or GaN photodiode, the photoactive area is often reduced to below 1 mm^2 and barely fills the sensor housing. So far, the top sensitivity offered by the semiconductor has been sacrificed for a competitive

Tab. 5.12 Compilation of UV photodiodes

Type & manufacturer	*Active material & construction*	λ_{peak} [nm]	$\lambda_{>10\%}$ [nm]	S_{peak} [mA/W]	*Visible blindness* $S_{peak}/S_{400\ nm}$	*Active area* [mm^2]	*Price indication* (*)
Hamamatsu S1226-BQ	UV-enhanced Si	720	190–1000	360 (130 in the UV)	–	1.2	**
Perkim Ellmer DF-300	UV-enhanced Si + interference filter	300	290–310	120	>1000	20	****
Infinion SFH 530	UV-enh. Si + filter + lens + amplifier	320	280–340	110 (10^9 V/W)	1000	1, lens aperture 11	**
Hamamatsu G5842	GaAsP + filter	370	260–400	60	15	0.58	*
EPIGAP EPD-365-0/1.4	GaP + UG11 filter	355	260–380	70	200	1.2	***
ifw JEC1	SiC	275	210–380	130	40	1	****
ifw JEC1B	SiC + filter	315	280–325	80	>1000	1	****
ifw JECF 1 I-DE	SiC + filter, resembles erythema curve	285	230–310	2	10000	1	****
SVT Assoc. GaN-0.8D	GaN	360	200–375	160	2000	0.5	****
SVT Assoc. AlGaN	AlGaN	280	190–290	30	1000 $S_{peak}/S_{300\ nm}=80$	0.5	****
twlux TW30SY	*poly*-TiO_2	300	220–385	20	100	15	*
Hamamatsu UV-TRON	gas discharge tube with Mo cathode	200	170–260	reacts to 0.1 pW/cm^2	$S_{peak}/S_{300\ nm}>1000$	–	***

* Price indication for quantities below 10 pcs. *: <20 €; **: 20–40 €; ***: 40–100 €; ****: >100 €.

Fig. 5.45 A twlux PFD306B visible-blind UV photodiode based on polycrystalline TiO_2. The chip with an active area of 2.9×5.4 mm^2 is mounted on a TO-39 header. The sealed cap has an entrance window made from Schott's UV-transparent glass that transmits down to 240 nm.

price. Developing a photodiode with, say, 10× larger photoactive area and a quarter of the sensitivity would be much more worthwhile.

In the near future, UV photodiodes made from polycrystalline wide band-gap semiconductors may fill the gap in the market. Although they have a lower sensitivity (photocurrent per area) they promise to have a better merit-rating in terms of photocurrent per sensor costs. The other major drawback of polycrystalline photodiodes, the risetime of micro- to milliseconds, is not relevant for household applications. Fuji Xerox Laboratories in Japan are developing visible-blind UV photodiodes made from polycrystalline GaN [12], while twlux AG in Berlin, Germany is developing visible-blind UV photodiodes made from polycrystalline titanium dioxide [13]. A prototype is shown in Fig. 5.45.

5.4.6
References

1 M. Razeghi and A. Rogalski, J. Appl. Phys. **79**, 7433 (1996).
2 J.-Y. Duboz, Phys. Stat. Sol (a) **176**, 5 (1999).
3 F. Omnes et al., J. Appl. Phys. **86**, 5286 (1999).
4 S. Yagi, *Appl. Phys. Lett.* **76**, 345 (2000).
5 C. Rost et al., Appl. Phys. Lett. **75**, 692 (1999).
6 "A reference action spectrum for ultraviolet induced erythema in human skin", *CIE Journal* 6, 17-22 (1987).
7 "Health Issues of Ultraviolet A Sunbeds used for Cosmetic Purposes", Health Physics **61**, 285, (1991).
8 "Reference action spectra for ultraviolet induced erythema and pigmentation of

different human skin types", CIE Technical Collection 1993/3 (1993).
9 European Standard EN 298, CEN, Bruxelles (1998).
10 European Standard EN 230, CEN, Bruxelles (1991).
11 "UV-Desinfektionsanlage für die Trinkwasserversorgung – Anforderung und Prüfung". Technische Regel W294, DVGW, Bonn (1997).
12 S. YAGI and S. SUZUKI, Proc. Int. Workshop on Nitride Semiconductors, IPAP Conf. Series **1**, 915–918 (2000).
13 More detailed information can be found under www.twlux.com.

5.5 Displacement Sensors in Washing Machines

E. HUBER

Displacement sensors are used in a very wide field of applications in research, development, quality inspection, automation, machinery and process control. Many physical parameters can be reduced to a displacement or change of distance, and can be measured with highest precision. Displacement sensors detect physical parameters like bending, deflection, deformation, diameter, eccentricity, elongation, gap, length, play, position, revolution, roundness, shift, stroke, thickness, tilt, tolerances, vibration, wear and width.

5.5.1 Contact and Non-contact Displacement Sensors

There are two product categories of displacement sensing technology. Whenever the moving target is mechanically connected to a sensor, it is referred to as a 'contact' displacement sensor. The physical principles adapted for contact displacement sensors are primarily:

- inductive
- magnetostrictive
- potentiometric

If the object to be measured cannot be mechanically attached to the moving target, does not permit contact or any external force, a non-contact displacement sensor has to be used. The physical principles for non-contact displacement sensors are primarily:

- laser-triangulation
- eddy current
- capacitive
- camera vision systems
- ultrasonic

Selection of the measuring principle is mainly determined by the requirements of the measuring task. A comparison of the specifications for linearity, resolution, band with and temperature stability will qualify one or the other principle. In combination with the sensor technology, the sensor design and electronics and can be selected to obtain the desired performance, measurement accuracy and life span.

5.5.2
Measuring the Load of Washing Machines

Washing machines are responsible for a substantial part of the total power consumption in a household. By improving their efficiency, major energy savings can be achieved. Optimal detergent dosing in connection with fine-tuned minimal water and energy use in washing machines is becoming increasingly important from an economic and environmental perspective. According to EU directives concerning labeling of household appliances, washing machine manufacturers are legally bound to declare and to label the energy consumption of the machine, making it clearly visible to the consumer.

To achieve the best washing performance, the drum should be loaded to its maximum tolerable weight. A half-loaded machine requires nearly the same energy, water and detergent as a fully loaded machine. In order not to overload the machine and to wash less laundry more economically and environmentally friendly, washing programs have to be adjusted to the actual load. For an optimum adjustment, this load has to be measured prior to the start of the program. The washing machine therefore has to become "intelligent".

The basis for this latest development was a displacement sensor designed by Micro-Epsilon. The sensor combines unbalance detection and measuring of the load in a single sensor module. With the information provided by the sensor, the electronic controller of the machine detects the laundry load and suggests to the user the required amount of detergent. Moreover the necessary amount of water is controlled. By sensing the unbalance of the drum, the number of revolutions

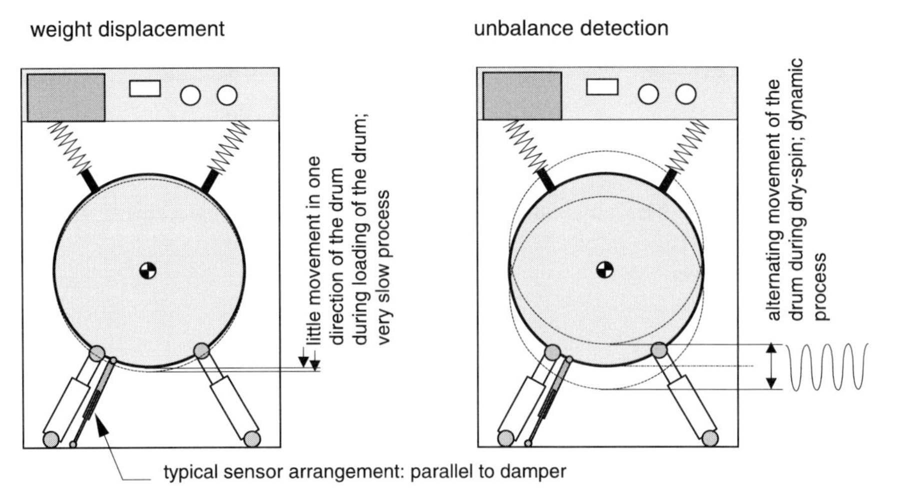

Fig. 5.46 Sensor arrangement.

Tab. 5.13 Requirements of the sensor system.

Functional requirements	– Measuring range 50 mm – Non-linearity: ≤±1% FSO (Full Scale Output) – Sensor resolution: 11 BIT – Operating voltage: 5 V – Operating current: 20 mA
Connector	Integrated standardized plug according to RAST 2.5 standard
Shock and vibration:	
Vibration:	DIN IEC 68-2-6
Shock:	DIN IEC 68-2-27
Environmental conditions:	
Storage temperature	–40 °C...+70 °C
Operating temperature	+5 °C...+70 °C
Air humidity	95% (non-compensating)
Electronic protection	Protection class II, VDE 0700
Material	UL 94, V-0

during the spin-dry cycle can be regulated. Smooth running and high spin-dry efficiency can be achieved. The stress on the bearings is thus reduced, and the lifetime of the machine will be extended.

5.5.3 Displacement Sensors in the Washing Machines

The drum of the washing machine is coupled to the housing via springs and dampers. During loading, washing and spinning, the drum moves in relation to the housing. Displacement sensors in the washing machine detect the weight and unbalance of the drum. The sensor therefore must be able to detect static and dynamic displacement.

In order to satisfy both requirements of load displacement and unbalance displacement with one sensor system, the sensor must have:

- high dynamic resolution of the total measuring range and
- high sensitivity to resolve slight drum movements during the loading process

5.5.4 Sensor Design and Measuring Principles

One special feature of the 'sensor parallel to damper' design is that the sensor can be used with any damper system with a defined neutral position (Figure 5.46).

Fig. 5.47 Sensor mounted parallel to damper, photo: courtesy of MIELE.

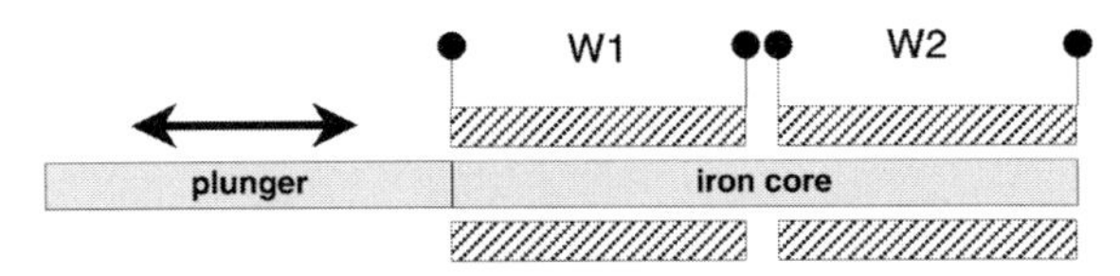

Fig. 5.48 Inductive measuring principle.

5.5.4.1 Inductive Measuring Principle

The displacement of the drum is detected by a rod connected to a magnetic iron core. The core moves inside a coil with two separate windings. The measurement is based on magnetic coupling between the two windings. This also means that the resolution of transducer is infinite. Dependant on the amount of iron inside the coil, the voltage induced in the windings changes. A bridge circuit measures the change of the induced voltage and transforms the movement of the core into a linear electrical signal.

5.5.4.2 Displacement Sensor Integrated into Damper

To reduce noise emission during the washing and spinning process, an improved spring – friction damper – sensor system has been developed jointly by Micro-Epsilon and SUSPA. During high speed rotation the dynamic force of the drum is transmitted to an integrated spring of the damper. During this process, the friction liner does not move, thus a smooth movement with less noise can be achieved. Less spinning energy is needed, and the drives can be made smaller and at a lower cost.

Special features of the 'sensor integrated into damper' design:

- sensor is protected against environmental influences by damper housing
- combined mounting of the damper and sensor simplifies installation

Fig. 5.49 Sensor integrated in damper, photo: courtesy of B/S/H and SUSPA.

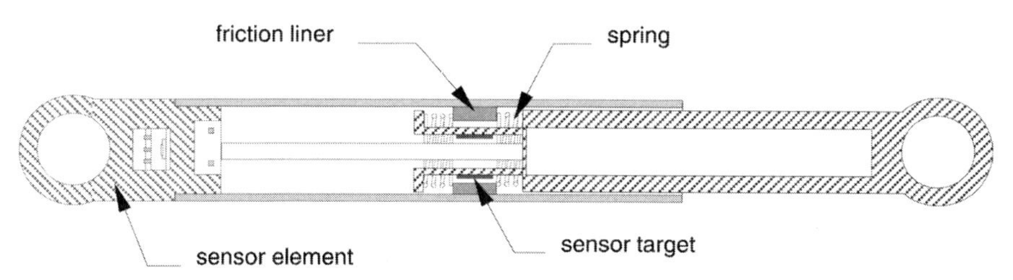

Fig. 5.50 Function of sensor integrated into damper.

- less mounting effort and lower costs
- the sensor body is used for rigid mounting of the damper to the housing of the washing machine
- electrical connection is integrated into the friction damper with an industry standard connector

5.5.4.3 **Inductive–Potentiometric Measuring Principle**

The sensor body consists of a coil with two connections and an electrode with a tapping point. An aluminium measuring ring is attached to the object to be measured. The measurement is based on the eddy – current principle. The non-conductive layer between the core and the coil produces a capacitive circuit. The output signal is related to the position of the target.

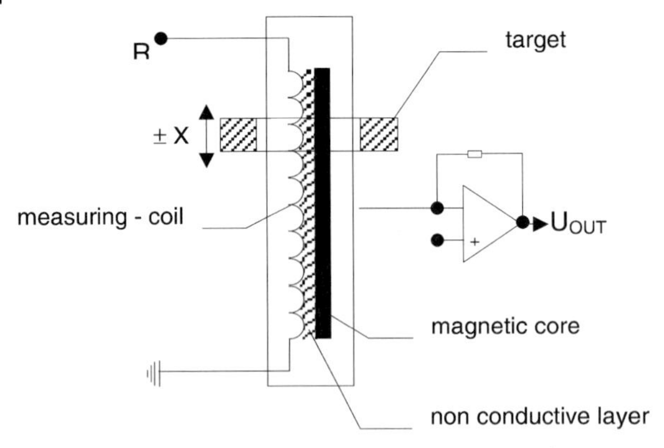

Fig. 5.51 Inductive-potentiometric measuring principle.

This patented sensor principle allows the integration of the sensor inside a compact friction damper. Compared with the limited space inside the damper, a maximum usable measuring range could be achieved.

5.5.5 Sensor Control and Signal Evaluation with Discrete Electronics

1. oscillator
2. sensor driver with complementary output
3. sensor with electrical connections
4. synchronous demodulator
5. amplifier

The components of the discrete electronics can be easily integrated into the control unit of the washing machine. This enables the user to define layout and performance of the electronics.

For direct control and signal processing by a micro-controller the sensor coil is primarily fed by an oscillating digital input signal. The sensor output is then demodulated, filtered and amplified. The secondary digital signal processing is achieved

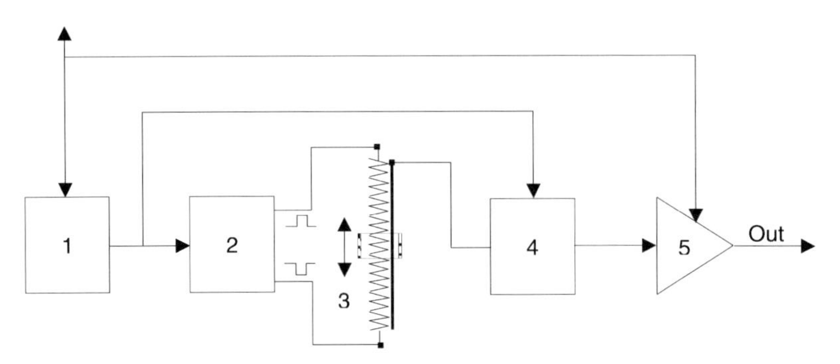

Fig. 5.52 Sensor control and signal evaluation with discrete electronics.

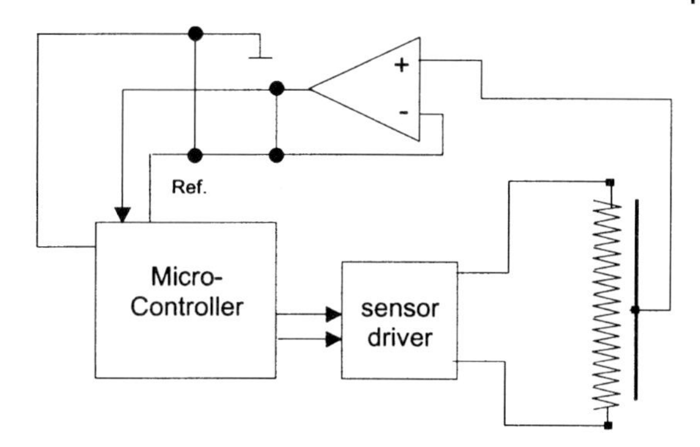

Fig. 5.53 Sensor control and signal evaluation with micro-controller.

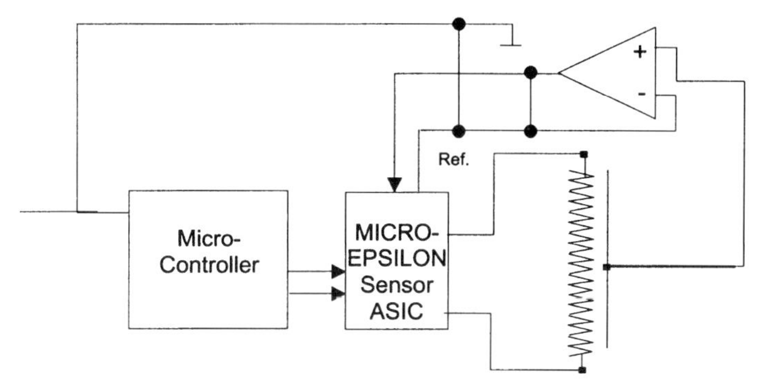

Fig. 5.54 Sensor control and signal evaluation with MICRO-EPSILON sensor ASIC.

with an integrated A/D converter of the micro-controller. The use of direct control and signal processing from a micro-controller, makes analog signal processing superfluous. The amount of necessary electronic components is reduced to a minimum.

To separate the processes of the integrated micro-controller of the washing machine from sensor control and signal processing, a specific sensor ASIC is used. The ASIC drives the sensor, performs a pre-processing of the sensor output signal and then digitises the signal. The sensor ASIC can be integrated into the sensor housing.

5.5.6 Summary

By sensing the displacement of the drum, considerable advantages for the manufacturer and end users of washing machines are achieved:

- the direct controlling of the unbalance of the washing machine allows the reduction of vibrations and increased the lifespan of the washing machine

- the necessary power for the electrical drives can be reduced
- automatic adaptation of the washing process according to the load
- the requested amount of the detergent is indicated, thus saving detergent, energy and water as well as increasing cleaning efficiency
- further developments permit automatic dosing of the detergent, dependant on the selected washing program

Integration of displacement measurement in more and more washing machines can be expected in the future.

5.6 Low-Cost Acceleration Sensors in Automatic Washing Machines

R. Herden

5.6.1 Imbalance in Automatic Washing Machines

Automatic washing machines are subject to the greatest demands when spinning the washing. In order to reduce water consumption, modern washing machines spin the washing not only at the end of the program but also after the main wash and after the rinsing cycles. During the final spin, ever faster spin speeds remove more and more water from the washing to save time and energy during the subsequent drying process. As a result, modern machines spin much more frequently in all wash programs, and reach speeds of from 1600 to 1800 rpm during the final spin. These high speeds subject on the washing to accelerations of up 800 times the acceleration due to gravity, thus imposing corresponding forces on the washing machine.

The washing is not always evenly distributed around the drum, resulting in imbalances. The magnitude of these imbalances depends, among other things on the type, quantity, size and weight of the individual pieces of washing. The position and extent of the imbalances are distributed statistically, leaving only limited scope to influence the imbalance, depending on how the washing is positioned (Fig. 5.55).

During the spinning cycle, water is withdrawn from the washing, thus normally also reducing the imbalance. In other words, the imbalance is reduced with increasing spinning time and speed. This only applies to imbalances affected by the withdrawal of water, but other items can be also washed in a washing machine, such as rubber boots, rain coats, foils, mats, etc. In such cases it is possible for water to accumulate inside the items, which cannot then be drained away. The imbalances thus persist, possibly right through to the top spinning speed.

It is also possible for washing to be virtually evenly distributed when it is placed in the machine so that it spins with only a very small imbalance. However, if the batch of washing includes an item which drains poorly, if at all, and the rest of the washing drains well, then here it is possible for imbalances to occur only when the spinning cycle has reached higher speeds.

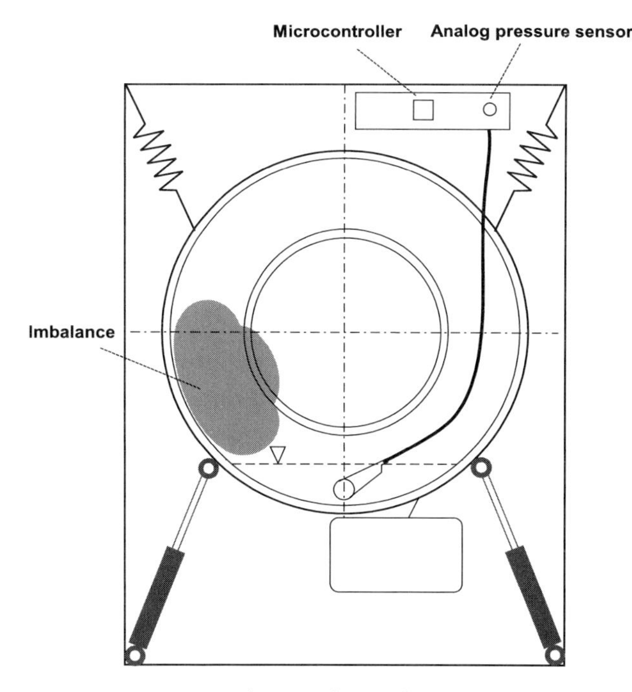

Fig. 5.55 Automatic washing machine with imbalanced washing.

The situation described here illustrates how important it is to know the washing imbalance value with the greatest possible accuracy throughout the whole spinning process. In this way, the particular imbalance situation can always be taken into full consideration and the washing can be drained to the best possible degree. The cycles are thus much gentler on the washing machine, which can then achieve top performance ratings and do not have to be rated for imbalance situations that only occur very rarely.

5.6.2
Normal Procedure for Detecting Imbalance in Automatic Washing Machines

The spinning procedure starts with the positioning phase. The speed is increased slowly, increasing the centrifugal acceleration on the washing until at approximately 100 rpm it is all spread completely around the drum wall, distributed as evenly as possible. Normally an imbalance will have developed at this stage.

The drum is driven by a belt from a motor.

The speed of the drive motor and thus of the drum is normally measured by a multipolar tachogenerator and controlled to a defined value with the assistance of a microprocessor.

If an imbalance has developed at this stage, it will also affect the acceleration due to gravity. As a result, the speed of the drum is accelerated when the imbal-

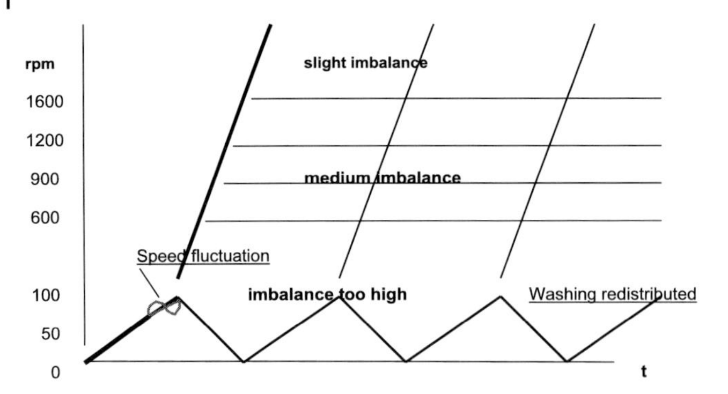

Fig. 5.56 Detecting imbalance from speed fluctuations.

ance moves downwards, and it decreases when the imbalance has to be lifted, resulting in periodic fluctuations from the nominal speed. The amplitude of this fluctuation is an indication of the imbalance in the drum. The microprocessor evaluates this fluctuation. On the basis of defined limit values, it decides which maximum speed can be allowed, or indeed whether this imbalance can be spun at all. If the imbalance is too large, the washing is redistributed (Fig. 5.56). Most washing machines implement this process with control drives.

However, this system has considerable drawbacks. If, along with the imbalance, there is a still larger quantity of evenly distributed washing in the drum, the greater moment of the inertia of this mass stabilises the speed of the drum and gives the impression of a lower imbalance. This means that frequently only half of the actual imbalance is detected.

At higher speeds, the effect of acceleration due to gravity compared with centrifugal acceleration is so slight that it no longer causes any measurable fluctuations in speed. In other words, it is no longer possible to observe the drainage behavior of the imbalance at higher speeds so that the worst case scenario of a non-draining imbalance (rubber boots) always has to be presumed. As a result, the maximum speed is restricted, although a higher speed would normally be possible after draining away the imbalance.

However, this procedure actually fails to detect any imbalance formed by uneven drainage of water from the washing, which only emerges at higher speeds. So the aim is to find a procedure which can detect an imbalance at both low and high speeds.

5.6.3
New Procedures for Detecting Imbalance in Automatic Washing Machines

In the most frequently used type of domestic washing machine, the drum containing the washing turns in a suds container. The suds container is spring-mounted in a housing with shock absorbers connecting it to the bottom of the housing in order to attenuate vibrations (Fig. 5.55).

The critical resonance speeds of this spring mass system are between approximately 120 and 350 rpm. The imbalance in the rotating drum not only produces the speed fluctuations described, but the resulting forces and counter-forces also move the whole suds container in its spring-mounted suspension. In the sub-critical and super-critical speed range, this movement is an indication of imbalance in the drum. During progress through the resonance range, the movement of the suds container is not defined.

The easiest way of measuring movement of the suds container is with distance sensors. Chapter 5.5 describes some typical examples. The advantage of these sensors is that they are also used to monitor the washing load. Their drawback is the high price.

Movement of the suds container always results in a change in length of the springs, thus the same information can also be obtained by measurement of the force at the suspension points of the springs. Up till now, this type of solution has not been used for cost reasons.

As a result of sensor development in the automotive industry, low-priced acceleration sensors are now available. Acceleration sensors are fundamentally also suited to observing excursion (Fig. 5.57) of the suds container caused by imbalance. However, static measurement of the weight of the washing, as achieved with a distance sensor, is not possible with acceleration sensors.

In order to be able to measure imbalance well into the sub-critical speed range at around 100 rpm and also at higher speeds of around 1600 rpm, it must be pos-

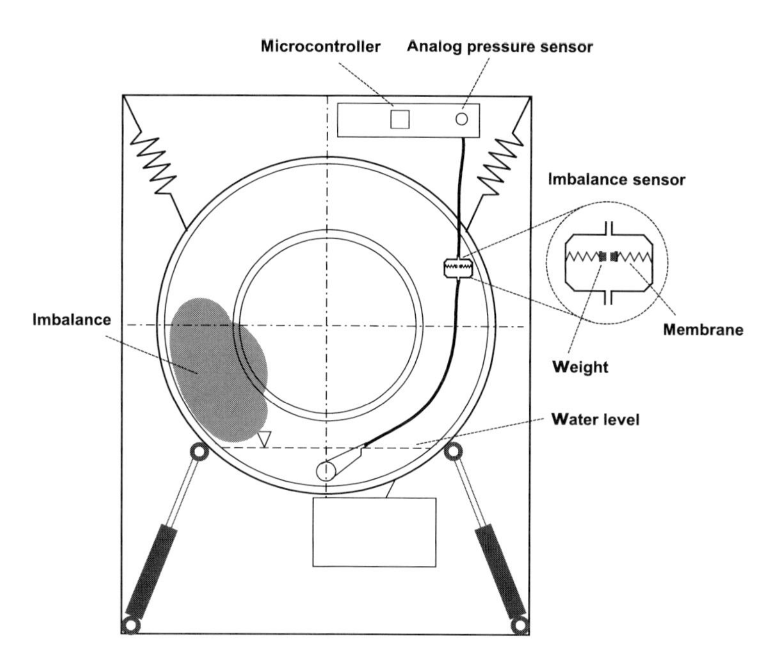

Fig. 5.57 Washing machine with level sensor and additional imbalance sensor component.

sible to detect acceleration from about 0.6 to 160 m/s^2. This large range of measurement is a problem for many sensors. In addition, the relatively low-priced sensor elements also require electronic circuitry, housing, wiring and signal processing, which then equal or exceed the costs of distance measurement as described above. Thus, these are the reasons why such acceleration sensors have not been put into wide spread use up till now.

However, very economical measurement of acceleration is possible when the washing machine is equipped with a high-resolution analog pressure sensor with microprocessor control. These pressure sensors come as a separate part or are integrated directly on to the PCB of the washing machine control. They work as differential pressure sensors and measure the static pressure via a hose in the lower part of the suds container, thus measuring the water level in the suds container (Fig. 5.57).

During the spinning cycle, the suds pump is switched on constantly and the pressure sensor does not detect a level.

An additional component, fitted in the hose pipe between the pressure sensor and the suds container and fastened to the suds container, is capable of generating a pressure signal from the dynamic movement of the suds container. This is then sent via the pressure sensor to the microprocessor for evaluation (Fig. 5.57).

This component consists of a housing with two fittings for the hose connections to the pressure sensor and to the lower suds container. The inner volume of the component is divided by an elastic membrane with a weight in the middle. A bore in the weight allows for the exchange of air between the chambers. Measuring of the water level is not affected because the bore always allows for pressure compensation.

During the spinning cycle, the mass inertia of the weight with the elastic membrane generates a pressure signal when the suds container moves. As a result, extremely low-priced, robust acceleration detection is possible in these washing machines.

This principle does not apply to low acceleration rates at sub-critical speeds, hence sub-critical imbalance detection methods based on speed fluctuations, as described above, are used here (Fig. 5.58).

Further cost reductions are possible if the pressure sensor can be fastened to the vibrating suds container instead of requiring an additional component. Pressure sensors normally measure the force exerting pressure on a defined surface. Depending on the principle of the particular pressure sensor, it will require an accelerating mass which, when accelerated, produces an additional force and the resulting acceleration signal. Three possible examples are described below.

Fig. 5.59 shows a section through a Piezo-resistive silicon pressure sensor. The ambient pressure is applied from above, while the pressure being measured is applied from below. A silicon membrane that deforms under the pressure is applied to a silicon carrier structure. Piezo-resistive structures are fitted in the membrane, which then change their resistance accordingly when the membrane deforms. A bridge circuit generates an electrical output signal which is proportional to the difference in pressure.

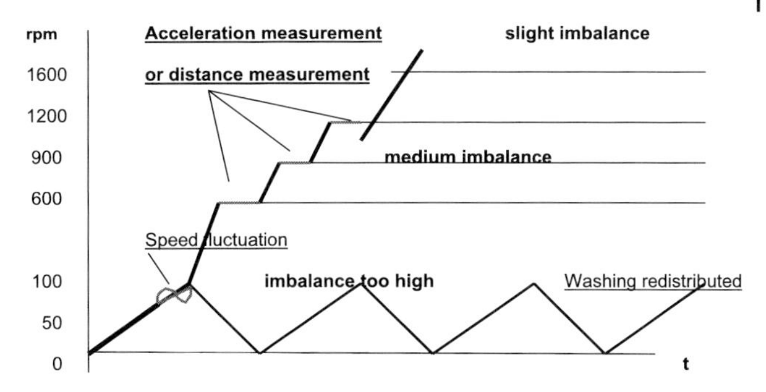

Fig. 5.58 Detecting imbalance with acceleration or distance sensors.

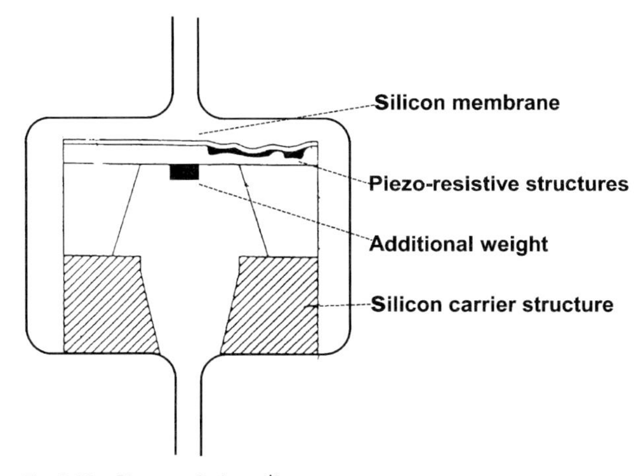

Fig. 5.59 Piezo-resistive silicon pressure sensor.

In order to measure the acceleration, the membrane mass is increased by an additional mass. In this way it is possible for this sensor to detect both the water level and the acceleration of the suds container.

The sensor shown in Fig. 5.60 is based on ceramic bending beam technology. A membrane divides the sensor housing into two chambers. The pressure is applied from below through a connecting fitting. A ball fitted to the membrane transfers the pressure to the ceramic bending beam, which deforms according to the amount of pressure applied to the connecting fitting. The deformation changes the values of the resistors mounted on the ceramic bending beam. An electronic evaluating device also fitted on the ceramic bending beam converts the change in resistance into a voltage that is proportional to the pressure and places this at the output pins. An additional mass is mounted on the ceramic bending beam to detect acceleration as well as pressure.

Fig. 5.61 shows a section through a capacitive pressure sensor. It consists of a metallic housing that is divided into two chambers by the electrically conductive

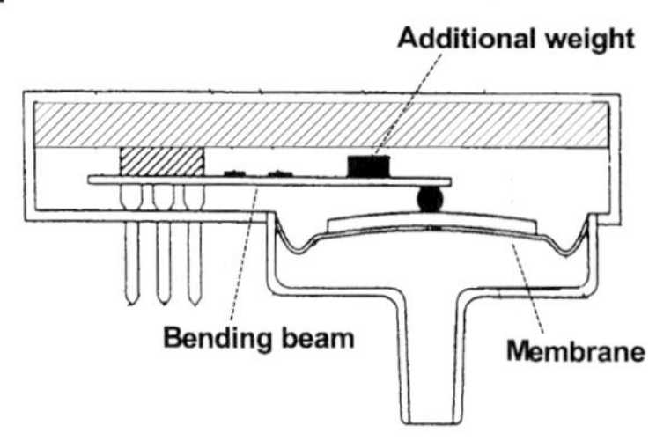

Fig. 5.60 Pressure sensor with ceramic bending beam.

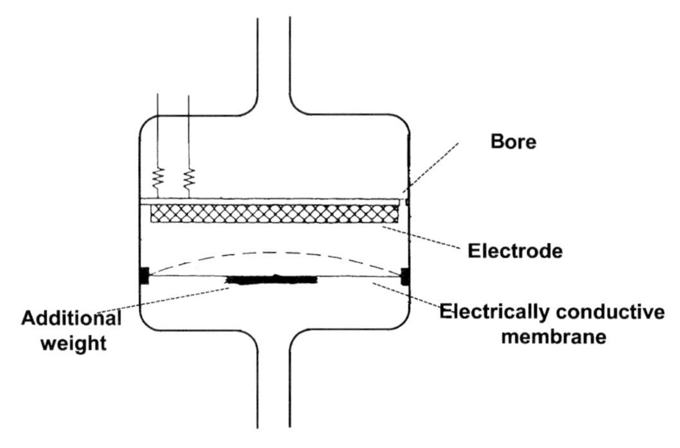

Fig. 5.61 Capacitive pressure sensor.

membrane. The membrane is welded to the housing. An electrode bracket is fitted in the chamber. Electrodes are vaporized onto the bracket: the potential of these electrodes is conveyed via the leads from the housing to the microprocessor control. The electrode bracket is drilled through on one side so that it is permeable to gas. In order to measure the differential pressure, the pressure being measured is applied to the connecting fitting. The atmospheric air pressure increases on the reference pressure side in the chamber. The increasing pressure moves the membrane toward the electrode. The membrane is connected to the housing so that the shorter distance results in an increase in capacity between the housing and the electrode. A measuring bridge converts this into a voltage signal, which is evaluated as a differential pressure by the microprocessor control.

The membrane is also reinforced in the middle. This reinforcement increases its mass. If acceleration affects the sensor housing, the inertia of the membrane with its mass causes a relative movement between the membrane and housing. This in turn alters the distance between the membrane and the electrode and thus alters the associated capacity change, so that this sensor can also measure acceleration. This means that the static pressure can be detected by mean of capacity, and the acceleration as the subsequently modulated capacity change.

5.6.4
Summary

The use of sensors can produce considerable reductions in the high mechanical load on washing machines. Optimum spinning results together with a long service life are possible without necessarily changing the design of the washing machine. Extremely simple and low-cost sensors can be used successfully as certain components such as analog pressure sensors and efficient microcomputers are already widely available.

5.7
Fuzzy and Neurofuzzy Applications in European Washing Machines

H. Steinmueller

5.7.1
Introduction

The washing of clothes and other textiles has become an everyday procedure as part of human hygiene. Whilst the components of the washing process – water, detergent, duration, temperature and agitation – remain unchanged, the technical procedure has developed dramatically. This type of development can be roughly subdivided into the stages shown below.

Before the washing machine entered the scene, doing the laundry was "feared" because of the hard work that was usually necessary to perform the cleaning. There were only a few devices to help with the process of cleaning, rinsing and removing as much excess water as possible. A step towards a new level was effected by the introduction of, initially, the semiautomatic, and then later on the fully automatic, washing machines. Manual operations became increasingly unnecessary. As time went by, washing devices were clearly improved due to technical advances. The specific aim of this effort was the economical use of the raw materials required for washing, especially water. The water consumption for each filled wash drum has decreased by more than 70% (Fig. 5.62).

However, the amount of water was virtually independent of the laundry weight and therefore the washing of smaller loads became ineffective. New techniques, such as the "Mengenautomatik", created a correlation between the wash load and the water requirement. A consumption of 70 l of water for 5 kg of laundry and 50 l for 1 kg was achieved. This is clearly a noticeable reduction, but there was still a wide gap between the specific water consumption of large and small laundry loads. With a full load it was possible to wash with 14 l/kg, while a 1 kg load still required 50 l/kg (Fig. 5.63). The "Mengenautomatik" was therefore a solution that in time could not keep up with the growing environmental awareness of both manufacturers and consumers.

To achieve further improvements to the washing process, an option was integrated with which it was possible for the user to reduce the water level in the

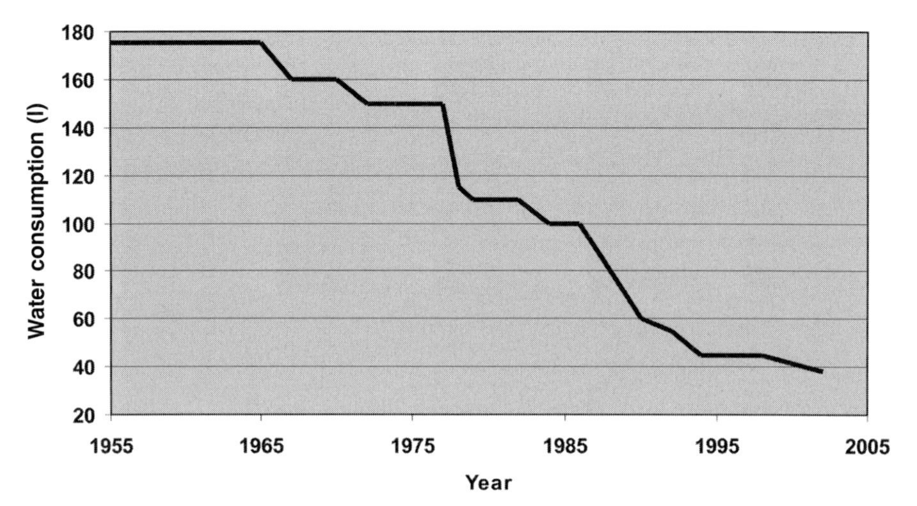

Fig. 5.62 Water consumption for 5 kg IEC 60456 standard load.

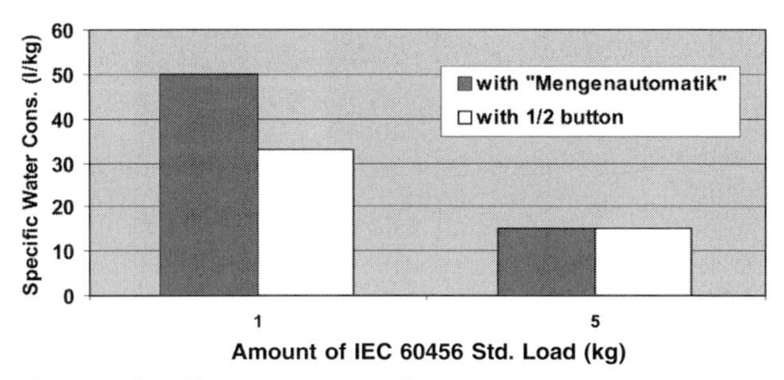

Fig. 5.63 Specific water consumption.

drum with the push of a button (1/2-button) and thus to decrease the supply of water for half loads. With this method, the water consumption for 1 kg of laundry was decreased to 32 l (Fig. 5.63), provided however, that the user was able to make the decision that the actual load meets with the corresponding requirement of half a load or smaller (Fig. 5.64).

When the washing machines with conventional controls are optimized further, more and more decisions have to be left to the user, even though the consumer is not in the situation to check whether or not these decisions are right. The users are only able to evaluate whether the wash result meets with their demands or not, they cannot prove how the same result could have been reached with a reduced use of raw materials. To avoid such a problem, AEG have moved on to the third phase of the development of washing techniques. With the help of the concepts of FUZZY LOGIC, a washing machine has been invented that adapts its wash processes to the demands of the laundry in order to offer the most in "easy operation" and ecology.

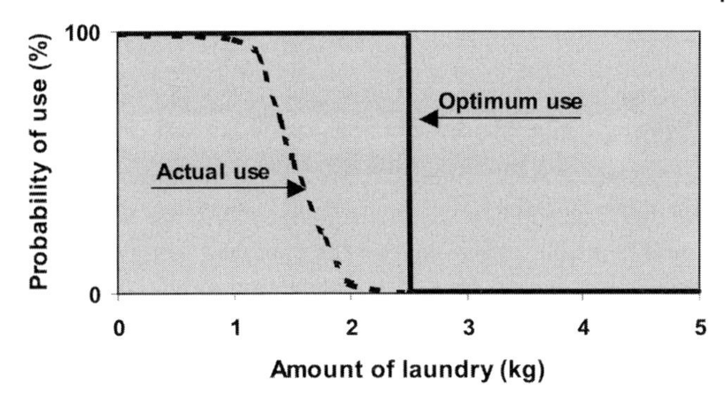

Fig. 5.64 Use of ½ button.

5.7.2
Explanation of the Conventional Wash Process

In order to understand the application of FUZZY CONTROL in the development of washing processes, a short explanation of "conventional" washing techniques will be necessary.

The washing cycle divides into three major stages (Fig. 5.65). In the main wash, the required cleaning is achieved with the use of time, temperature, detergent and the rotation of the drum. At the end of the first stage, the soiled water is pumped out, and a large part of the remaining water in the items of laundry is removed by spinning. The second part of the process is the rinse cycle. Using fresh water and mechanical action introduced by the rotating drum, the remains of the detergent are removed from the laundry. The spinning cycle concludes the washing process, thus reducing the amount of absorbed water in the wash load. To design the optimum process control, regarding the water consumption, several criteria have to be considered from the conventional wash cycle.

1. The absorbed water is a primary function of the laundry type and amount.
2. The energy consumption in the main wash stage is proportional to amount of water introduced.

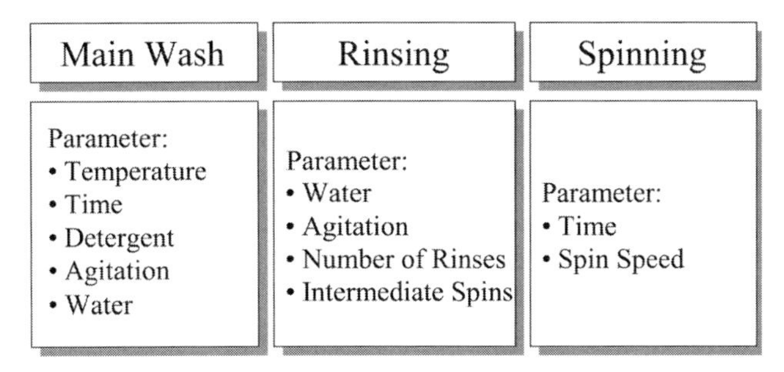

Fig. 5.65 Conventional washing cycle.

3. Sufficient water has to be present in the wash and rinse as early as possible to achieve the desired results.
4. The total amount of required rinse water relates to the amount of absorbed water in the main wash cycle.

Added together, the criteria define the following "FUZZY CONTROLLER" profile: At the beginning of the main wash stage, the right amount of water must be made available depending on the type and amount of laundry. Too much water would shift the energy consumption to undesirable levels while not enough water would result in a poor washing performance. The amount of rinse water is pre-determined by the water consumption of the main wash cycle in order to keep the rinse short and effective.

5.7.3 Engineering the "FUZZY RULE SET"

Several experiments have shown that the "Suction Speed" and the amount of water taken in over a certain time period closely matches the water demand of the laundry introduced. By placing an unknown load of laundry in the washing machine drum and adding a defined water level for its disposal, the laundry absorbs, according to its composition, different amounts of water if the drum is moved in a specially developed spin rhythm (Fig. 5.66). After a given time T_1, the amount of absorbed water is checked. The resultant value, called "Suction Speed", defined as the amount of water absorbed during T_1, defines the average specific absorbing ability of this wash load. This means that a laundry load with a fairly high specific absorbing ability will correspond to a high Suction Speed and *vice versa*.

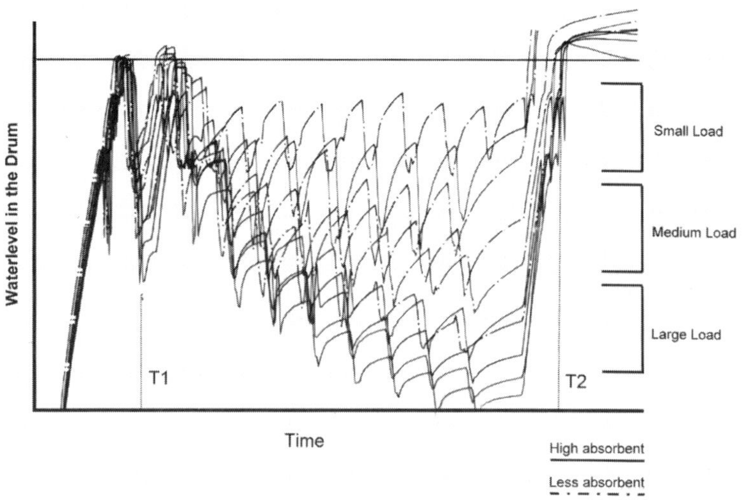

Fig. 5.66 Recorded behavior of the waterlevel as fct (load size, absorbent ability).

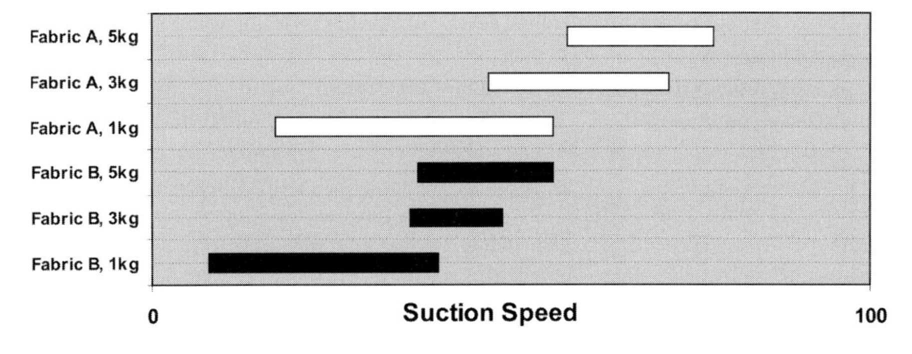

Fig. 5.67 Suction speed per type and amount of load.

At the end of an additional time T_2 the total quantity of absorbed water is measured. This term "Suction Amount" is defined as the total absorbing capacity of the introduced load. Through this, a large amount of absorbed water corresponds to a large load, a small amount of water to a small load. The results of the experiments show that the type and size of a washing load is related to the "Suction Speed" and the "Suction Amount". Apart from this recognized tendency (Fig. 5.67), the results are somewhat "fuzzy" and thus justify the use of "FUZZY CONTROL".

The combination of the two values Suction Speed and Suction Amount (Fig. 5.68) allows a relatively precise value of the expected water demand in the main wash to be determined. From experiments this derived amount of wash water ("First Water Demand") is added shortly after the end of the measurement. In cases where the laundry needs still more water the water level will eventually

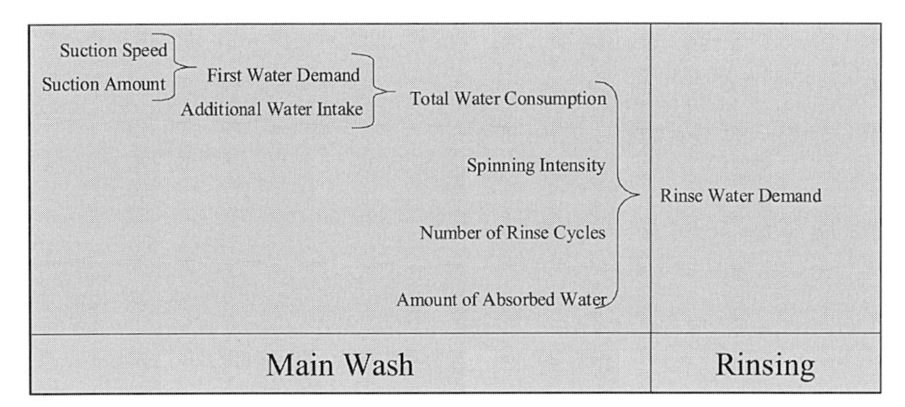

Fig. 5.68 Structure of the First Fuzzy System.

drop and extra water will be added. These additional water intakes in the main wash cycle are summed up and combined with the "First Water Demand" to form the "Total Water Consumption" thus describing the manner of water intake in the main wash.

The resulting value of "Total Water Consumption" and the result of the summed up value "on-times", defined as the "Amount of Absorbed Water", form the foundation for the attachment of the "Rinse Water Demand".

However, the amount of rinse water results not only from the amount of water used in the main wash, which now has to be removed, but also depends on the number of rinse cycles that are to follow on and the spinning intensity at the beginning of each rinse cycle. The sum of all these influencing factors finally leads to the required amount of rinse water that is put in at the beginning of each rinse cycle.

The relationships described form the concept (Fig. 5.68) of the FUZZY RULE SET. The membership-functions of the linguistic variables are derived from the experiments that have already been mentioned; however, the plausibilities of the RULE SET have not been touched on so far.

Initially, one expert tried to fix the plausibility aspects, but in the inspection the resultant RULE SET, with a total of 58 rules, showed a large number of weak points. Furthermore, the application of the plausibilities by this expert proved itself to be time consuming due to the large amount of data and the complexity of the structure.

To make the tuning procedure of the RULE SET more efficient and to reduce the ratio of errors a step was taken to introduce a NEURONAL NET to fix the plausibilities. The required examples to provide a learning experience were through experimentation data transformed by the expert. The input data described the accumulated knowledge regarding the laundry characteristics ("Suction Speed", "Suction Amount") and the output data corresponds to the appropriate conclusions ("First Water Demand" for main washing, "Rinse Water Demand"). Each learning cycle ends with results falling short of the committed error limit, which had been previously laid down. In the next step the plausibilities tuned by the NEURONAL NET were inspected. A recognized error area in the RULE SET was limited in this structure and the "error" plausibilities were optimized with a local learning procedure.

The RULE SET received, consisting of 159 rules, proved to have an error quota of less than 2% and was designed in 3 days.

5.7.4
Setting up the Actual Control Unit

Since it is not necessary to permanently recalculate the coherence between the Suction Speed, the Suction Amount and the required amount of rinse water, implementation of a special FUZZY PROCESSOR with the **AEG Öko-Lavamat** washing machines was abandoned. Instead, the perceived knowledge was registered in a conventional microprocessor as look-up tables. The values were taken from the instructed “FUZZY CONTROLLER”. Theoretically the assignment between the amount of water demand in the main wash and the necessary rinse water would occur in graduated steps, but since the number of table values chosen was rather high, a continuous connection appears in practice.

5.7.5
Summary

The very first AEG washing machine to include this technology was the **Öko-Lavamat 6953** and from the consumption data achieved the effectiveness of FUZZY CONTROL in the engineering process and the wash cycle regulation has been proved. With the wash and rinsing results remaining at a known level it has succeeded in reducing the water consumption according to the laundry type and amount. For a full load, less than 11 l/kg are required. At the same time, the specific water consumption for 1 kg of laundry dropped from 32 l/kg to 27 l/kg.

Since then FUZZY LOGIC has taken over more and more functions. Today the methods have been refined and improved: the machine detects the type and amount of laundry and adapts the whole washing process from water consumption to the steepness of the spin speed acceleration accordingly. Thus the water consumption dropped to 39 l for 5 kg (Fig. 5.69). To conclude, the use of FUZZY LOGIC in the **Öko-Lavamat** series is not only environmentally friendly but also removes the burden from the user as to which program would be the most appropriate for a particular washing load. On considering an average household, we learned that about 20% of the required water for washing can be saved by the new AEG washing machines. Furthermore, the energy consumption is minimized due to the reduced water consumption.

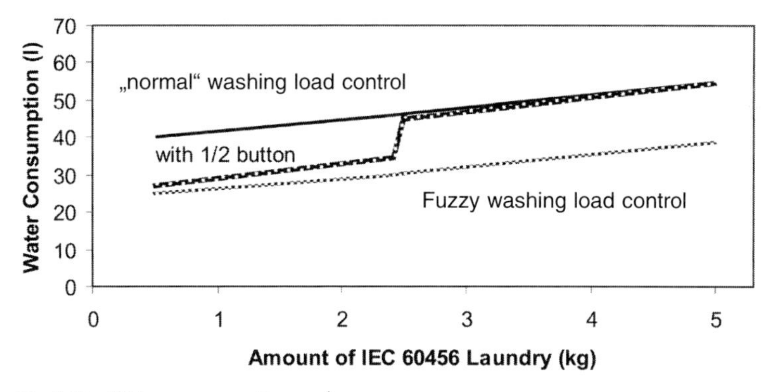

Fig. 5.69 Water consumption cycle.

5.8
Cutting-edge Silicon-based Micromachined Sensors for Next Generation Household Appliances

F. Solzbacher and S. Bütefisch

5.8.1
Brief Introduction to MEMS-based Sensors

Current market trends indicate a gaining momentum of corrosion and temperature-resistant MEMS devices for the use in mass market applications. Silicon-on-insulator material properties make way for a whole new range of uses and applications while being available in sufficiently large quantities to satisfy large volume production requirements. A breakthrough of these devices can be expected through automotive applications, if they can meet the price pressure.

Whereas integrated circuits only require planar structures of a few Nanometers to Micrometers, micromechanical devices usually exhibit thicknesses of several micrometers to millimeters. The general approach towards the processing of the devices however remains largely unchanged to most other semiconductor-based technologies: functional layers and structure are created by the succeeding deposition/implantation and lithographic structuring of layers and bulk material.

5.8.1.1 Microsystems Technology (MST) – A Complex Game of Materials, Technologies and Design

Microsystems is a term first introduced in Europe the late 1980's and early 1990's referring to a new generation of miniaturized micro systems consisting of sensing, processing and actuation components. Europe (MST), the USA (MEMS-Micro-ElectroMechanicalSystems) and Japan (MicroMachines) are the key drivers in the development and market introduction of MST products with Europe taking the lead in the commercialisation of MST products.

For the most part, new materials suited for low cost mass production purposes and micro machining technologies are used to form three basic building blocks of every micro system:

- Sensing/actuation element
- Package
- Signal Processing.

The driving force for the use of new technologies is the market need: an enhancement of functionalities (i.e. also addressing applications that could formerly not be addressed), stability and reliability or a reduction in fabrication cost (i.e. through batch processing, higher yield, cheaper processes, etc.). New technologies only succeed, if significant enough profit (safety, reliability, financial, etc.) can be gained, leading to some of the key challenges faced by this new generation of products [1]:

- Price – new technologies usually have to match or outperform existing technologies at 10–20% lower price.

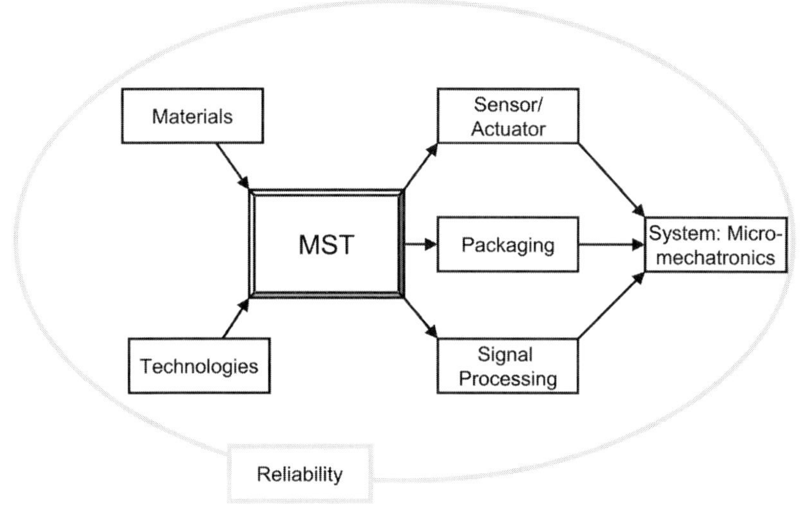

Fig. 5.70 Microsystems technology: key components and issues [1].

- Media compatibility – the majority of the high market growth applications today require high media compatibility, due to the fact that the processes to be controlled frequently involve aggressive liquid or gaseous media (e.g. hydraulic oils, etching gases, body fluids (biomedical applications)).
- High temperature compatibility – a large proportion of the key applications in high volume markets are found in automotive or industrial control systems with high temperature requirements (e.g. engine management, mechatronics systems, etc.).
- Integration of multiple functionalities – use of Microsystems can bring about a significant cost reduction compared to conventional solutions; logistics and quality control can frequently be simplified when integrating several components in one system: e.g. multiple sensor arrays.
- Reliability – failure modes and reliability of microsystems are already well investigated in many areas

The market penetration of MST products is far deeper than commonly known, many of which have become standards, such as automotive sensors (acceleration sensors for airbag ignition, pressure sensors for engine management, break and mechatronics control, etc.), biomedical sensors (e.g. blood pressure sensors for catheter applications or artificial respirators, micro-probes for nerve signal measurement or stimulation) and sensors for white goods appliances (e.g. sensors for water and foam level measurements in washing machines and dish washers), inkjet printheads in office printers and many more. This development is but the start of a new generation of products. The white goods industry, because of the smaller value chain and lower cost of the finished system, leaving less room for costly sensor systems than a car, as well as the lower innovation pressure is still somewhat dragging behind more innovation-sensitive fields. The predominant position of the European white goods industry in the market is in the high price level due to technical innovation and excellence. This will require the increasing use of sensors and electronics if the European industry is to prevail at an international level.

Environmental concerns and the necessity to withstand the market pressure by keeping the technology leadership however has triggered a new wave of sensor and electronics development for European white goods manufacturers.

5.8.1.2 Materials

Device, package, electronics and the assembly and packaging technology constitute the core components of the MST product. These rely heavily on the advances in materials science and technology.

Progress in semiconductor processing has evolved in a number of substrate materials, pre-destined for the use in micro structured devices, such as Silicon, Silicon-on-Insulator (SOI), Silicon Carbide and Gallium Arsenide [1].

- Si: Silicon substrate and processing technology and research has reached a level of perfection unmatched by any other material. Inherent material limitations are frequently overcome by continued optimisation and progress in processing technology.
- SiC: Silicon carbide is one of the most promising materials for ultra high media resistance and high temperature compatibility. Material is available as bulk or layer/coating material. Its processing technology is only just developing.
- SOI: Silicon on Insulator material is today's most advanced Silicon-based substrate technology: it is basically Silicon material with an integrated insulation layer underneath a single crystal Si layer. It combines most of the advantages of the most well established Silicon technology with high radiation, media and temperature compatibility.
- GaAs: Gallium arsenide, a so called wide-bandgap semiconductor is well suited for high temperature applications. However, little processing technology has been developed in terms of micromachining processes.

Pricing as well as reliability considerations have led to an almost exclusive use of Si-based (i.e. Si and SOI) micro machined devices. Packaging and assembly has focused on ceramics (Al_2O_3, AlN, Low Temperature Co-fired Ceramics LTCC), Printed Circuit Board (PCB-) and Surface Mount Device (SMD-) technology and multichip modules (MCM's).

5.8.1.3 3D Processing Technology

3D-machining techniques of semiconductor materials are amongst the core processes leading to the miniaturized structures (sensors, actuators (e.g. valves, relays, etc.) and passive components (e.g. micro channels)).

Bulk and surface micro machining constitute the main 3D-semiconductor processes. Wet etchants such as KOH, EDP or TMAH or alternatively, a dry etching

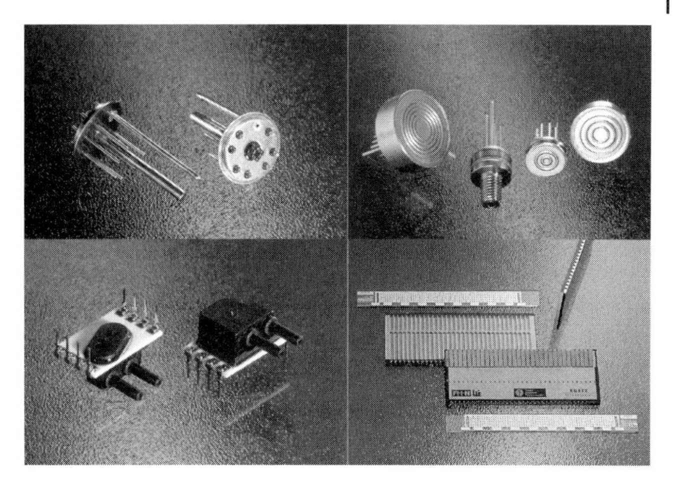

Fig. 5.71 Examples of MST-based sensors and actuators in various packages [1].

process using e.g. SF_6 gas is used to etch 3D structures directly into the Si-wafer in bulk micromachining. Structures of up to wafer thickness can easily be realised. In surface micro machining, sandwiches of sacrificial and functional layers are used on top of the Si-wafer. Removal of the sacrificial layers results in free standing structures with thickness of typically about 1–5 μm.

5.8.1.3.1 **Sensors, Actuators and Passive Components**

The core components of a complete microsystem are the integrated sensing, acting or passive micromechanical devices. In most cases, a naked chip manufactured in bulk or surface micromachining is used for the detection of a physical or chemical quantity or some actuation principle, like the dosage of ink droplets in inkjet printheads. A complete microsystem can consist of a complex set of these devices.

5.8.1.3.2 **Bulk Micromachining Technology**

The majority of micromechanical devices require 3D machining of the bulk silicon material with etching depths of up to wafer thickness. Generally, three basic etching process types can be distinguished:

- Isotropic etching (i.e. etching of bulk material with equal etch rates in all material/crystal directions).
- Anisotropic etching (i.e. etching of bulk material with etch rates depending on material/crystal orientation, used in single crystal material in order to determine clear features and geometry aspect ratios).
- Electrochemical etching (i.e. etching using an electrochemical potential for the control and stopping of the etching process).

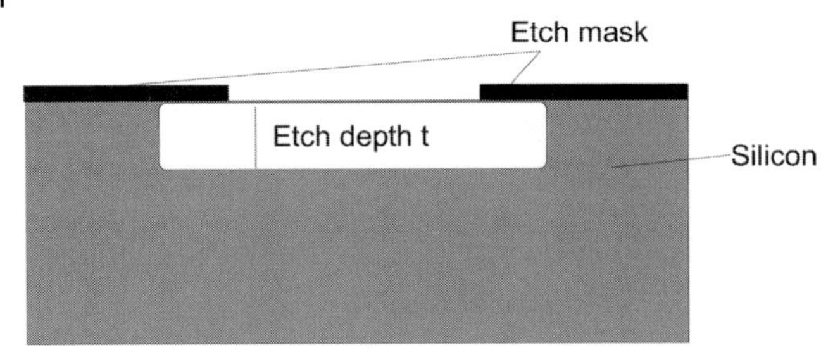

Fig. 5.72 Underetching of etch mask during isotropic etching [2].

Isotropic Etching

In single crystal semiconductor material, isotropic etching processes etch the bulk silicon material equally in all crystal directions. $HF/HNO_3/H_2O$ is amongst the most frequently used wet etchants. Masking layers can be e.g. Si_3N_4 or Au/Ti layers. Underetching of the masking layers with extensions about equivalent to the etching depth can be expected.

Alternatively, in dry etching processes, a gaseous etchant is combined with ion bombardment thereby physically removing material. These so called RIE (Reactive Ion Etching) processes use e.g. SF_6/O_2 as etching agents. Recent years have also shown the upcoming of a new generation of Deep Reactive Ion Etching processes and equipment (DRIE) which allows the etching of deep trenches with high aspect ratio with low underetching [2]. DRIE today is most commonly done using STS reactors.

Anisotropic Etching

In single crystal semiconductor material, anisotropic etching processes exhibit etch rates depending on the crystal orientation of the material. A large proportion of today's micromechanical devices and structures make use of this type of processing technology. Only basic etchants can be used for this purpose: examples are potassium hydroxide (KOH), sodium hydroxide (NaOH), Ammoniumhydroxide (NH_4OH/TMAH) and organic solutions such as Ethylenediamine-Pyrocatechol (EDP). Today, KOH is used almost exclusively, for safety and health reasons.

Etch rate and homogeneity and anisotropic characteristics are the predominant factors in determining the resulting micro system device properties. Temperature and concentration of the KOH solution as well as the doping concentration of the silicon material have the largest impact on these properties and have to be thoroughly controlled.

Crystal Orientation of Etch Rate in Anisotropic Etching

For good aspect ratio and device geometry, anisotropic etching agents have to exhibit a strong difference in etch rate between crystal directions. In a typical KOH solution for single crystal Si, about a two orders of magnitude smaller etch rate in

(111) direction of the crystal compared to the (100) and (110) direction can be expected. As a rough model, the number of free dangling bonds in the respective crystal directions influence the bonding energy of Silicon atoms thereby determining the etch rate. Thus, in (100) and (110) material structures limited by (111) surfaces can be created. These surfaces are oriented to each other at characteristic angle of 54.7° (100-material) or 90° (110-material) [2].

Protection and Passivation Layers

Protection of the Microsystems devices has in the past primarily focused on the protection of the implemented electronic semiconductor devices (i.e. transistors, resistors, memory cells, etc.) from humidity, or material diffusion processes during operation or on masking layers during processing. During KOH etching so called passivation or masking layers are used. These are almost exclusively Si_3N_4 or SiO_2 layers. The higher etch rate of SiO_2 in KOH in most cases leads to the use of Si_3N_4 for deeper etching structures. The masking layers are deposited via Low-Pressure-Chemical-Vapor-Deposition (LPCVD) or Plasma-Enhanced-CVD (PECVD) with LPCVD layers being the more dense and KOH resistant. Today, more and more new material combinations like e.g. PECVD and LPCVD deposited SiC-layers are used for the protection of the finished device against mechanical, chemical or temperature impacts during operation. These are either applied as protective "coating" or even constitute the active device part (e.g. as piezoresistors for high temperature pressure sensors).

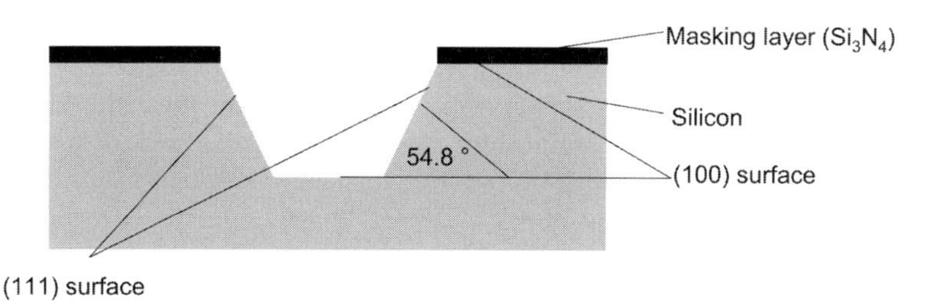

Fig. 5.73 Anisotropic etching using KOH in (100) silicon material [2].

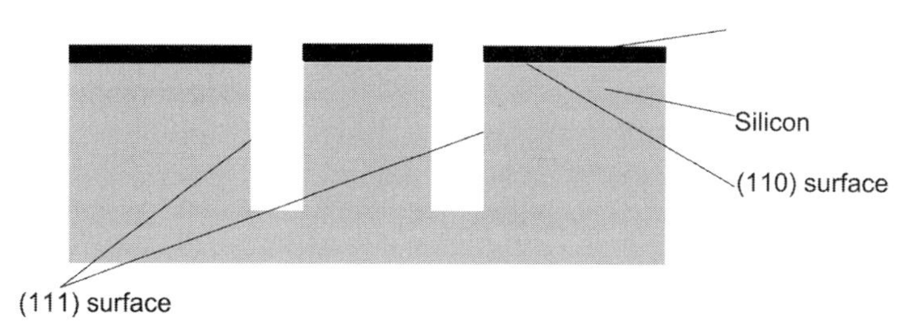

Fig. 5.74 Anisotropic etching using KOH in (110) silicon material [2].

Problems of Anisotropic Bulk Micromachining: Convex-corner Underetching

Depending on the micromechanical structure's geometry, convex corners are likely to occur in the system design and thus the masking layer layout. These convex corners expose crystal surfaces with high etch rate to the etchant (e.g. KOH). The resulting high underetching of the masking layer in this case depends strongly on type of etchant, etch time and depth as well as the surface characteristics of the substrate material (silicon wafer).

A variety of strategies has been employed to prevent this effect from destroying the desired geometrical shape. Predominantly, compensation structures at the convex corners are used. These structures are being consumed during processing, protecting the original intended shape and geometry of the structure. Alternatively, additives like Isopropylalcohol (IPA) can be added to the KOH, changing the etchant composition and etching mechanism.

Electrochemical Etching of Silicon

Electrochemical etching is one way of controlling the etch rate and determine a clear etch stop layer when bulk micromachining Silicon. In this case, the wafer is used as anode in an HF-Electrolyte. Sufficiently high currents lead to oxidation of the silicon. The resulting oxide which is dissolved by the HF-solution. Since lowly doped silicon material is not exhibiting a notable etch rate, it can be used as an etch stop.

Silicon-based pressure sensors are amongst the most common devices making use of this process. A thin low-n-doped epitaxial layer on the wafer determines an etch stop depth and thus the thickness of e.g. the pressure sensor membrane.

In addition to HF solutions, this process can also be carried out in anisotropic solutions like KOH/H_2O. In this case the so called pn-junction (essentially the etch stop layer) resulting from the epitaxial layer being placed on top of the wafer is put into reverse bias voltage with the complete potential difference occurring across the pn-junction. The etching process runs like a normal KOH-etching process until reaches the junction, where the potential changes resulting in a passivation layer stopping the etching process.

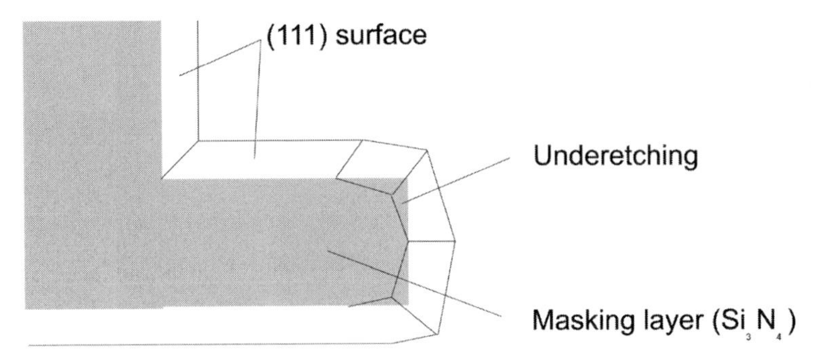

Fig. 5.75 Convex corner underetching [2].

5.8.1.3.3 **Surface Micromachining of Silicon**

Surface micromachined structures are mainly used in high volume production of acceleration sensors and gyroscopes (i.e. automotive and some aerospace applications), some micro optics and micro motors as well as gears and are so far of little interest for the use in white appliances. These 3D-structures have a feature thickness of only a few micrometers. The structures are created by succeedingly depositing and etching sandwiches of active layers and separating sacrificial layers. Removing the sacrificial layers leaves the desired 3D structures on the wafer surface.

5.8.1.4 **Summary**

When designing Microsystems components, the choice of materials and processing technologies has to be considered carefully. It depends strongly on the application field of the MEMS product to be manufactured. KOH wet etching processes are considered to be amongst the cheapest production processes available, allowing batch processing of large quantities of wafers. Surface micromachined pressure sensors or gyroscopes as well as a number of acceleration sensors have achieved good commercial success, in business areas outside white goods. Easy integration of signal processing (amplification, temperature compensation, etc.) in surface micromachined pressure sensors lead to fully functional low cost sensor systems and may well lead to an increasing use in white appliances.

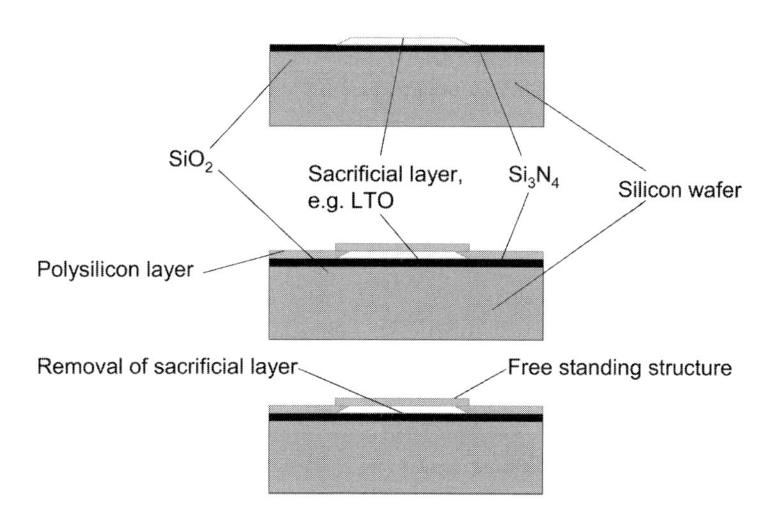

Fig. 5.76 Principle schematic process flow of surface micromachined structures [2].

5.8.2
Comparison of Standard vs. Customized Solutions

Unlike the automotive industry, the white goods industry has in the past always considered sensor components to be exchangeable standard items for two main reasons:

1. The simplicity of the application use allowed employing standard sensors after an internal qualification or slight modifications without the need for a new development.
2. Cost considerations based on the low total system price for a household appliance did not allow for the customer-specific development of semiconductor components. The high development cost would either have to be split over a sufficiently large production volume or result in an unacceptably high unit price. Both are not given in the European market.

Recent developments towards modularisation of processing and device technology now allow for a cost efficient customisation of sensor devices for white goods appliances.

No general guideline can be given as to whether standard or customized solutions are the best choice for white goods appliances. Each application case has to be considered individually. Nevertheless, the key issues considered are summarized in Tab. 5.14.

5.8.3
Examples of Application Use

Recent developments in sensor technology as well as the aforementioned increasing pressure towards technological leadership of European white goods manufacturers have led to continuous spread of the use of sensors in a variety of applications. Some of them, like temperature sensors, level sensors and switches have been in use for quite a while. Others, like optical, chemical and displacement sensors, have only been introduced more recently. Furthermore, more and more information is being gained from sensor signals, frequently exceeding their initial purpose. In this chapter, only a few highlighted examples will be given to illustrate the increasing variety sensors currently in use in household appliances.

Pressure sensors are being used in all water consuming household appliances (e.g. dish washers, washing machines), but also for filter (differential pressure) applications. Washing machines usually feature only one pressure sensor yielding at the same time:

- Level measurement for water in washing machines – this can be used in order to optimise the water use depending on the type of fabric and weight of the laundry.
- Foam level measurement – European and American detergent and appliance manufacturers follow different philosophies and use changing chemical detergent compositions resulting in varying soap foam quantity.

Tab. 5.14 Comparison of standard and customized sensor solutions

Property	***Standard product***	***Customized solution***
Cost		
Production price	World market level, depending on type of product	Has to be considered in context with other systems components (e.g., integrated microprocessor may reduce overall system cost and allow for higher price)
Development cost	None	Between 10 000 € (modification of standard product) and 500 000 € (complete development of sensor system)
How much may a sensor cost?	e.g., Pressure sensor: $\sim$1.50–2.00 € at >100 k pcs/yr	
Qualification cost	Low, depending on the no. of suppliers qualified, frequently with products already qualified for other white goods manufacturers	High, tests usually have to be planned and projected specifically for the new product, first application of product in specific field
Synergy effects	None	Allows for a re-design of the complete "system" with significant performance and cost advantages rather than relying on open market products
Performance		
Suitability for the application	"Does the job", frequently however not well suited without modification	Perfect fit, device is designed for one sole purpose
Measurement range	Limited by ranges available on the market, electronics have to adhere to the sensor characteristics	Free ranges within technological limits allow modification of the sensor characteristics to suit electronics requirements → performance optimisation
Properties (physical/chemical) measured	Limited by products on the market, but still covering a wide range already: • Temperature • Pressure (also: level measurement and force) • Flow • Chemical composition/concentration (gaseous/liquid) • Position/displacement • Optical properties • Acceleration	

Tab. 5.14 (cont.)

Property	*Standard product*	*Customized solution*
Match with other system components	Weak, depending on system design	Excellent, specifically designed for one purpose/system
Lifetime/stability requirements	10 years/>10000 hours → 15 years/>20000 hours	
Reliability	Has to be guaranteed by statistical tests, available low-cost sensors have to be protected in order to meet the requirements, many years of experience and data available	No extra protection necessary, inbuilt media separation and device design for high stability and reliability possible
Logistics/strategic considerations		
Procurement	Standard procedures, product/pricing mix possible if quantities allow for it, product can usually easily be exchanged	Long-term contract with one supplier necessary in order to guarantee constant supply and support, also a guarantee for supplier's risks
Dependency on suppliers	Usually low, flexibility by switching standard products	High, no immediate replacement product available by other suppliers

- Determination of the balance/moment of inertia of the drum as input for the electric motor control during spinning, thereby reducing the need for heavy counterweights, thus leading to lighter machines.

Tumble dryers use pressure sensors for filter measurements by measuring the differential pressure across the lint filter signalling the user when to clean the filter.

Refrigeration equipment (refrigerators, freezers, A/C systems) can make use of pressure switches or sensors in order to control the refrigerant compressor and thus the cooling power of the system. The use of new cooling agents like R788 (CO_2) instead of R134a leads to higher temperatures and pressures (about 150–220 °C, 80–160 bar for CO_2, about 90–125 °C, 15–40 bar for R134 a) increasing the requirements for the sensors in use. The majority of systems in use however still relies solely on a built in temperature sensor or switch, triggering the compressor. The advantage of the pressure sensor use lies in the improved temperature control time constant and reduced power consumption, ergo better environmental compatibility.

Position/Displacement sensors can be used in active damping systems of industrial washing machines in order to compensate for vibrations coming from unbalanced drums/laundry. These can work much like a simple automotive shock absorber.

Temperature sensors are amongst the most common sensors in use in household appliances. They range from sensors for cooking equipment (e.g. ovens, hotplates, kettles, ...) to refrigeration equipment. Their design ranges from simple bi-metal stripes and switches to thermocouple and PT-sensors.

Chemical sensors are amongst the "youngest" sensors available for the use in industrial applications. The Figaro™ sensor based on the Taguchi sensing principle is one of the most commonly available sensors. In recent years, also new technologies (e.g. metal oxide thickfilm sensors on ceramic by companies like UST, Gschwenda) have been introduced. The stability and selectivity of these devices is only moderate, in particular in harsh environments, e.g. in contact with detergents (alkaline solutions, etc.). Significant effort is being made to use these sensors to measure the composition of detergents during laundry and dish washing cycles.

One of the key factors in achieving satisfactory washing performance lies in the combination of the detergent composition and the washing cycle. Thus, both have to be adjusted to each other, which imposes a troublesome and continuous optimisation process on the appliance manufacturers, since the chemical composition of the detergents is frequently changed without sufficient information. A sensor analysing the composition and concentration of the detergent would allow an automatic adjustment of the washing cycle to the detergent "quality" and composition.

Humidity sensors are often technologically closely linked to chemical sensors and can be used e.g. to determine the drying cycle length in tumble dryers and dishwashers, by determining the remaining air humidity caused by wet laundry or dishes.

5.8.4 Perspectives for Future Developments

Future developments will see a variety of changes in technology and will require a change in doctrine by the white appliance manufacturers.

One of the key trends of sensor technology will lie in modularisation of the technology steps in order to provide "quasi-customized" solutions without the need for a complete development. A modularisation of the sensor hardware in combination with standardization attempts are also being followed, but are deemed too expensive for the use in many applications, without a significant customisation advantage.

Europe's white goods suppliers are technologically leading the world, supported by environmentally conscious and technology-open clients. Their success lies in the offering leading products in high price ranges. The race for the low-cost market has for years been won by huge manufacturers mostly from South-East-Asia like e.g. Haier Corp. in QingDao, PRC. Thus, the European manufacturers will only survive if they keep the innovation and technology leadership, allowing the extra-cost for extra-performance.

In order to achieve this, a change in purchasing and development strategy will be necessary. The white goods suppliers, like 10–15 years ago the automotive industry will have to take over the responsibility to push RnD on the component and subsystem level forward by jointly developing technology and applications roadmaps with their suppliers, instead of using market standards.

Both the suppliers and the appliance manufacturers have to co-operate closely in order to determine the next decades requirements and technological solutions in order to stand a chance to hold up their unique market position.

5.8.5 References

1 F. Solzbacher, "Microsystems technology – a new era leading to unequalled potentials for automotive applications", AMAA Yearbook 2002, Springer, 2002.

2 F. Solzbacher, "Fabrication Technologies for 3D-microsensor structures", AMAA Yearbook 2002, Springer, 2002.

6 Influencing Factors – Today and Tomorrow

6.1 Future Developments, Roadmapping and Energy

G. Tschulena

6.1.1 General Trends

There is strong international competition in the area of domestic appliances, partly due to market saturation for many household appliances, as described in Chapter 2. The number of newly built houses and flats, the growing age of the population and the trend towards an increasing number of single households are major influencing factors for buying decisions of domestic appliances. Especially in Europe, it is not only the price that determines buying decisions, but also benefit to customers and their social environment need to be taken into account. Many trends in the technical improvement of household appliances were consumer-driven. In Europe in particular, customers wanted:

- Increased user comfort
- Reduced energy consumption
- Reduced water consumption
- Reduced washing detergent consumption
- Noise and vibration reduction
- Autonomous and perhaps remote operation.

Nearly all domestic appliances are using more and more electronics. Mechanical functions are replaced by electronic devices, mainly sensors and electromechanical actuators (such as pumps or electromotors) and microelectronic control systems.

In Fig. 6.1, an attempt is made to show to what extent sensors have been penetrating the appliance market over the past years, a trend which is set to continue in the next decade. In the beginning, there were relatively simple sensors for temperature, pressure, flow, etc. Over the last years, non-contact measuring devices have attracted much attention, such as non-contact temperature monitoring for toasters or for hair blowers. The introduction of more complex sensor systems, such as water quality sensors or multi gas sensing artificial noses is imminent.

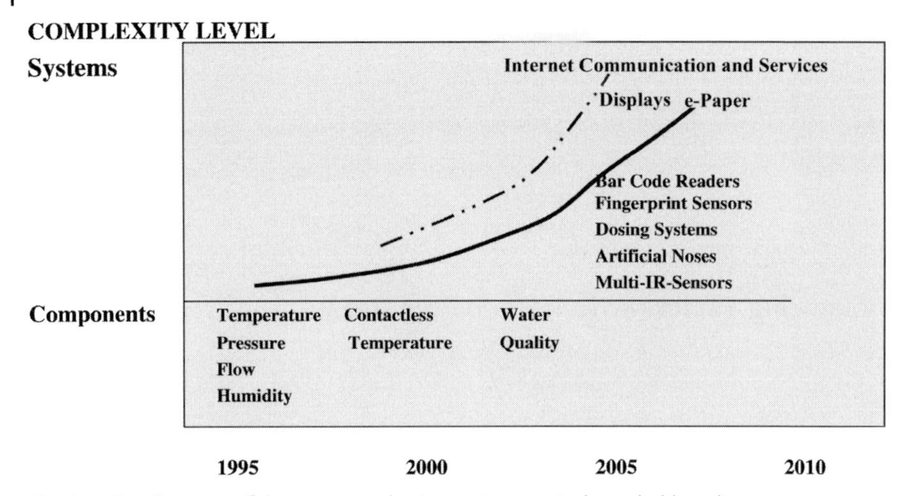

Fig. 6.1 Development of the sensor and microsystem use in household appliances.

Simple bar code readers or fingerprint sensors may also be introduced into household appliances in the next decade. Visionaries even expect the introduction of displays in the kitchen or bathroom within the next decade.

However, the pressure to keep down the costs of all parts used in domestic appliances is omnipresent and also applies to the area of sensors and microsystems. The lifetime of sensors, sometimes under adverse conditions, is critical, and a lifespan between 8 and 15 years should be expected. Commercially viable prices for sensing modules are rather low, depending on the individual functions, but less than 5 Euro for a complete module are typical [1]. Also the financial risk involved in serial production has to be considered. Just one faulty unit would cost the producer nearly 100 Euro to repair, even if this unit costs only 2 Euro. If the factory output is 2000 units per day, the loss may amount to 20 million Euro if a flaw is detected by one of the customers after 100 days of production.

In the following chapters we describe the functional trends for large and small appliances and their current need for sensors as well as the need anticipated for the near future. The final chapter deals with "Visions for Appliances" and the sensors that could help turn these visions into reality.

6.1.2
Functional Trends for Large Household Appliances

The main functional application trends in household appliances can be summarized as follows:

- Increased User Comfort
- Reduced Energy Consumption
- Reduced Water Consumption
- Noise Reduction and Vibration Reduction

Some typical figures for the reduction of energy, water consumption, detergent and salt consumption of recent decades for washing machines, dish washers, electric cookers and refrigerators are given in Fig. 6.2. These trends are expected to continue over the coming years. Further reductions, however, cannot be achieved by conventional changes in the construction of the appliances, which is why washing, cooking or cooling processes will undergo further sophistication in the near future. Such further process improvements must be based on detailed information about the main process parameters, i.e. the number of sensors is going to rise, and their functionality and accuracy will improve steadily. We will give some examples of sensors that are already in use and of others that could be introduced.

In the present situation and in the near future, better sensors and improved control systems would require improved measurement functions, and new parameters could be defined. More and more electronics will be used in domestic appliances. Mechanical functions are being replaced by electronic devices, mainly sensors and electromechanical actuators (such as pumps or electromotors) and microelectronic control systems.

Major trends in the technical development of washing machines and dishwashers as well as vacuum cleaners within the next 5 to 10 years are shown in Tab. 6.1.

Further examples of potential future functions in appliances that use improved or new MST devices comprise:

- Cleanliness determination in washing machines
- Humidity status in dryers
- Dirt content of clothes for automatic wash detergent control
- Tagging (and non-contacting reading) of washing goods according to the type of material

6.1.3 Sensors in Large Appliances – Current and Future Trends

About 10 years ago the main sensors in large appliances were temperature control and level control systems. Now the number of sensors has increased and their quality improved:

- NTC and PTC resistors for temperature determination are built into nearly all large appliances
- Conductivity sensors for residual humidity
- Tacho generators for motor rotational speed in washing machines and dryers
- Capacitive sensors for push-button switches.
- Motor overload indication through integrated temperature sensors.

In washing machines and dish washers new and improved microsensors have been introduced over the last years, such as:

- Pressure sensors for water level switches, or, in a more sophisticated form, also for foam content surveillance in washing machines and dryers.

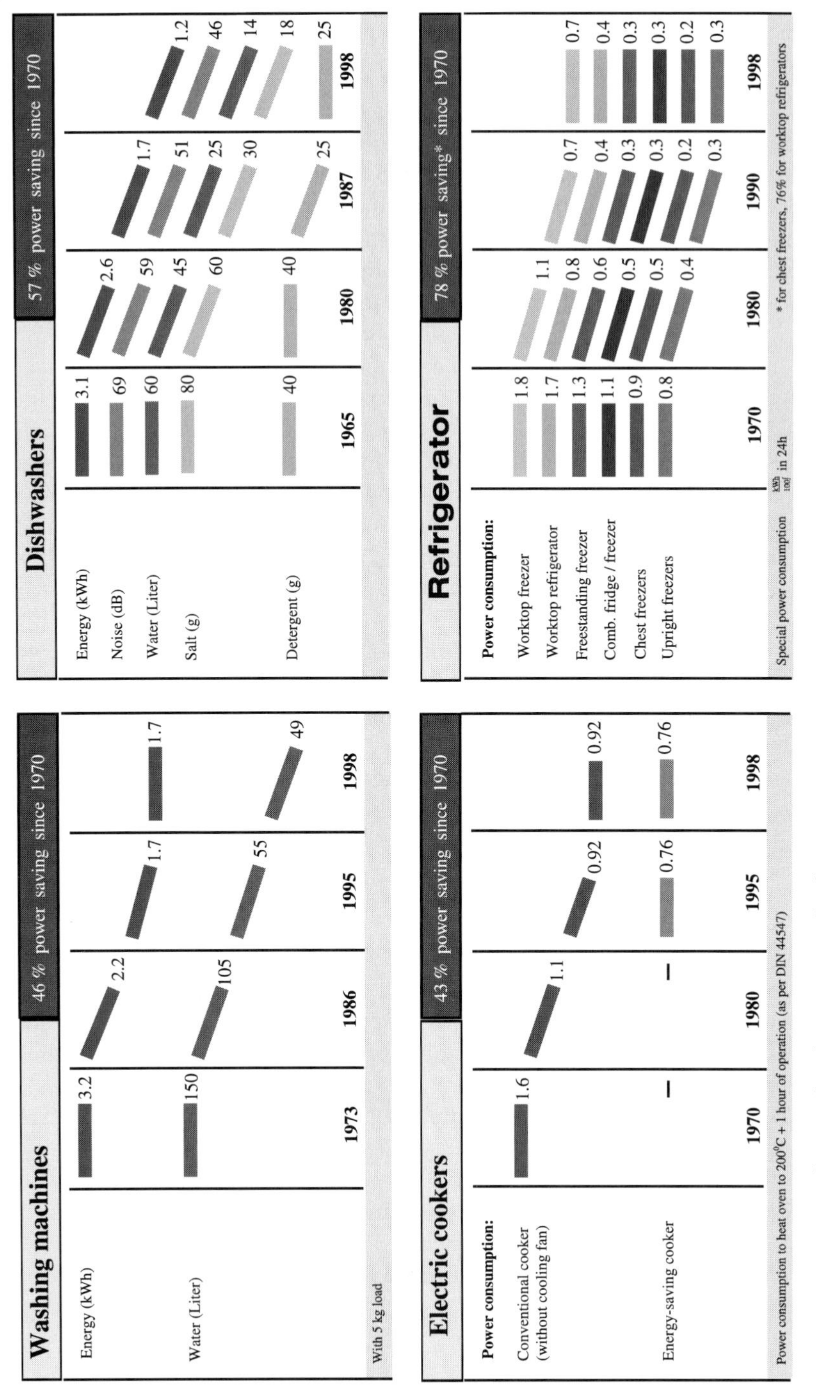

Fig. 6.2 Consumption figures of appliances (examples from [2]).

Tab. 6.1 Development of functional activities in washing machines and dish washers for the next about 5 years (preliminary).

Future	*Currently*	*In 5 yrs*	*Comments*
Legal requirement			
Energy consumption	1.7 kWh 0.95 kWh	–20%	For boilers, 95 °C For 60 °C
Industrial requirement			
Water consumption	49/39 l for 5 kg	–	Rinsing vs. detergent consumption Keep performance
Medium temperatures	–	Reduction	
Noise reduction	–	–3–6 dB	In spinning, from 600 → 1400 rpm
Detergent consumption	–	–30%	"Right amount"

- Chemical sensors for water quality determination. The parameters measured include turbidity, color, surface tension, detergent concentration, pH-value etc.
- Optoelectronic systems for monitoring the turbidity of washing water allow the number of flushing cycles to be adapted to the actual need. (Aqua-sensor system).
- Magnetic sensors for controlling the movements of the spray arms in dish washers.

A more general description of sensors for some selected large appliances is given in Tab. 6.2

Examples of MST Products, which will be improved over the coming years in their technical performance, lifetime and cost reduction, include

- Flow meters
- Pressure sensors
- Optical sensors
- Chemical sensors for liquids (like pH, surfactant concentration, water hardness)
- Gas sensors and artificial noses
- Temperature sensors, both contacting and non-contacting
- Humidity sensors
- Weight sensors
- Surface tension sensors

Such sensors should be combined with microelectronic control systems and microsystems.

The technical development of all types of washing machines and their sensor and control systems depends to a large extent on the use of appropriate detergents, their chemical composition and physical qualities.

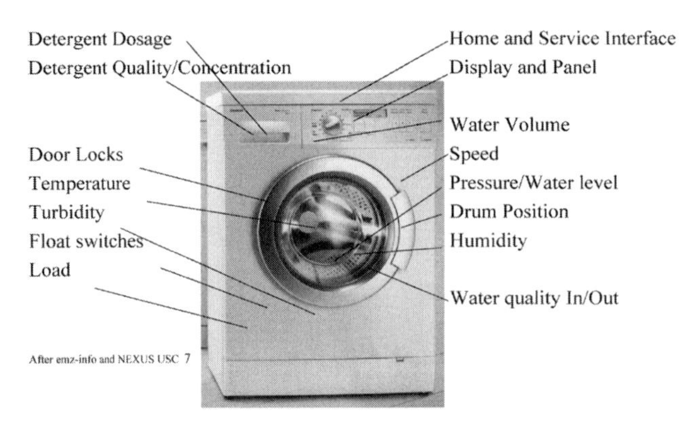

Fig. 6.3 Sensors in washing machines [3].

Tab. 6.2 Examples of sensors for large appliances [4, 5].

Appliance	*Variable measured*	*Type of sensors used today*
Washers	Temperature	NTC, Klixon
	Rotation	Tachometers
	Level	Pressure sensors
	Weight	Inductive sensors
	Position	Reed switches
	Unbalance	Tachometer, pressure sensors
Dryers	Temperature	NTC, Klixon
	Humidity	Conductivity sensing
Dish washer	Temperature	NTC
	Salt level	Reed contact + swimmer
		Density sensors
	Cleanser level	Reed contact
	Spray arm rotation	Magnetoresistive sensors
	Flow	Rotary water meters
		Pressure sensors
Freezers	Temperature	
Oven	Temperature	Expansion capillars, Pt 100
Microwave ovens	Temperature	NTC
	Humidity	Ceramic sensors
	Gases	

6.1.4 Functional Trends in Small Household Appliances

The sensor application trends in small household appliances are similar to those for large appliances. Additionally, the following features are required:
- smaller size
- low power consumption, as many of the devices are hand-held.

In several household appliances remote or autonomous operation will be introduced. So the first autonomous cleaning robots will be available, mainly for industrial purposes. Once the price of these autonomous devices have dropped, mass production for various applications can be expected.

Examples of small appliance features due to new or improved sensors include:
- Automatic baking control, e.g. by introduction of intelligent multi-gas sensors (artificial noses) in combination with non-contacting temperature distribution recognition
- Remote non-contacting temperature, humidity and color detection for hair care equipment
- Integration of medical sensors into dental care equipment
- Improved UV sensors for tanning appliances as a protection against overexposure to UV radiation
- "Smart shoes" are another new application that will first focus on athletes before the technology is introduced to the public at large. Such smart shoes will include e.g. acceleration sensors for the detection of foot movement and a possible calculation of distances for runners and walkers.
- Distance and pressure sensors could be introduced into shavers and ensure continuous adjustment of the blades.
- In many kitchen appliances, such as fryers, some sensors could be introduced for "auto-cooking functions" with the aim of
  a) Determining the condition of the food product
  b) Monitoring potential health hazards, e.g. the condition of oil used in a fryer
- Remote operation of vacuum cleaners

Tab. 6.3 Example of the development of functional trends in vacuum cleaners and in the next about 5 years.

Function	*Now*	*In 5 yrs*	*Comments*
Power consumption		–20%	Low power Consumption will be a differentiatng feature
Suction power			In Western Europe (except France)
Efficiency	30%	45–50%	

6.1.5
Sensors for Small Appliances – Current and Future Developments

Many sensors can be used in appliances in order
- to improve the quality of the processes,
- to improve the user-friendliness of the application
- to realize complex smart functions

In small appliances, a range of specific sensors are used, as shown in Tab. 6.4. There has been an increased use of microsensors in recent years.

- Thermopiles have been used for non-contacting temperature measurement in hairdryers, to prevent damage to the hair and to speed up the drying process.
- Thermopiles are also used in new ear thermometers or in forehead thermometers to measure the infrared radiation emitted from the skin. This allows quick and reliable temperature measurement and is easy and comfortable to use.
- A new application of thermopile sensors can be found in toasters, ensuring a more reliable control of the browning process, which makes toasting a healthier process.
- In irons, acceleration sensors are used that cause the iron to switch off when in an unwanted position or stationary.

Below, further examples of microsensors are given, the performance of which will certainly improve within the next few years, as will production costs.

Tab. 6.4 Examples of sensors for small appliances [4, 5].

Appliance	*Variable measured*	*Type of sensors used*
Hair dryers	Temperature Contactless temperature Humidity Air flow	NTC IR Thermopiles
Toasters	Temperature Contactless temperature	NTC IR thermopiles
Irons	Temperature Humidity Position, movements	Bimetal switch, NTC IR sensors, thin film sensors Tilt sensors, acceleration sensors
Vacuum cleansers	Air flow Particles	Film sensors Microoptic sensors
Air cleaners	Gases Particles	Ceramic sensors Microoptic sensors

- Gas sensors and artificial noses
- Temperature sensors, both contacting and non-contacting
- Humidity sensors

6.1.6
On Energy Consumption in Appliances

As a consequence of recent technical developments as well as legal and policy requirements, overall energy consumption for domestic appliances in Europe is expected to fall gradually. The total energy consumption of washing machines, dryers and dishwashers is estimated to be around 60 TWh in Europe, equivalent

- to a consumer expenditure of approximately 8 Billion Euro, or
- CO_2 emissions of around 30 million tons.

The energy consumption of wet appliances in Europe has been investigated and is shown in Fig. 6.4. Two different projections of future energy consumption have been made, one of which can be described as the "BAU – business as usual"-scenario, the other as the "ETP" scenario, utilizing the "full economic and technical potential" and resulting in a forecast 28% decrease in energy consumption until about 2010. The latter scenario takes both technical improvements and changes in consumer behavior into account, as brought about by the new EU energy label for wet appliances in the late nineties [6].

In 1995 the energy consumption of average European machines was expected to drop over the coming years

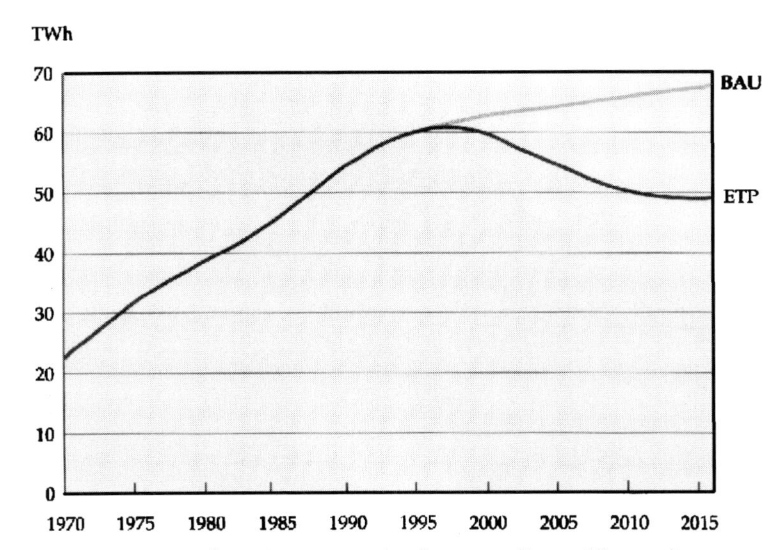

Fig. 6.4 European electricity consumption for wet appliances [6] according to two different scenarios (business as usual, and utilisation of the full economic and technical potential).

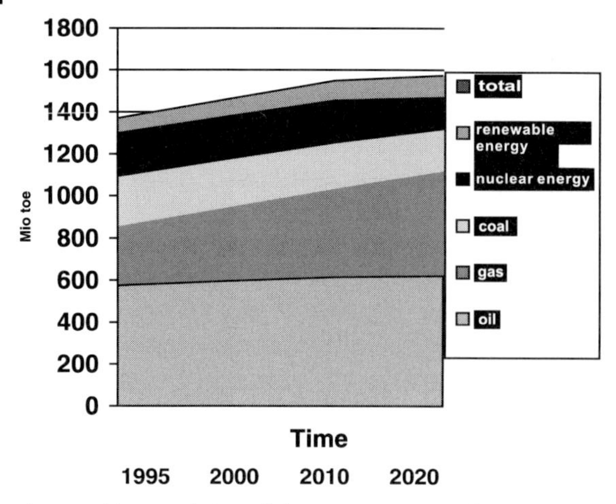

Fig. 6.5 Time evolution of the raw energy supply in Europe [9].

- by about 25% for washing machines
- by about 33% for dish washers, and
- by about 10% for dryers.

In recent years, energy reduction has begun in earnest, due to governmental initiatives such as the energy label for new appliances as well as by informing and educating customers.

Another important change will take place in the next decade regarding the raw energy supply in Europe. We have currently a mixture of

- coal energy, which will be reduced in the future because of its environmental incompatibility and high cost,
- nuclear energy, the share of which will also be reduced by political consensus,
- gas,
- natural renewable energy from wind, water and solar plants a small, but growing share.

In the next decade, only the gas sector will see major growth in Europe. This will probably lead also to an increased use of gas-powered appliances in the future, This seems to be no problem for cooking appliances where burners could still be made more efficient, and more control and safety-related sensors could be introduced. Gas-powered washers and especially dryers should be made more widely desirable.

6.1.7
Heating and Climate Control

Heating and climate control are essential for comfort, and people have always been looking for improvement in this field. In large rooms, such as offices or public buildings (e.g. concert halls), automatic temperature control has become increasingly important. The creation of a comfortable climate is far more complex

Tab. 6.5 Desired improvements in modern gas appliances [7].

New technologies in gas appliances	
Current	***Future***
Thermoelectric flame failure detection	Electronic safety pilot
Analog burner control systems	Electronic burner control systems
Safety temperature cut-out	Electronic cut-out with NTC
Mechanical pressure switch	Electronic pressure sensor/transmitter
Mechanic/pneumatic gas-air-ration control	Electronic gas-air-ration control with ionisation signal or O_2 sensor
Thermoelectric flame supervision	Ionisation flame supervision
Thermal combustion products, discharge safety devices	Electronic combustions product discharge safety device

than it may seem, depending on many physical and chemical parameters not only on temperature, but also on air flow, gas composition, smell, light and even infrared radiation, and more. We are currently using all types of heating systems and especially in southern countries – climate conditioning. Heating systems use oil, gas or electricity as raw energy source. In the field of domestic heating there is a growing demand for improved technical solutions, because of

- lower heating requirements owing to the improved insulation of buildings
- higher efficiency combustion processes with reduced emission
- better safety features
- a need for cost reduction.

For gas heaters, intelligent firing systems are the way ahead, which means improved combustion control due to automatic monitoring of the combustion process and closed-loop control at all times.

Gas heating can be more economical than electric heating if conditions are right. This largely depends on policy decisions over the coming years. Gas supply will be deregulated – for example in Germany within the next few years, and subsequently also in other European countries. This means that natural gas will come from a variety of sources and thus vary in composition (methane, hydrocarbons, sulfur, hydrogen,...), which, in turn, has an impact on heating values. It is therefore important that burning value and gas composition can be monitored and burners adjusted accordingly.

6.1.8 Climate Control Current and Future Sensor Developments

Reliable gas sensor systems that are small and inexpensive and do not need re-calibration over a lifespan of at least 10 years are required for

- Ventilator operation and efficiency control systems
- Heating and climate control systems with integrated gas sensors to improve comfort

For building control and safety, the presence, concentration and flow of gases are crucial parameters. We distinguish the following categories:

- Toxic gases such as CO, CO_2, exhaust gases, smoke
- Combustible gases such as CH_4, C_2H_6 (gas detection via flame detection (Europe), fire detectors, caravans with gas detectors)
- Gases that influence air quality such as CO_2, VOC (volatile organic compounds), humidity, ozone, radon, particles

National safety regulations determine to what extent toxic gases are measured. For example, every Japanese home has to have sensors that monitor toxic and explosive gases.

In order to reduce energy and improve comfort in high-end commercial buildings optical detection of air quality is becoming more and more important.

Cheap, very sensitive pressure sensors will be used to monitor the behavior of ventilation systems. These will be based on silicon sensors with built-in signal conditioning.

Tab. 6.6 Sensors for heating and climate control.

Task	*Measurements performed*
Comfort, room climate, air conditioning	Temperature Pressure Air quality like CO_2, VOC (volatile organic compounds), Humidity Ozone Particles Air flow Water flow IR-Radition Luminous intensity Presence detection
Energy saving	Temperature Air flow Water flow Luminous intensity Presence detection
Safety and security	Temperature Toxic gases like CO, CO_2, exhaust gases, smoke, ... Combustible gases like CH_4, C_2H_6 (gas detection via flame detection (Europe), fire detectors, caravans with gas detectors) UV sensors Radon sensing

The use of gas sensors is an important area which is expected to grow strongly in the future. Their introduction has only been making slow progress for a while because in the past there was a lack of selective, sensitive sensors with high long-term stability, but this has now changed, and the following Tab. 6.7 lists some gas sensors that may be used.

Further optimization of the combustion and heat transfer process in burners can be expected within the next years. Therefore an integration of actuators for the optimal air-gas/fuel ratio may be useful. This control of the air-fuel ratio can be achieved by controlling

- the gas flow, or
- the air flow.

Tab. 6.7 Gas sensors for potential use in climate control [8].

Sensing parameter	*Device*	*Technology used*
CO	Resistive sensors: pellistors, metal-oxide sensors Optical sensors: micro-spectrometer, IR-sources, IR-detectors, IR-filters	Hybrid or integrated, surface micro-machining SnO_2 sintered thick film (Figaro, FIS, ...), SnO_2 thin and thick film on silicon (MiCS, Microsens) IR spectroscopy (Vaisala, Honey-well, ...)
CO_2	Micro-spectrometer, IR-sources, IR-detectors, IR-filters	Hybrid or integrated, surface micromachining IR spectroscopy (Vaisala, Honey-well, ...)
H_2O, humidity	Capacitive polmyer sensors and ceramic sensors	Thin film technology and thick film technology
O_2	Oxygen-sensors, metal-oxide sensors	Hybrid or integrated
O_3, Ozone		Hybrid or integrated
Exhaust gases	Resistive sensors: pellistors, oxygen-sensors, metal-oxide sensors	Hybrid or integrated
Smoke		Hybrid or integrated
Combustible gases: CH_4, C_2H_6	Resistive sensors: pellistors, metal-oxide sensors Optical sensors: micro-spectrometer, IR-sources, IR-detectors, IR-filters	Hybrid or integrated, surface micro-machining SnO_2 sintered thick film (Figaro, FIS, ...), SnO_2 thin and thick film on silicon (MiCS, Microsens) IR spectroscopy (Vaisala, Honey-well, ...)
VOC (volatile organic compound	Work-function sensors, micro-spectrometer, IR-sources, IR-detectors, IR-filters	Hybrid or integrated, surface micro-machining

Additionally gas quality sensors are required, in order to allow natural gas from a variety of sources to be utilized as efficiently as possible.

Also, several types of gas quality sensors can be utilized, such as:

- Solid state resistive gas sensors
- Flame ionization sensors
- Semiconductor sensors
- Optical sensors
- Pellistors.

For all sensors, and gas sensors in particular, long-term stability (more than 10 years lifetime) under realistic conditions is crucial.

Household appliances can also benefit from improvements in other areas. For example, oxygen sensors that measure the O_2-concentations in exhaust gas have been developed that combine a Nernst type lambda gauge (which can measure only the O_2-concentration at one lambda-point) with an amperometric O_2-pumping cell.

6.1.9
Visions for Household Appliances

While it is quite difficult to describe one or several "visions" for the future of household appliances – and only a short overview can be given here – it can also be very inspiring for users as well as for the developers of appliances and sensors. This why we want to present some of them to our readers and describe the sensors required.

It is not easy to develop such an overall picture, as we are trying to include a wide range of appliances, each of which has its own specific developments and technical requirements. To what extent these visions can be realized is still a matter for debate.

6.1.9.1 The Automated Kitchen of the Future

Several new and innovative features can be expected in kitchens of the future. Efficient and inexpensive sensors are the key component.

Food could be stored in refrigerators or freezers, which can recognize the type and date of the goods stored. If some goods reach their use-by date, it is brought to the attention of the user. When food is taken out of the freezer, a reminder will be displayed to buy more, or, if the user has subscribed to a re-stocking service, it may even automatically order from the supplier.

This can only work with adequate tagging systems. readers must be installed in the refrigerator or freezer, and all the goods must be tagged. Technical solutions have to be investigated, and the system needs to be standardized to be introduced on a large scale. Optical systems, bar code readers, rf-systems or magnetic systems could be used.

Kitchen scales, mixers or refrigerators could be equipped with food content analysis systems that would not only determine the weight e.g. of fruit or vegetable, but also provide data on freshness and nutritional information, e. g. about energy level, carbohydrates, proteins, vitamins, fat, or of other main food constituents.

The chemical analytical systems or biochemical sensors (up to lab-on-the-chip) required have to be investigated.

Many features of ovens, ranges, grills, microwave ovens can be automated to a larger extent than today. This means not only the predetermined timing of all the functions, but also a better and nearly automated control of the food processed according to the (non-contacting) temperature and the temperature distribution at the surface and throughout the food, the color, the humidity of the food, and the gases emitted.

This can be realized with modern sensors such as infrared thermopiles, thermopile arrays, microspectrometers and color sensors, several types of humidity sensors, artificial noses and multi-gas sensors.

The position of the pot or pan in the oven could be determined and heat distribution adapted. Dough rounders with better control of many functions could be developed, including the determination of dough viscosity, humidity, color, particle size and some constituents.

Some of these functions could be monitored with improved sensors, instruments and microsystems, like microspectrometers and color sensors, thermopiles, artificial noses, etc. Also some dosing and mixing functions (e. g. of herbs and spices) could be controlled by microfluidic systems.

The kitchen of the future will have a computer or internet access – perhaps with a large display on the freezer. It will come up with recipes and suggestions on how to use the ingredients stored in the fridge and adapt the suggestions to the taste of the cook.

This requires low cost and flexible displays, some of which are a sort of microsystem. Competition may come from OLEDs or from different types of LCD displays.

Extractor hoods could be controlled and adapted to air quality, taking into account air parameters like humidity, specific gases, and also gas mixtures.

This can be done with gas-sensing microsystems, ranging from simple gas sensors to artificial noses and multi-gas sensors.

6.1.9.2 **The Laundry of the Future**

The laundering process will become more energy-efficient and use less water and adequate low amount of detergent. It will be more or less automatically adapted to the types of textiles, the washing program, dirt content, wash load, to temperature levels accepted and the detergents used, using sophisticated electronic control systems.

Several new sensors will have to be introduced alongside the current ones – several types of temperature sensors, level sensors, load and weight sensors, humidity sensors etc.

Water quality could also be monitored, both for the inlet and outlet water quality (sud monitoring).

This requires other kinds of sensors, including those for inlet water hardness, water conductivity, outlet water turbidity in order to control the efficiency of the washing and rinsing process, water flow and water level, foam level, or even chemical and biochemical sensors. Turbidity sensors were the first type to be used for suds monitoring. Thus, residual dirt, detergent type and concentration, fluff and other parameters can be determined.

The automatic determination of the type of laundry, e.g. by reading the textile care information automatically into the washing machine, seems to be somewhat more difficult.

Adequate and widely accepted tagging systems, with readers in washing machines or dryers and low-cost tags on the clothes, are a prerequisite. Technical solutions have to be investigated, and the system needs to be standardized to be introduced on a large scale. Optical systems, bar code readers, rf-systems or magnetic systems could be used.

Alternatively non-contacting measurement systems may be considered for distinguishing between the different textile families, e.g. based on analyses of the optical or infrared optical spectra of the textiles. However, difficulties will arise, considering the variety of textiles in use.

Close co-operation between machine producers and detergent manufacturers is absolutely necessary because the quantity and type of detergent used has a considerable impact on the washing process.

In the future, monitoring of surface tension could be an attractive option for measuring surfactant concentration. Automatic dosing systems could also be introduced for a controlled supply of concentrated detergents.

As washing detergent manufacturers keep developing new products, washing processes also need to be adapted several times over the lifetime of a washing machine. New programs could be transferred to the machines every so often, either by a technician through an infrared interface or conventional PC interface, or simply by a download from the Internet.

Interfaces may also be very useful in machine repair. The service technician can obtain detailed information about the machine status via an appropriate diagnostic program and then carry out the repair. This can be done through direct contact, e.g. with electrical or infrared interfaces, or in a remote way via internet access.

6.1.9.3 **Service Robots**

Robots, already well-established in the industrial environment, are expected to become part of private households. These "service robots" could be used as

- Cleaning robots, which can already be found on the market. Their functionality is expected to increase, while prices are going to come down, as they are competing with existing hand-operated systems.
- Safety and security robots, which can be used for fire detection, for security, environmental monitoring or building management.
- Lawn mower systems for gardening purposes are expected to be introduced. They could be powered by solar cells.
- Medical mobile transport systems for the transport of medications, supplies, meals, and equipment, e. g. in the hospital between departments and wards. ro-

bots for personal hygiene could also be considered, e.g. for tasks like washing face, hands, and body, and cleaning teeth for the elderly and disabled.
- Care robots could offer multimedia communication, in combination with operation of home electronics, reading books or playing games, and could guide, assist or support the elderly and disabled.

Many sensors have to be integrated into such mobile robotic systems, with a major share of miniaturized, low-power microsystems. Self-guiding features are used in such robots, as well as autocharger docking system features.

6.1.9.4 Services for Health

New application areas, new services and new devices will find their customers, partly by the integration and improvement of currently existing devices, partly by development of new devices, e.g. in telemedicine or sport medicine or in the care for the elderly where more safety, security and services could be offered.

For example, kitchen or bathroom scales could be equipped with fingerprint recognition systems, thus giving access only to a limited number of users. Bathroom scales could also have a voice recognition system with a limited vocabulary, and for limited applications.

A home medical box in the bathroom could be introduced, comprising simple medical tools, such as a wrist blood pressure monitor or a (non-contacting) electronic thermometer, and a stethoscope microphone. Additionally an interactive medical encyclopedia could be accessible, with in depth explanations and animated models. Also a connection to a medical doctor could be integrated with wireless video links.

Sport devices are expected to contain more integrated sensing functions. For example, sport shoes could have an integrated pressure sensing and telemetry unit, incorporating also some power-generating features. Thus, a specially designed wristwatch could display the number of steps, the distance or the amount of energy burned, or other health-related functions.

6.1.9.5 Home Automation

One likely trend will be the integration of several functional features in the house, such as integrated control of illumination, safety, heating and ventilation, domestic appliances and audio and video function. This will be called "home automation". A whole range of manufacturing areas is targeting this market, including:
- Intelligent domestic white goods,
- computing,
- home entertainment, with audio and visual electronics,
- telecommunication and especially all internet based functions,
- building controls (lightning, security),
- climate control, heating and ventilation,
- security and safety related functions and devices, etc.

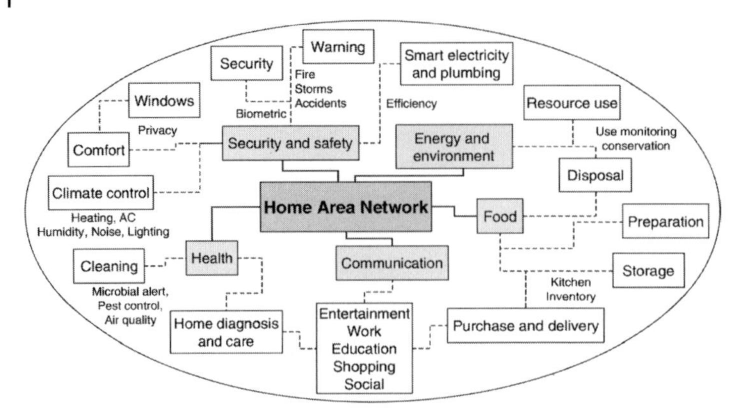

Fig. 6.6 Potential integration of several functions toward home automation.

Communication plays an increasingly important role. Several technical possibilities of common protocols for such communication networks are under investigation, and some common standards are being developed, such as the EIB, the European Installation Bus system. Also communication access via power lines (PLC Powerline Communication, at frequencies in the 10 to 150 kHz band) may be introduced, where house automation services and energy-related services (like remote access to the current counters) may be linked.

There is much technological progress in each of these sectors, but the future lies in their integration. The decisive factors that will give systems the edge over their competitors are

- Cost
- Ease of installation
- Functionality, flexibility and level of integration
- Service provision
- User-friendliness

6.1.10
References

1 Lahrmann, A: Smart domestic appliances through innovations, 6th International Conference on Microsystems 98, Potsdam, Ed. H. Reichl, E. Obermeier, VDE-Verlag, Berlin (1998).
2 Bosch Siemens Hausgeräte AG, Berlin, company information.
3 emz Elektromanufaktur Zangenstein Hanauer GmbH, Nabburg, company information.
4 NEXUS Market Analysis for Microsystems II 2000–2005, Ed. R. Wechsung.
5 G. Tschulena, mst news 2 (2001).
6 Energy Consumption of Washing Machines, Dryers and Dish Washers in Europe, Final Report, European Energy Network, 1955.
7 Gas Wärme Institut, Essen, communication 1999.
8 The NEXUS Technology Roadmap for Microsystems (2000), Ed H. Zinner.

6.2 Smart Buildings – Combination of Sensors via Bus Systems

K. Abkai

6.2.1 Introduction

The evolution of technology and society go hand in hand. Every generation has sought to gradually improve their living conditions, which is reflected in the differences in the buildings and cities our society has inhabited over the centuries. Technical breakthroughs and new design concepts have contributed to its functionality and comfort. Continual innovation and technological progress will dramatically change lifestyle and interaction patterns in our society. The smart building concept integrates new technologies from a range of areas, e.g. computer automation, space age materials, and energy management. The resulting structure is more than just a modern building – it is an independent system and entity. Intelligent buildings may look like any other kind of building from the outside, but what makes them so special is the use of highly advanced computer technology in order to make all familiar aspects of the building (i.e. heating, appliances, lights, etc.) more convenient and efficient.

6.2.2 What Is Behind the "Smart Building" Concept?

The concept of high technology buildings was first developed in the United States in the early 1980s in the context of facility management. In the mid-eighties this topic was also introduced in Europe where the concept was slightly modified. For Americans, smart buildings are places that permit the smooth running of business almost instantaneously and at all hours. They are user-friendly, easily accessible to telecommuters and business owners alike. They provide a safe and secure working environment.

In Europe, a smart building is seen as an object, and no difference is made between private or public usage. The emphasis is on more efficient and comfortable buildings, be they public or private. Administration and services around a building are targeted.

As technologies are developed and integrated into large public buildings, the acceptance of these products will grow, and the production of higher quantities will lower prices. This will benefit the public as well as the private sphere.

And the development will be go further, first of all because investors are rapidly pumping vast amounts of money into intelligent building research. According to Fortune magazine, one of the new companies which are developing intelligent building technology research, had no problem raising 30 million dollars in venture capital.

Secondly, prices of information technology are falling rapidly. While automated home systems were already available two decades ago for $ 20,000–$ 200,000, better systems can now be purchased for between $ 3500 and $ 5000 [Business Week].

On the hardware side it is possible to manufacture very small sensor systems (microsystems) with computing and communication options. These sensors could get down to the size of a penny and cost less than $ 1 each.

Furthermore, due to modern software architecture, it is possible to run small platforms with a tiny operating system that gives the sensor computing power and makes it "smart".

In fact, it seems plausible that intelligent buildings will make good financial sense in the near future. By turning off unnecessary lights and not heating unoccupied rooms, for instance, these buildings can reduce utility bills by 20–30%. According to research done by the University of Berkeley, small changes in our ability to monitor and control energy consumption can dramatically reduce the waste of energy as well as expenses. A network of small wireless sensors capable of monitoring the lighting and temperature of a building and communicating with an energy management system has the potential to cut for example California's energy costs by as much as $ 7 billion a year.

The European markets for home automation are fragmented and suffer from a lack of standardization. Currently this market is divided into different sectors:

- telecommunication and internet services
- home entertainment
- computing
- domestic white goods
- heating and ventilation
- building controls

In this new market, manufacturers as well as consumers, have to define their aims. By lowering costs and targeting potential customers directly, manufacturers are able to increase their market share considerably. A market report of Frost & Sullivan estimates the volume of the European market for home automation with $ 432.1 million in 2007 (see Table 6.8).

A wide range of control systems is used in household appliances. A standard control loop consists of sensors, control units and actuators. The appliances become more powerful and efficient, as the technology is developed and integrated into microsystems. Matchbox-sized sensors can be equipped with wireless radio transceivers and their own miniature operating system to transmit continuous data to the facility manager.

Unfortunately, most of these applications are designed for their specific tasks only. There is currently no software architecture that integrates them into a network that would enable intelligent interaction between them. This is where the future lies. For example, a sensor could recognize the opening of a window and make the heating control of a radiator shut down. The same information about the window could also be built into a security system that would then check what caused the window to open. An alarm would be set off if the person who opened it is not recognized.

Tab. 6.8 Market report of Frost and Sullivan.

Year	*turnover (in Mil. US $)*	*Rate of growth (in %)*
2000	94,3	–
2001	132,3	40,3
2002	177,4	34,1
2003	226,6	27,7
2004	281,4	27,7
2005	334,8	19,0
2006	381,6	14,0
2007	432,1	13,2

There are many strategies and concepts in designing and implementing home automation solutions, but there is very little cooperation between suppliers and no standards have been agreed upon. To make the step from smart products to intelligent home systems a connection between the physical layer and the data managing space needs to be made. This requires simple, flexible and open systems for monitoring, measuring and controlling. Standard specifications would all sensors or actuators to communicate directly with one another. Communication via building bus-systems is the first step in this direction, but the exchange of information between different products and bus systems can also cause problems.

Another important aspect from a practical point of view is the entire installation process. During an installation many different working groups come together and have contact via more or less specified interfaces. In Germany, the installation activities are usually divided into three sectors:

- electrical installation
- sanitation
- heating and air conditioning installation

Each of these sectors use their own specific products and favorite bus systems. Diversity in data transmission rates and different power requirements makes it unlikely that one standard bus system will develop. Products with interfaces for different bus-systems are also available, but production costs are too high for small companies. In recent years the border between the traditional sectors has become more and more flexible and transparent. Standards in bus systems like EIB and LON have been defined and specified.

All special information from such a network of different bus systems would have to be coordinated by one single management system. Facility management means coordinating the physical workplace with the people and work of the organization. It integrates the principles of business administration, architecture and the behavioral and engineering sciences. The ultimate goal is to integrate these operating

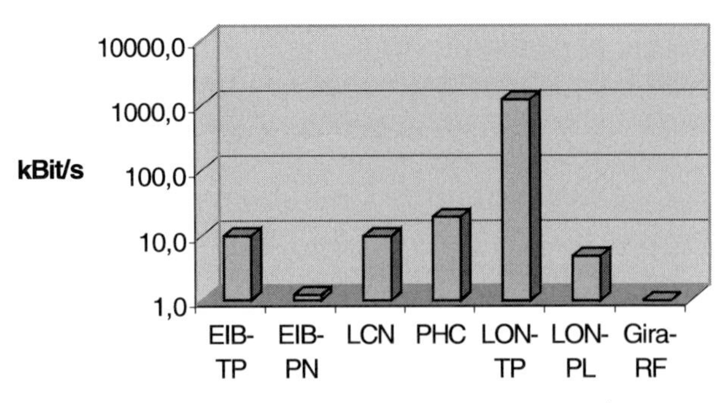

Fig. 6.7 Data transmission rates [Source: FH Dortmund].

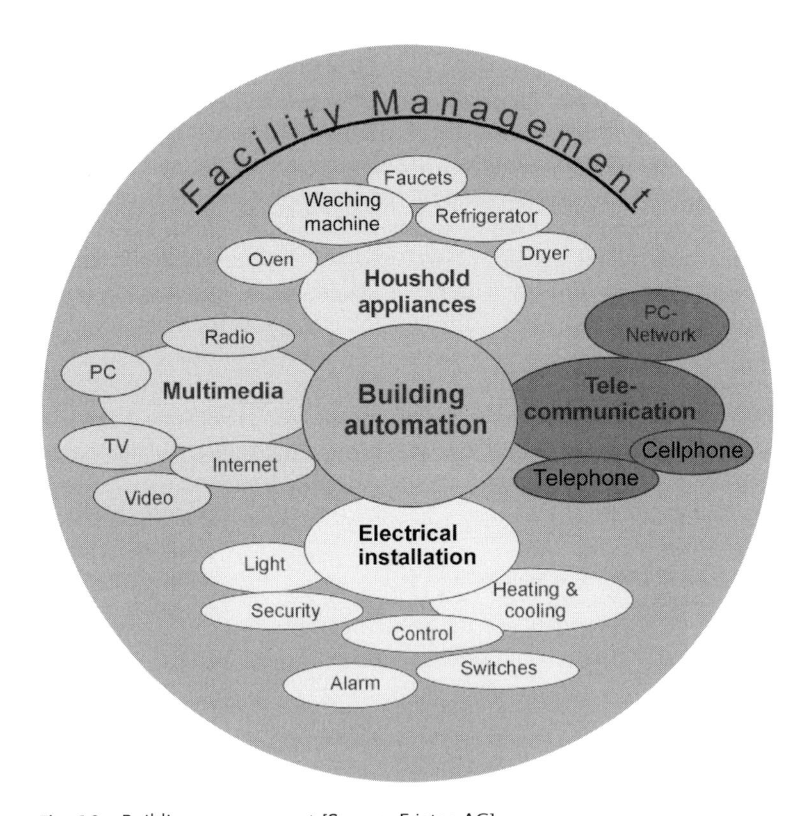

Fig. 6.8 Building management [Source: Friatec AG].

areas into one single computerized system with all hardware and software furnished by one single supplier or partners who use compatible resources and equipment.

Smart building technology can be seen from three operating aspects:

- efficiency
- security
- communication systems

6.2.3 Efficiency

The aim is to reduce the consumption of energy and water to the bare minimum. Computerized systems are used extensively. Such systems have different names: Building Automation System (BAS), Energy Management System (EMS).

These systems would involve:

- Programmed start/stop
- Automatic demand control
- Adaptive control
- Heat & cold optimization
- Light control
- Warning and alarm monitoring
- Optimal energy sourcing

Even simple tasks like closing the curtains when the sun shines into a room or lowering the temperature when nobody is in the house can lead to substantial saving of 20–30% on the utility bill. Electronic faucets are not only more hygienic but also save water through the use of "time out" settings and low-flow aerators (see Fig. 6.9).

6.2.4 Security

Security in a smart building involves the use of high technology to maximize the performance of security systems and comfort while minimizing costs. This can include:

Fig. 6.9 Touchless electronic faucet [Source: Oras Ltd.].

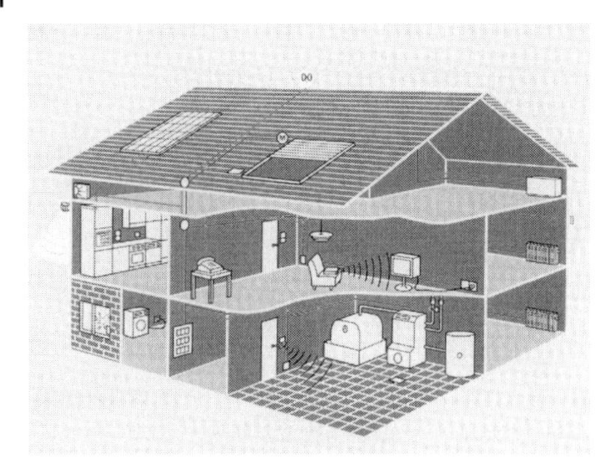

Fig. 6.10 Smart home [Source: Siemens].

- Smoke detection
- More comfort for aged and handicapped people
- Closed-circuit television
- Intrusion alarms
- Emergency control, HVAC
- Universal Power Supply Systems (UPS)

Multi-sensor fire detectors provide faster detection with fewer false alarms. These sensors are capable of monitoring the environment for multiple purposes, e.g. carbon monoxide concentration, concentration of flammable gases and indoor air quality, too.

6.2.5 Communication Systems

Voice and video communications in a smart building involves offering the users many sophisticated data and telecom services and features at a considerably reduced cost. Some of these features are:

- High-speed Internet access
- Local and long distance services
- Cablevision
- Videotext
- Electronic mail
- Direct access to satellite communication
- Enlarged riser capacities
- Videoconferencing facilities
- Emergency electrical back-up
- Disaster back-up and recovery access

For example, the system could mute the volume of the television when the phone rings to make a comfortable conversation possible.

Currently, research is being carried out all over the world to study the behavior of the people using these new technologies. In projects such as the "in Haus-Project" of the Fraunhofer IMS or the "SmartHouse" of ZVSHK the functionality of different home systems in one object are investigated and observed. The "SmartHouse" has been presented on the ISH fair 2001 at Frankfurt for the first time.

As a partner in the "SmartHouse" project, Friatec AG presents their new products in this field. Flushing systems with EIB interfaces enable a connection to managing systems higher up in the hierarchy. Information from various parts of the system can be read and used for further control steps. The status of flusher and fauct valves, water consumption, remote control and vandalism alarms are some of these features. This very fast integration of different technologies in a young product segment was only possible through cooperation with our partners in this field. Only the know-how of electronic design from our partners and our own experience in the installation of sanitation products have made this implementation possible.

6.2.6 Conclusion

The concept of the Smart Building is new, but the tradition it follows is not. Since the time when humans lived in caves, there has been a steady evolution of the structures in which society lives, works, and interacts. The basic concept of the Smart Building will most likely change and adjust as new technologies become available. Most technologies are still hypothetical and will not be implemented for many years.

The facility management aspect of smart buildings deals with the automation of every function of a building, such as lighting, heating and cooling, as well as communications. By controlling these areas through a main computer, the building becomes not only more functional, but also much more energy-efficient.

It takes time for new products to find their place in the world market. People are skeptical about new ideas, even if they are likely to yield straightforward benefits in the future. However, some aspects of Smart Buildings will be finding its way into reality very soon. The sheer range of possibilities intelligent technology opens in private and public buildings, across the various communication systems and installation sectors are very exciting. It is time now for manufacturers to establish themselves as industry leaders with a strategic mix of cooperations, alliances and mergers that offer end-to-end solutions for consumers.

6.3
Smart Homes – A Meeting with the Future Even Now

C. Kühner and U. Koch

Everyone who is involved with the potential of modern techniques is surely fascinated by the unbelievable developments that have taken place in this particular sector within the last 100 years. Knowledge, abilities and the possibilities for mankind as a whole, and specifically the individual, have also seen dramatic evolutionary changes over the last few decades and reached a level our ancestors would never have even dreamed of. So from the present point of view, it is only a question of time before our future technology will be that which today's science fiction authors are writing about.

Many of these authors have explicit and detailed views on what the lives and way of life of future generations will look like, and how intelligent systems can and will make our lives easier, more secure and more comfortable. A first step towards realizing this is being achieved by the German eBuilding GmbH. Their project "Living in an Intelligent House" is combining aspects of the future with today's capabilities and compatibility with future developments. The realization of this project will be in a south-western German city within the next two years and combines sensorial and technical possibilities with a high quality of life. A key point of the eBuilding philosophy is to provide the maximum level of user friendliness for a single human user through today's high level sensor and computer techniques. This is because the techniques are used to make life easier and not to control our every day lives.

During the project 1200 households to be built in a suburban housing area should be linked with a high-speed data connection. Eventually a central service- and security-center will collect data from all households.

To provide easy access to the central-service center, fiber-glass data-lines will be installed together with the other supply lines from the outset. These data lines will also provide individual maximum-speed internet connections for all households.

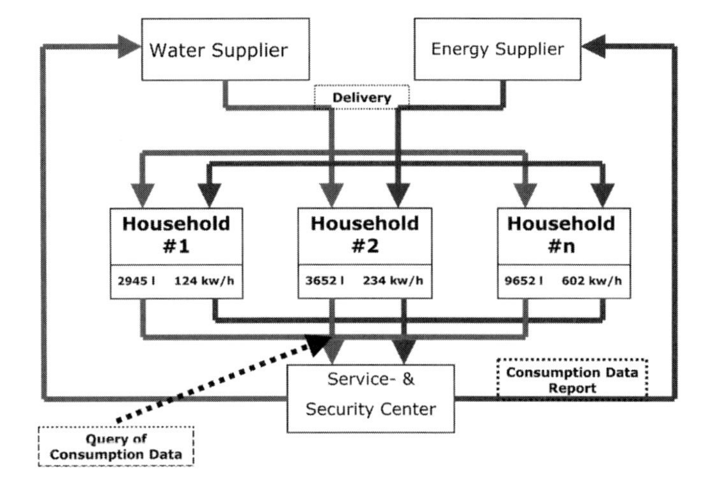

6.3.1
Comfort and Security as Central Aspects

As compared with today's so called "Internetcommunities", which rely on regular internet data transfer to the owner, the eBuilding concept is one of linking the houses and all connected appliances directly to the central-service center. This ensures that all reports of irregularities are collected and dealt with. The owner will only be contacted if the problem reaches a specified alert level.

The ambitious goal of eBuilding is to maximize the comfort of the house owner at a minimum cost level. This will be achieved by using sensors in all sensitive areas, for instance, constant control of the air-conditioning or to cut off cost intensive peaks. Also sensors in smoke-detectors, windows, doors and other appliances will take care of the house security.

Even though the modules Comfort and Security operate independently of each other, their combination will reduce the overall operating costs of the house. This particular aspect is surely becoming more and more important these days. However, the most obvious of all the questions raised is "What is the advantage of an intelligent property?" The answer from eBuildings Managing Director Ulrich Koch is quite simple: "An intelligent real-estate is innovative and cost-friendly".

Giving some examples will indicate the advantages the inhabitants gain from living in an intelligent house. Light sensors control the illumination inside and outside the house corresponding to the luminous intensity and to the season. So energy is saved during the twilight hours because the illumination is not then working at 100%. The heating and air-conditioning level is also controlled by temperature sensors, so the house or even single rooms are always at their defined temperature level. Costly extreme fluctuations in temperature are equalized because of these measures.

The blinds are also controlled by sensors to reduce the cost of running the air-conditioning and heating as well as the total-energy costs. During the night and while nobody is in the house, specific energy-intensive devices are reduced to a minimum or completely shut off. So standby functions do not affect the energy costs.

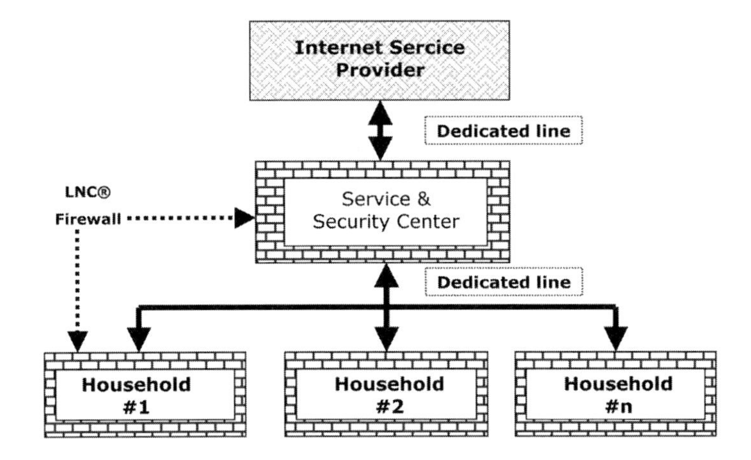

Not only comfort and service but most of all security is the main advantage of an "intelligent real-estate", because not only does everyone want to live comfortably but most of all they want to live in a save area. As eBuilding judges security as definitely the most important point, they offer a high standard on their security package. All of these measures are accounted for by numerous different precautions inside and outside a single house. These manpower, software and hardware components offer security against a variety of disasters. The common burglar alarm is just a small key feature in the various portfolios of eBuilding. Particularly new and comfortable are the in-house and child security features.

As in the field of comfort and service, sensors in combination with high-speed connections provide overall security for peoples homes. Sensors at doors and windows are not only used as primary control devices against burglary while the owner is absent, they also provide information on the air-conditioning while somebody is at home. Depending on the season, temperatures can be raised or lowered if a window or door is opened from the inside. Outside weather sensors also provide data to control the blinds against strong winds or the air-conditioning to provide pre-determined temperature. In addition, inside and outside lights can be activated. Sensors at supply lines report electrical shorts, gas leaks, water damage and cut off immediately. Upon demand, video and digital cameras inside and outside the house can also be used to transfer pictures to the service center or to the owners mobile phone, office PC or laptop anywhere in the world. These pictures can subsequently be used as evidence if necessary.

The implementation of an automatic power shut-off when leaving the house turns off any sockets, where standby appliances waste energy or, for example, the iron may have been forgotten and could cause a fire.

A separate energy line secures the functioning of the whole system. In case of any failure the service center will send someone for immediate action and notify the owner if required.

However, the security package of eBuilding is not just a one-way vehicle. If the owner has to stay away longer than planned any type of change can also be made by mobile phone or via the internet so that the presence of someone in the house is simulated.

6.3.2 The Service and Security Center (SSC)

This center is designed to control and manage the connected houses on the specified demand. Depending on the type of alert or action the owner will immediately be informed by SMS or email while the SSC will already be taking action.

However, the SSC does not just report on security situations. The center also collects data provided by the appliances in the homes. These data are processed and given to the owners in the form of monthly reports.

To transfer the data to the local internet or intranet, the Gateway LNC® is used. LNC® is a Hardware-Firewall, which guarantees that no one from outside the SSC system can interfere. Because of the SPS technique this firewall guarantees absolute security. Messages and information are transferred to the SSC via the

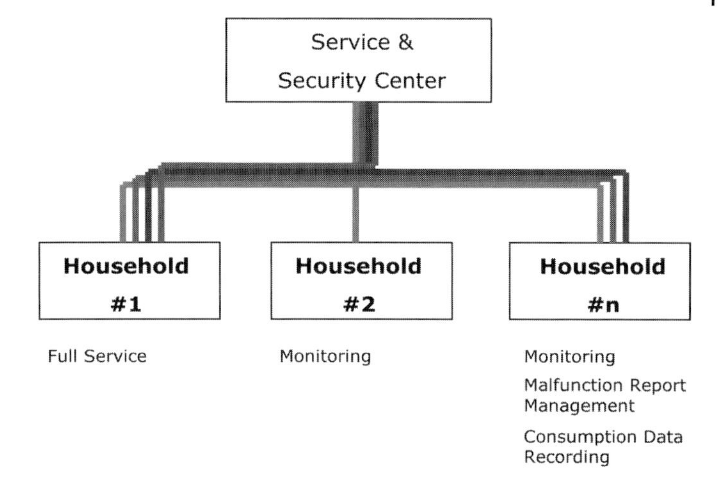

LNC®. LNC® has been tested for efficiency, capacity and security over a long period of time in large office buildings, hence it will be used in this project to link the houses and the SSC.

To cope with growing demand, the SSC is designed in modules. This ensures that it keeps up with advancing technology and the connection of new real-estate nearby. This way the provider and user attain the optimum performance in the acquisition and processing of data. All data will be secured in a central unit so that an individual record of each property is available. The data from these records can be used in standard programs such as Microsoft® or Apple®, so that individual processing is granted. Because of this system the functions of the SSC are guaranteed the whole way through. Thus the inhabitants of the individual households do not need to waste a thought on the prospect of "sitting in the dark", because service, comfort and security are fail safe at any time.

The SSC will take care of lots of different actions and measures to make life and living in an "intelligent house" as easy and comfortable as possible and to offer a high level of safety at the same time. As a bonus the SSC provides specific statistics on how to optimise the environmental friendliness of the individual buildings and the project itself. Hence architects and urban developers could gain a great deal of important hints and information by analysing the whole eBuilding pilot-project data.

However, the SSC is more than a collector of information and data: it is designed to make peoples lives easier and to serve the individual needs of the house owners. The key point of the philosophy of eBuilding is that each household can put together an individual package of services they desire without interfering with the concept as a whole. Thus as times and needs change the packages can be changed as well. So everybody receives exactly the amount of service, comfort and safety they wish to enjoy in their personal lives.

Appendix:
Examples of Commercial Sensors for Household Appliances

MEDER electronic AG
Robert-Bosch-Strasse 4
D 78224 Singen
Tel.: 07731-83990
Fax: 07731-839932
eMail: info@meder.com
Internet: www.meder.com

type: MK12-1C90C-500W

part number: 9123903054

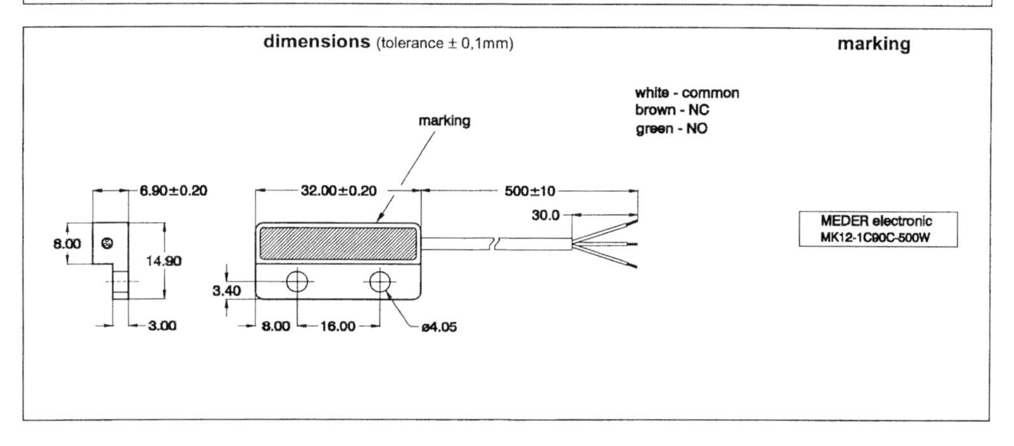

contact data 90 (FormC/Dry)	condition	Min.	Typ.	Max.	Unit
pull-in energization contact unmodified		15		20	AT
pull-in energization contact modified		58,97		95,14	AT
test coil		KMS-31			
contact material		Rhodium			
rated power	each combination of the switching voltage and current must not exceed the given rated power			3	W
switching voltage				175	VDC
switching current				0,25	A
carry current				1,2	A
static contact resistance	starting values measured with 1,4 x $AT_{pull-in}$			150	mΩ
insulation resistance	RH Ω 45%	10^9			Ω
breakdown voltage		200			VDC
operate time incl. bounce	measured with 1,4 x $AT_{pull-in}$		0,7		ms
release time			1,0		ms
capacitance	without test coil			1,0	pF

general data					
contact resistance incl. cable	measured with 1,4 x $AT_{pull-in}$			250	mΩ
shock	½ sine wave, duration 11ms			50	g
vibration	10 – 1000Hz			30	g
operating temperature		-20		70	°C
storing temperature		-20		70	°C

washability		fully sealed
material of case		Glass fibre reinforced polybutylene terephtalate (PBTP) self-extinguishing Self-extinguishing V-0 according to UL94
sealing compound		polyurethane
cable		round cable LIYY 3 x 0,14 mm^2, grey colour of wire: white, brown and green cable end with approx. 5 mm tinned leads

remark

The switching distance can be decreased by mounting the MK12 on iron.
When mounting the sensor, magnetically conductive screws must not be used.

MEDER electronic AG
Robert-Bosch-Strasse 4
D 78224 Singen

Phone: 07731-83990
Fax: 07731-839932
eMail: info@meder.com
Internet: www.meder.com

type: MK6- 5 - C

part number: 2206050002

dimensions (tolerance ± 0,1mm) **marking**

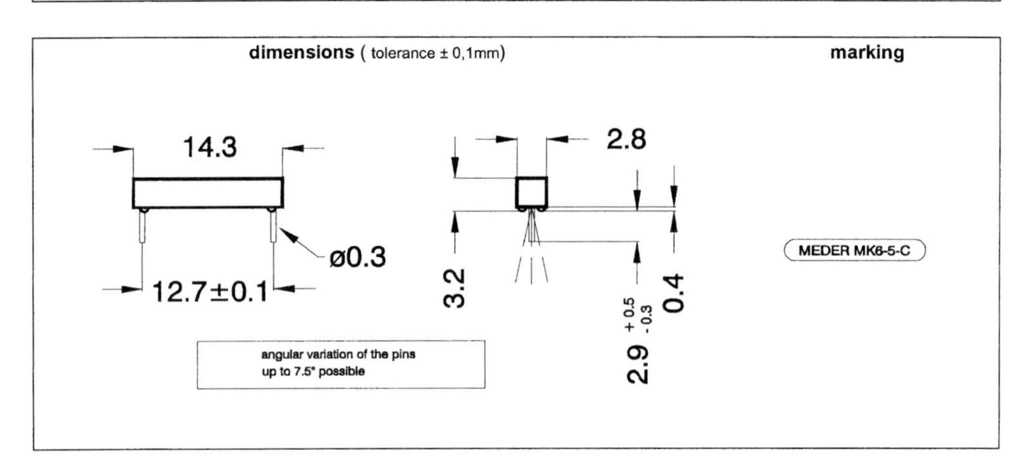

contact data 87 (Form A/Dry)	conditions	Min.	Typ.	Max.	unit
pull-in energization contact unmodified	measuring ramp 10 mA / s	15		20	AT
pull-in voltage contact modified	measuring ramp 10 mA / s	10,8		18,3	VDC
test coil		KMS-16			
contact material		Rhodium			
rated power	each combination of the switching voltage and current must not exceed the given rated power			1	W
switching voltage				24	VDC
switching current				0,1	A
carry current				0,3	A
static contact resistance	measured with 1,4 x $AT_{pull-in}$			200	mΩ
insulation resistance	RH Ω 45%	10^9			Ω
breakdown voltage		150			VDC
resonant frequency			7500		Hz
operate time inclusive bounce	measured with 1,4 x $AT_{pull-in}$		0,5		ms
release time			0,3		ms
capacitance	without test coil			0,2	pF

shock	½ sine wave, duration 11ms			30	g
vibration	10 – 1000Hz			20	g
soldering temperature	4 sec. at			360	°C
operating temperature		-20		85	°C
storing temperature		-35		85	°C
material of case		polyamid 6 black			
sealing compound		epoxy resin			

MEDER electronic AG
Robert-Bosch-Strasse 4
D 78224 Singen

Phone.: 07731-83990
Fax: 07731-839932
eMail: info@meder.com
Internet: www.meder.com

type: MK14-1A71C-500W

part number: 9143711054

dimensions (tolerance ± 0,1mm) / **marking**

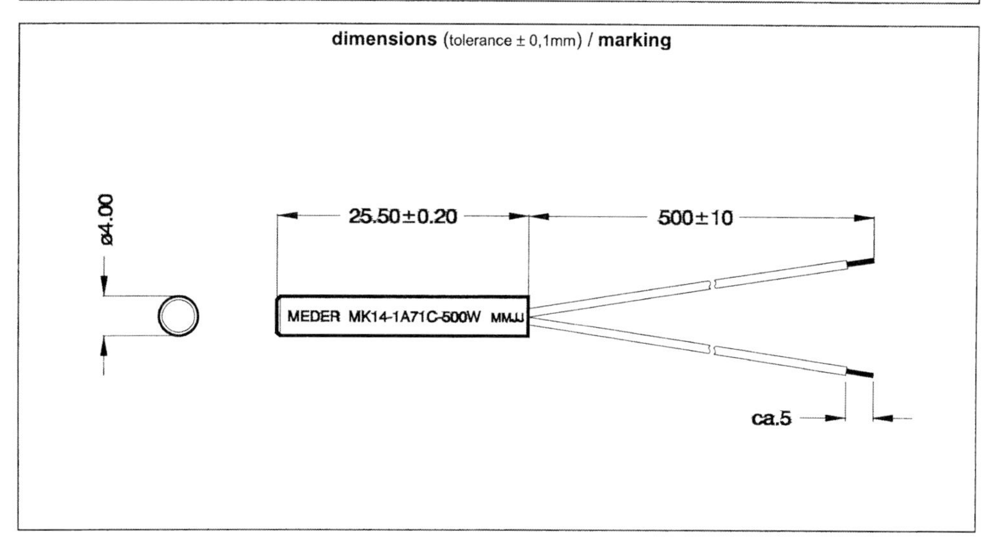

contact data 71/7 (FormA/Dry)	condition	Min.	Typ.	Max.	Unit
pull-in energization contact unmodified		15		20	AT
pull-in energization contact modified		28,31		49,06	AT
test coil		KMS-13			
contact material		Ruthenium			
rated power	each combination of the switching voltage end current must not exceed the given rated power			10	W
switching voltage				180	VDC
switching current				0,5	A
carry current				1,5	A
static contact resistance	starting values measured with 1,4 x $AT_{pull-in}$			150	mΩ
insulation resistance	RH Ω 45%	10^{12}			Ω
breakdown voltage		200			VDC
resonant frequency			6700		Hz
operate time incl. bounce	measured with 1,4 x $AT_{pull-in}$		0,5		ms
release time			0,1		ms
capacitance	without test coil			0,3	pF

general data					
contact resistance incl. cable	measured with 1,4 x $AT_{pull-in}$			300	mΩ
shock	½ sine wave, duration 11ms			150	g
vibration	10 – 2000Hz			10	g
operating temperature		-20		85	°C
storing temperature		-35		85	°C

washability		fully sealed
material of case		non-reinforced, semi-crystalline polybutylene terephtalate (PBTP)
sealing compound		epoxy resin
cable		2 x single wire LIY 0,14 mm^2 colour of wire: blue with 5mm tinned leads

MEDER electronic AG
Friedrich-List-Strasse 6
D 78234 Engen-Welschingen

Phone.: +49(0)7733/9487-0
Fax: +49(0)7733/9487-32
eMail: info@meder.com
Internet: www.meder.com

type: MK4-1A71B-500W

part number: 2242711054

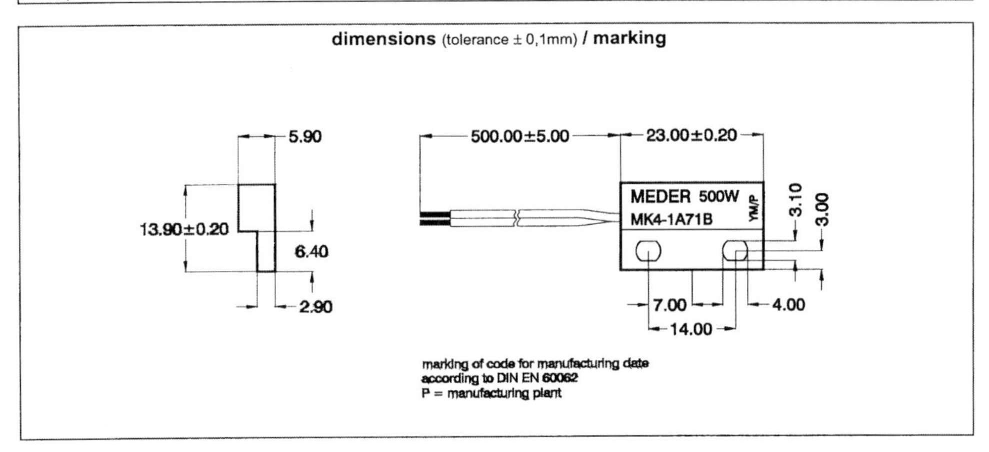

contact data 71 (FormA/Dry)	condition	Min.	Typ.	Max.	Unit
pull-in energization contact unmodified	measuring ramp 10 mA/s	10		15	AT
pull-in energization contact modified	measuring ramp 10 mA/s	23,15		43,52	AT
test coil		KMS-14			
contact material		Ruthenium			
rated power	each combination of the switching voltage end current must not exceed the given rated power			10	W
switching voltage				180	VDC
switching current				0,5	A
carry current				1,5	A
static contact resistance	starting values measured with 1,4 x $AT_{pull-in}$			150	mΩ
insulation resistance	RH Ω 45%	10^{12}			Ω
breakdown voltage		200			VDC
resonant frequency			6700		Hz
operate time incl. bounce	measured with 1,4 x $AT_{pull-in}$		0,5		ms
release time			0,1		ms
capacitance	without test coil			0,3	pF

general data					
contact resistance incl. cable	measured with 1,4 x $AT_{pull-in}$			250	mΩ
shock	½ sine wave, duration 11ms			150	g
vibration	10 – 2000Hz			10	g
operating temperature		-5		70	°C
storing temperature		-25		70	°C

washability		fully sealed
material of case		Glass fibre reinforced polybutylene terephtalate (PBTP) self-extinguishing Self-extinguishing V-0 according to UL94
sealing compound		polyurethane
cable		flat cable LIYZ 2 x 0,14 mm^2, white cable end with approx. 5 mm tinned leads

remark

The switching distance can be decreased by mounting the MK4 on iron.
When mounting the sensor, magnetically conductive screws must not be used.

modifications in the sense of technical progress are reserved   index 02   11.12.2001

MEDER electronic AG
Friedrich-List-Strasse 6
78234 Engen-Welschingen

Phone.: +49 (0) 7733 / 9487-0
Fax: +49 (0) 7733 / 9487-32
eMail: info@meder.com
Internet: www.meder.com

type: MK17-E-2

part number: 9171000025

dimensions (tolerance ± 0,1mm)

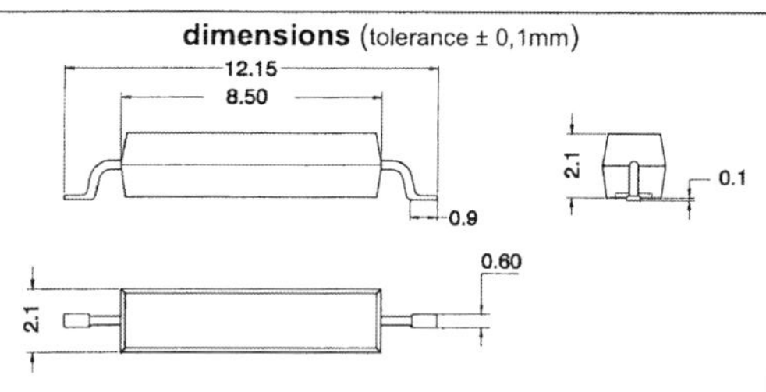

marking

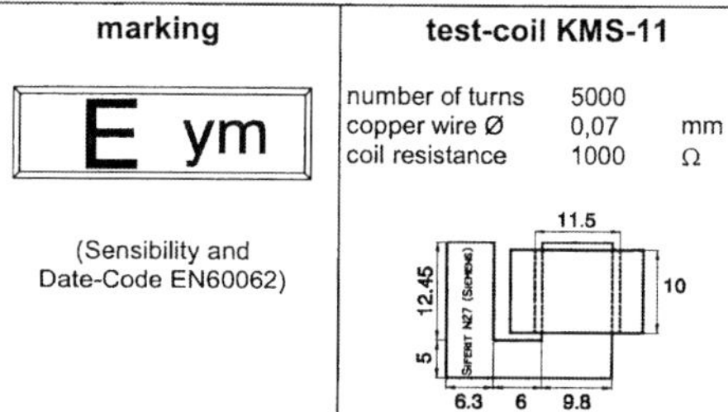

(Sensibility and Date-Code EN60062)

test-coil KMS-11

number of turns	5000	
copper wire Ø	0,07	mm
coil resistance	1000	Ω

reflow soldering conditions according to JEDEC norm JESD22-A113A

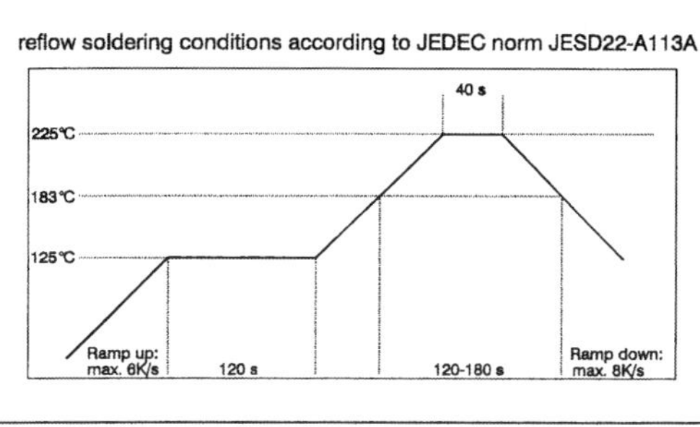

tape and reel packaging (IEC 60286-3)

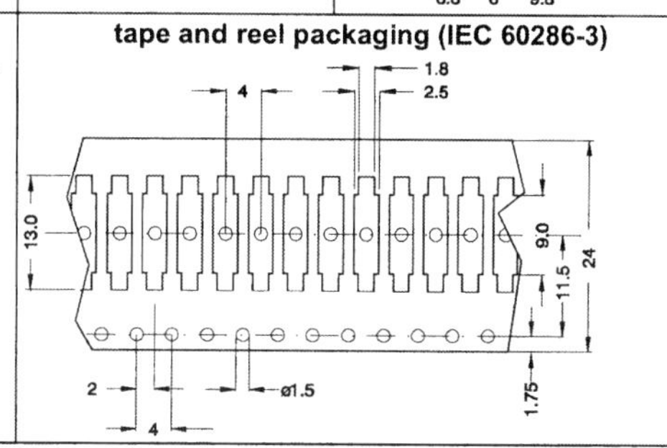

contact data (Form A/Dry)	condition	Min.	Typ.	Max.	Unit
pull-in energization contact unmodified	measuring ramp 10 mA/s KMS01	25		30	AT
pull-in energization contact modified	measuring ramp 10 mA/s KMS11	50		66	AT
contact material			Ruthenium		
rated power	each combination of the switching voltage and current must not exceed the given rated power			10	W
switching voltage				100	VDC
switching current				0,5	A
carry current				0,5	A
static contact resistance	measured with mit 1,4 x $AT_{pull-in}$			250	mΩ
insulation resistance	RH Ω 45%	10^{10}			Ω
breakdown voltage		150			VDC
resonant frequency			17900		Hz
operate time incl. bounce	measured with 1,4 x $AT_{pull-in}$		0,5		ms
release time			0,1		ms
capacitance	without test coil			0,2	pF

shock	½ sine wave, duration 11ms			30	g
vibration	10 – 2000Hz			20	g
operating temperature		-40		130	°C
storing temperature		-50		130	°C

general data					
soldering temperature	4 sec. at			360	°C
cleaning			fully sealed		
material of case			mineral-filled epoxy		

reel dimensions					
outer diameter			330		mm
outer width			30		mm
shaft diameter			12		mm

MEDER electronic AG
Robert-Bosch-Strasse 4
D 78224 Singen

Phone: 07731-83990
Fax: 07731-839932
eMail: info@meder.com
Internet: www.meder.com

Reed Sensor: Customer-designed Sensor

part number:

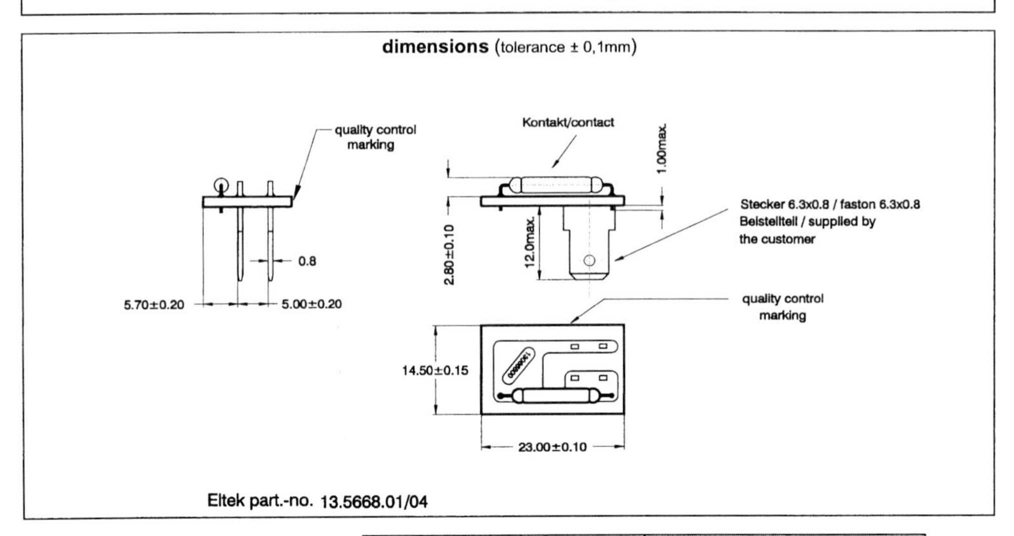

contact data (A/Dry)	conditions	87 Min.	87 Typ.	87 Max.	71/3 Min.	71/3 Typ.	71/3 Max.	Unit
pull-in energization contact modified	measuring ramp 10 mA / s	41		52	37		50	AT
hysteresys				0,8			0,8	
test coil		KMS-11			KMS-11			
contact material		Rhodium			Ruthenium			
rated power	any combination of the switching voltage and current must not exceed the given rated power			10			10	W
switching voltage				200			200	VDC
switching current				0,5			0,5	A
carry current				0,5			1,0	A
static contact resistance	measured with 40% overdrive			150			150	mΩ
insulation resistance	RH 45%	10^9			10^{12}			Ω
breakdown voltage		230			220			VDC
resonant frequency			7500			5500		Hz
operate time incl. bounce	measured with 40% overdrive		0,5			0,5		ms
release time			0,1			0,1		ms
capacitance	without test coil			0,2			0,3	pF
Environmental data								
shock	½ sine wave, duration 11ms			30			150	g
vibration	10 – 2000Hz			20			10	g
soldering temperature	4 sec. at			360			360	°C
operating temperature		-40		125	-55		125	°C
storing temperature		-40		125	-55		125	°C
remark								
switch marking		black			red			

modifications in the sense of technical progress are reserved   index 01   07.07.2000

Headquarter Europe	Headquarter USA		
MEDER electronic AG Friedrich-List Strasse 6 D-78234 Engen-Welschingen Tel.: +49(0)7733-9487-0 Fax: +49(0)7733-9487-32 eMail: info@meder.com Internet: www.meder.com	MEDER electronic Inc. 766 Falmouth Rd Mashpee, MA 02649 Phone: +1/ 508-539-0002 Fax: +1/ 508-539-4088 eMail: salesusa@meder.com	**Reed Sensor:** **Part Number:**	MK21-1A66B-500W 9212100054

Dimensions (tolerance ± 0,1mm) / **Marking**

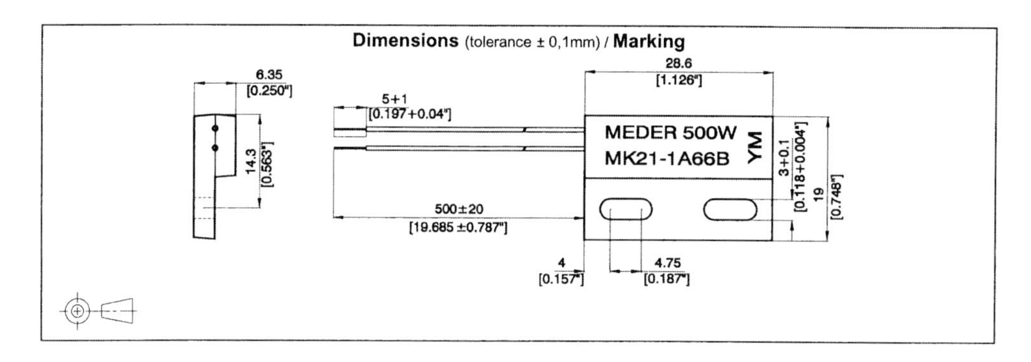

Magnetic Characteristics	Conditions at 20°C	Min.	Typ.	Max.	Units
Pull-In Switch unmodified	Testcoil SMC-01	10		15	AT
Pull-In Switch modified	Testcoil SMC-14	21		76	AT

Contact Data 66 (Form A/Dry)					
Contact Rating	Any combination of the switching voltage and current must not exceed the given rated power			10	W
Switching Voltage	DC or Peak AC			200	VDC
Switching Current	DC or Peak AC			0,5	A
Carry Current	DC or Peak AC			1,0	A
Static Contact Resistance (initial)	Measured with 40% overdrive			150	mΩ
Insulation Resistance	RH 45%	10^{10}			Ω
Breakdown Voltage		250			VDC
Resonant frequency			6700		Hz
Operate Time, including Bounce	Measured with 40% overdrive			0,5	ms
Release Time				0,1	ms
Capacitance			0,3		pF

Environmental Data					
Shock	½ sine wave, duration 11ms			150	g
Vibration	from 10 - 2000 Hz			10	g
Operating Temperature	10°C/min max. allowable	-30		150	°C
Storage Temperature	10°C/min max. allowable	-40		160	°C
Cleaning		fully sealed			
Material of Case		Mineral-filled-epoxy			
Cable	RADOX RXL 155 0,5mm² BU				
Contact Resistance with Cable	Measured with 40% overdrive			200	mΩ
Remarks	The switching distance can be decreased by mounting the MK21 on iron. When mounting the sensor, magnetically conductive screws must not be used.				

Customer / Customer part number	Standard product

Headquarter Europe
MEDER electronic AG
Friedrich-List Strasse 6
D-78234 Engen-Welschingen
Tel.: +49(0)7733-9487-0
Fax: +49(0)7733-9487-32
eMail: info@meder.com
Internet: www.meder.com

Headquarter USA
MEDER electronic Inc.
766 Falmouth Rd
Mashpee, MA 02649
Phone: +1/ 508-539-0002
Fax: +1/ 508-539-4088
eMail: salesusa@meder.com

Reed Sensor: MK3-1A71C-500W

Part Number: 2233751054

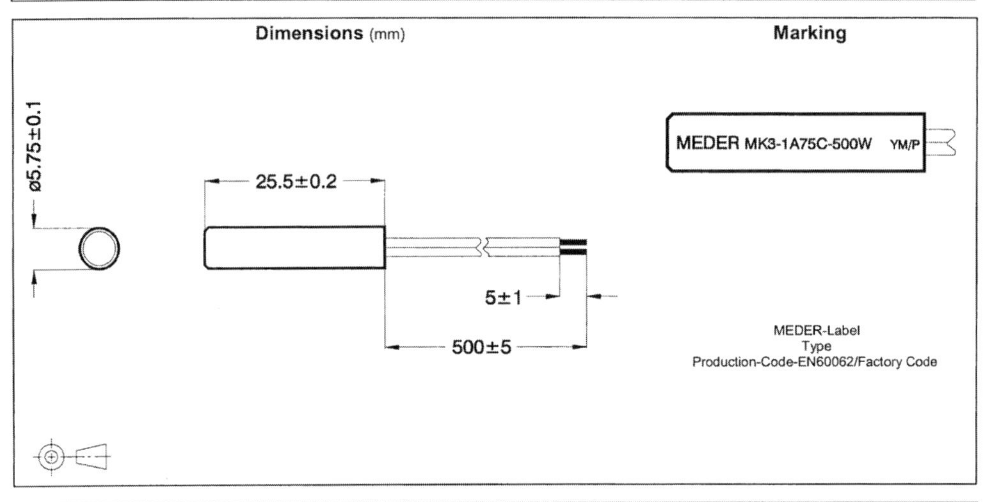

Magnetic Characteristics	Conditions at 20°C	Min.	Typ.	Max.	Units
Pull-In Switch unmodified	Test-coil KMS-01	15		20	AT
Pull-In Switch modified	Test-coil KMS-13	30		49	AT

Contact Data 71/7 (Form C/Dry)					
Contact Rating	Any combination of the switching voltage and current must not exceed the given rated power			10	W
Switching Voltage	DC or Peak AC			500	V
Switching Current	DC or Peak AC			0,5	A
Carry Current	DC or Peak AC			1,0	A
Static Contact Resistance (initial)	Measured with 40% overdrive			200	mΩ
Insulation Resistance	RH 45%	10^{10}			Ω
Breakdown Voltage		600			VDC
Operate Time, including Bounce	Measured with 40% overdrive			0,5	ms
Resonant frequency			4800		Hz
Release Time				0,1	ms
Capacitance			0,4		pF

Environmental Data					
Shock	½ sine wave, duration 11ms			30	g
Vibration	from 50 - 2000 Hz			10	g
Operating Temperature	10°C/min max. allowable	-5		70	°C
Storage Temperature	10°C/min max. allowable	-25		70	°C
Contact Resistance with Cable	Measured with 40% overdrive			320	mΩ
Cleaning		fully sealed			
Material of Case		non-reinforced semi-crystalline polybutylene terephthalate /PBTP)			
Sealing Compound		Epoxy resin			
Cable		Flat cable LiYZ 2 x 0,14 mm^2 Colour: white Ends of cable with 5 ±1mm tined leads			
Remarks					

Modifications in the sense of technical progress are reserved DESIGNED BY: Werner Kovacs APPROVED BY: Horst Wedele DATE: 17.12.2002
Page 1 of 1 [VRSp-02.doc] REVISION / Rev. No.: 01 REVISED BY: APPROVED BY: DATE:

Headquater Europa
MEDER electronic AG
Friedrich-List Strasse 6
D-78234 Engen-Welschingen
Tel.: +49(0)7733-9487-0
Fax: +49(0)7733-9487-32
eMail: info@meder.com
Internet: www.meder.com

headquater USA
MEDER electronic Inc.
766 Falmouth Rd
Mashpee, MA 02649
Phone: +1/ 508-539-0002
Fax: +1/ 508-539-4088
eMail: salesusa@meder.com

Reed Sensor: Customer-designed Sensor

Part Number:

dimensions (mm)

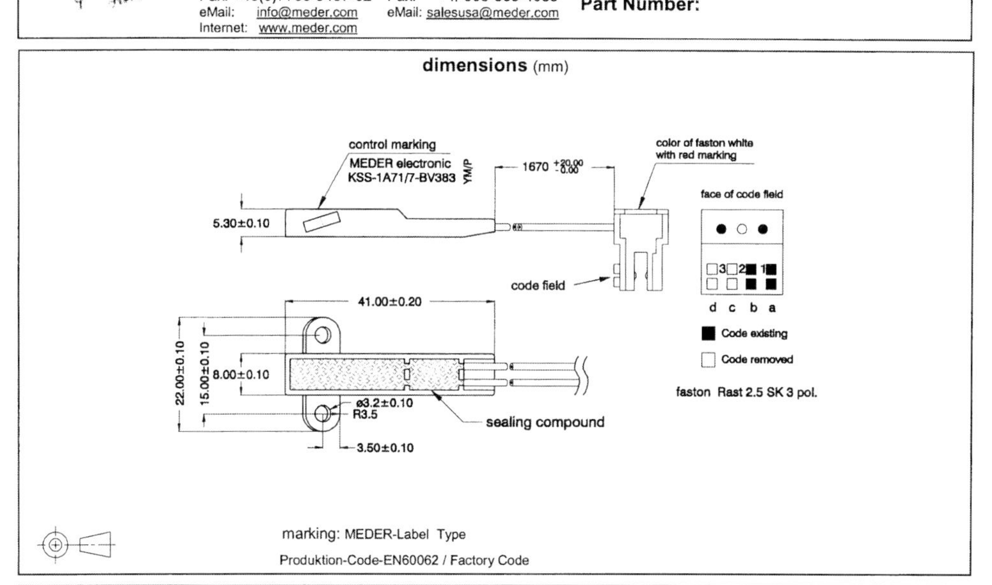

marking: MEDER-Label Type

Produktion-Code-EN60062 / Factory Code

Magnetic Characteristics	Conditions at 20°C	Min.	Typ.	Max.	Units
Pull –in Switch unmodified	Testcoil KMS-01	12,5		16,49	AT
Drop out Switch unmodified	Testcoil KMS-01	7,5		10,5	AT
Pull-in Switch modified	Testhead- MK4	25		50	AT
Drop out Switch modified	Testhead- MK4	18		40	AT

Contact data 71/7 (Form A/Dry)					
Contact rating	Any combination of the switching voltage and current must not exceed the given rated power			10	W
Switching voltage	DC oder Peak AC			180	VDC
Switching current	DC oder Peak AC			0,5	A
Carry current	DC oder Peak AC			1,5	A
Static contact Resistance (initial)	Measured with 40% overdrive			150	mΩ
Insulation Resistance	RH 45%	10^{12}			Ω
Breakdown voltage		200			VDC
Operate time, including bounce	Measured with 40% overdrive			0,5	ms
Release time				0,1	ms
capacitance			0,3		pF

Environmental Data					
Shock	½ sine wave, duration 11ms			50	g
Vibration	10 - 2000 Hz			20	g
Operating temperature	10°C/min max. allowable	-20		70	°C
Storage temperature	10°C/min max. allowable	-25		70	°C
Material of case	Glass fibre reinforced semi-crystalline polybutylene terephtalate (PBTP) self-extinguishing Self-extinguishing V-O according to UL94				
Sealing compound	polyurethan				
cable	Single wire FLRY 0,35mm black, ends of cable equipped with connector Rast 2,5 SK 3pol. Cable rolled to a ring of 100mm diameter and fixed with an 12mm wide adhesive tape 12 mm x 40 mm (Art.-Nr.: 4003009009)				
Contact resistance incl. cable	Measured with 40% overdrive			350	mΩ

sensor type: Non-Contact Capacitive Measuring System
physical principle: capacitive measuring principle
model range: Micro-Epsilon capaNCDT Series 620
typical applications: development, testing, in-process quality-control

			S601-0,05	S601-0,2	S600-0,2	S601-0,5	S610-0,7	S600-1	S601-1	S600-2	S600-3	S600-5	S600-10
Measuring range	el. cond. (metal)	mm	.05	.2	.2	.5	—	1	1	2	3	5	10
		inch	.002	.008	.008	.019	—	.039	.039	.078	.117	.195	.39
	Insulator [1]	mm (≈)	—	.4	.4	1	.7	2	2	4	6	10	20
		inch (≈)	—	.015	.015	.039	.027	.078	.078	.156	.234	.39	.78
Linearity	≤ ±0.2 % FSO	±μm	.1	.4	.4	1	1.4	2	2	4	6	10	20
Resolution	stat. (...30 Hz) ≤ 0.004 % FSO	μm	.002	.008	.008	.02	0.03	.04	.04	.08	.12	.20	.40
	dyn. (...6 kHz) ≤ 0.02 % FSO	μm	.01	.04	.04	.10	.14	.20	.20	.40	.60	1	2
Sensor outer diameter		mm	6	6	6	8	7	10	10	20	30	40	60
		inch	.23	.23	.23	.31	.27	.39	.39	.78	1.17	1.56	2.34
Weight		g	2	2	2.5	12	3.5	7.1	7.1	61	95	120	230
Active measuring area (diameter)		mm	1.3	2.3	2.3	3.9	2.5	5.5	5.5	7.9	9.8	12.6	17.8
		inch	.05	.09	.09	.15	.1	.21	.21	.31	.38	.49	.69
Guard ring width		mm	.8	1	1	1.4	.4	1.5	1.5	4	8.1	11.8	18.1
		inch	.03	.04	.04	.05	.015	.06	.06	.16	.32	.46	.71
Min. diameter of target	el. cond. (metal)	mm	3	5	5	7	—	9	9	17	27	37	57
		inch	.12	.2	.2	.27	—	.35	.35	.66	1.05	1.44	2.22
	Insulator	mm	—	7	7	10	8	12	12	24	36	48	72
		inch	—	.27	.27	.39	.31	.47	.47	.94	1.4	1.87	2.8
Temp. stability sensor	Zero	±μm/°C	.03	.06	.03	.06	.17	.17	.06	.17	.17	.17	.17
	Sensitivity	·ppm/°C	3	11	3	30	30	30	11	30	30	30	30
Temp. stab. electronics			≤ ±0.01 % FSO / °C										
Long term stability[2]			≤0.02 % FSO / month										
Sensitivity		V/mm	200	50	50	20	≈ 10	10	10	5	3.33	2	1
		V/inch	5080	1270	1270	508	≈ 254	254	254	127	83.58	50.8	25.4
Output	Voltage		0 - 10 VDC (max. 10 mA short circuit proof)										
Power supply	Model DT 620		±15 VDC (±5 %) / ±50 mA										
	Model DT 621		230 VAC / 16 VA or 115 VAC (with integral power supply)										
Band width			static .... 4kHz (-0.1 dB), ...6 kHz (-3 dB)										
Temp. range	Sensors		-50 to +200 °C / -60 to +400 °F (S600-1 max.150°C / 300°F)										
	Sensor cable		-50 to +150 °C / -60 to +300 °F										
	Electronics		+10 to +50 °C / +50 to +125 °F										
Air humidity	Sensor		5 to 95 % (non condensing)										
Electromagnetic compatibility (EMC)			EN 50081-1 Spurious emission EN 50082-2 Immunity to interference - uncertainty of measurement max. 2.6 %										
Protection class			Electronics and sensors: IP 40										

FSO = Full Scale Output 1 μm = 1 micron

1) The measuring range for insulators is approximate and depends on the relative dielectric constant of the insulator

2) At reference temperature 20 °C (68 °C) and steady state

sensor type: Non-Contact Eddy Current Displacement Measuring System
physical principle: eddy current principle
model range: Micro-Epsilon eddyNCDT Series 3300
typical applications: factory automation, machine monitroing, inspection and testing

Sensor model		ES04	EU05	EU1	ES1	ES2	EU3	ES4	EU6	EU8	EU15	EU22	EU40	EU80
Controller		DT 3300												
Measuring range	mm	0.4	0.5	1	1	2	3	4	6	8	15	22	40	80
Start of measuring range SMR	mm	0.04	0.05	0.1	0.1	0.2	0.3	0.4	0.6	0.8	1.5	2.2	4	8
End of measuring range EMR	mm	0.44	0.55	1.1	1.1	2.2	3.3	4.4	6.6	8.8	16.5	24.2	44	88
Linearity		≤ ±0.2 % FSO												
	µm	±0.8	±1	±2	±2	±4	±6	±8	±12	±16	±30	±44	±80	±160
Resolution* up to 25 Hz		≤ 0.01% FSO		≤ 0.005 % FSO										
	µm	≤0.04	≤0.05	≤0.05	≤0.05	≤0.1	≤0.15	≤0.2	≤0.3	≤0.4	≤0.75	≤1.1	≤2	≤4
Resolution* up to 25 kHz		≤ 0.2% FSO		≤ 0.1 % FSO										
	µm	≤0.8	≤1	≤1	≤1	≤2	≤3	≤4	≤6	≤8	≤15	≤22	≤40	≤80
Frequency response		kHz / 2.5 kHz / 25 Hz (-3 dB) selectable · option: 100 kHz (for small range senso												
Temperature compensation		10 ...100° C (50 ... 212° F) · option TCS: up to 150° C (302° F)												
Temperature range Sensors + cable		-50 ... 150 °C (-58 ... 302° F)												
Temperature range Controller		5 ...50° C (41 ...122° F)												
Temp.stability Sensors		≤±0.015 % FSO/° C (≤±0.008 % FSO/° F)												
Protect.class sensors + cable		IP67		IP65		IP67								
Max press. on sensor face		1450 psi [1)]		-		290 psi [2)]								
Sensor cable length		3 m (± 0.45 m) / option: up to 15 m (sensors ES04 and EU05 6 m max.)												
Signal output		selectable: 0 ... 5 V, 0 ... 10 V, ±2.5 V, ±5 V, ±10 V (or inverted) 4 ... 20 mA (load 350 ohm)												
Power supply DT 3300		±12 VDC / 100 mA, 5 VDC / 220 mA												
Electromagnetic Compatibility		acc. to EN 50081-2 / EN 61000-6-2												
Controller Functions		Limit switches, Auto-Zero, Peak-to-Peak, Minimum, Maximum, Average, Storage of 3 configurations (calibrations)												

FSO = Full Scale Output
Reference materials are Aluminum (non-ferromagnetic) and Mild Steel St 37, DIN 1.0116 /AISI 4130 (ferromagnetic)
Reference temperature for reported data is 20°C (70 °I *resolution data are based on noise peak-to-peak values
Resolution and temperature stability refer to midrange (MMI [1)] 10^7 Pa (100 bar) [2)] $2 \cdot 10^6$ Pa (20 bar)
Data of SMR, MMR, EMR are always related to the sensor face

sensor type: Inductive Displacement Sensors and Linear Gaging Sensors
physical principle: linear variable differential transformer principle
model range: Micro-Epsilon transSENSOR
typical applications: test stand, experimental station, mobile applications, OEM - sensor

displacement sensor with free moving plunger

Basic Model		DTA-1D-□-			DTA-3D-□-			DTA-5D-□-			DTA-10D-□-		
Connection		TA	CA	SA	TA	CA	SA	TA	CA	SA	TA	CA	SA
Linear Range	mm	±1			±3			±5			±10		
Linearity	0.3 % FSO	●	●	●	●	●	●	●	●	●	●	●	●
	0.5 % FSO	○	○	○	○	○	○	○	○	○	○	○	○
	0.15 % FSO	○	○	○	○	○	○	○	○	○	○	○	○
Excitation frequency	kHz	5									2		
Excitation amplitude	V_{eff}	5											
Sensitivity	mV/Vmm	133			85			53			44		
Termperature range	-20...80°C		●	●		●	●		●	●		●	●
	-20...120°C	●			●			●			●		
available options		W,P F	P,F H	P,F	W,P F	W,P F H	W,P F	W,P F	W,P F H	W,P F	W,P F	W,P F H	W,P F

Basic model		DTA-15D-□-					DTA-25D-□-				
Connection		LA	CA	CR	SA	SR	LA	CA	CR	SA	SR
Linear Range	mm	±15					±25				
Linearity	0.5 % FSO	●	●	●	●	●	●	●	●	●	●
	0.3 % FSO	○	○	○	○	○	○	○	○	○	○
Excitation frequency	kHz	1									
Excitation amplitude	V_{eff}	2.5									
Sensitivity	mV/Vmm	45					33				
Temperature range	-20...80°C	-20...80°C									
available options		W,P F,	W P,F H	W P,F H	W P,F	W P,F	W,P F	W,P F, H	W,P F, H	W P,F	W P,F

gauging sensor with sliding bearing and spring

Basic model	DTA-1G-□-		DTA-3G-□-		DTA-5G-□-		DTA-10G-□-	
Connection	CA	SA	CA	SA	CA	SA	CA	SA
Linear range	±1 mm		±3 mm		±5 mm		±10 mm	
Linearity	●	●	●	●	●	●	●	●
	○	○	○	○	○	○	○	○
	○	○	○	○	○	○	○	○
Excitation frequency	5 kHz						2 kHz	
Excitation amplitude	5 Veff							
Sensitivity	133 mV/Vmm		85 mV/Vmm		53 mV/Vmm		44 mV/Vmm	
Force in midrange (typical)	0.95 N		1.00 N		1.18 N		1.23 N	
Spring constant	0.22 N/mm		0.14 N/mm		0.12 N/mm		0.08 N/mm	
Temperature range	-20...80 °C							
Available options	Option V							

Options:
- W Waterproof housing
- P Pressure resistant sensor
- F Standard mounting flange
- H High temperature sensor up to 200°C
- V Pneumatic push

sensor type: High-Speed Laser-Micrometer
physical principle: laser based shadow measurement system
model range: Micro-Epsilon optoCONTROL 2500
typical applications: running conveyors, extrusion lines, drawing processes

Model		ODC 2500-35
Measuring range		34 mm (1.34 inch)
Distance lightsource - CCD-camera		300 mm (150-700 mm) 11.81 inch (5.91-28 inch)
Linearity [1]		<±10 µm
Resolution [2]		≤ 1 µm
Repeatibility		≤ 3 µm
Smallest diameter (detectable target)		0.5 mm (0.02 inch)
Sampling rate		2.3 kHz
Light source		semiconductor laser 670 nm, class II
Dimensions (W x H x	receiver	54 x 72 x 28 mm (2.13 x 2.83 x 1.10 inch)
	laser light source	110 x 72 x 28 mm (4.33 x 2.83 x 1.10 inch)
	ontroller (without connectors)	191 x 110 x 45 mm (7.52 x 4.33 x 1.77 inch)
Cable		2 m (option: extension 3 m / 8 m)
Protection class	receiver / light source	IP 64
	controller	IP 40
Operating temperature		0 °C to 50 °C (32 °F to 122 °F)
Storage temperature		-20 °C to 70 °C (-4 °F to 158 °F)
Output		analog 0 ... 10V, range -10 to + 10V digital RS 232 or RS 422 1 x error, 2 x limit, 2 x warning LC-Display, 3 x LED sync-out
Input		sync-in zero laser on/off
Supply voltage		24 Vdc (± 15 %)
Measuring programs		diameter, gap, position / edge, segment

[1] Valid for distance of the target to receiver 20 ±5 mm

[2] If output via integral LC-Display

sensor type: Optical Laser Displacement Measuring
physical principle: laser based triangulation principle
model range: Micro-Epsilon optoNCDT 2200
typical applications: vibration, thickness, deflection, dimension, stroke, dimensional testing

Model	ILD 2200-2	ILD 2200-10	ILD 2200-20	ILD 2200-50	ILD 2200-100	ILD 2200-200
Measuring range	2 mm (.08 ")	10 mm (.4 ")	20 mm (.8 ")	50 mm (2 ")	100 mm (4 ")	200 mm (8 ")
Start measuring range (SMR)	24 mm (0.94 ")	30 mm (1.18 ")	40 mm (1.57 ")	45 mm (1.76 ")	70 mm (2.72 ")	130 mm (5.06 ")
Reference distance midrange	25 mm (0.98 ")	35 mm (1.38 ")	50 mm (1.97 ")	70 mm (2.76 ")	120 mm (4.72 ")	230 mm (9.06 ")
End measuring range (EMR)	26 mm (1.02 ")	40 mm (1.58 ")	60 mm (2.37 ")	95 mm (3.76 ")	170 mm (6.72 ")	330 mm (13.06 ")
Linearity	1 μm	3 μm	6 μm	15 μm	30 μm	60 μm
	≤±0.05 % FSO	≤±0.03 % FSO				
Resolution 10 kHz	0.1 μm	0.5 μm	1 μm	2.5 μm	5 μm	10 μm
	0.005 % FSO					
Measuring rate	10 kHz					
Permissible ambient light	30,000 lx					
Spot diameter (midrange)	35 μm	50 μm	60 μm	80 μm	130 μm	1300 μm
Light source	1 mW laser, wavelength: 670 nm (visible/red)					
Laser safety class 2	DIN EN 60825-1/A1 12.99 / IEC 825-1/A1 12.99 / FDA					
Protection class	Sensor: IP 65 Controller: IP 50					
Operating temperature	0 to 50 °C (32 to 122 °F)					
Storage temperature	-20 to 70 °C (-4 to 158 °F)					
Output	Analog: ±5 V Digital: RS 485 / 691.2 kBaud					
Supply	24 VDC (±15 %), max. 500 mA					
Sensor cable	2 m (6.5 ') - integrated 5/10 m (16.5 '/33 ') - without additional calibration					
Controller	auto zero / signal averaging 143 x 145 x 52 mm (5.6 " x 5.6 " x 2 ") - without mounting clips					
Electromagnetic compability (EMC)	EN 50081-1 und EN 50082-2					
Vibration	2 g / 20 ... 500 Hz					
Shock	15 g / 6 ms / 3 Axis					

FSO = Full Scale Output
All specifications apply for a diffusely reflecting matt white ceramic target

continued on next page

Model		ILD 2200-2	ILD 2200-10	ILD 2200-20	ILD 2200-50	ILD 2200-100
Measuring range		2 mm (.08 ")	10 mm (.4 ")	20 mm (.8 ")	50 mm (2 ")	100 mm (4 ")
Reference distance	midrange	25 mm (1 ")	35 mm (1.4 ")	50 mm (2 ")	70 mm (2.7 ")	120 mm (4.7 ")
Linearity		1 µm	3 µm	6 µm	15 µm	30 µm
		≤±0.05 % FSO	**≤±0.03 % FSO**			
Resolution	at 10 kHz	0.1 µm	0.5 µm	1 µm	2.5 µm	5 µm
		0.005 % FSO				
Measuring rate		**10 kHz**				
Permissible ambient light		30,000 lx				
Spot diameter	midrange	140 µm	320 µm	45 µm	55 µm	60 µm
Light source		1 mW laser, wavelength: 670 nm (red)				
Laser safety class II		DIN EN 60825-1 03.97 / IEC 825-1 11.93 / FDA				
Protection class	sensor controller	IP 65 IP 50				
Operating temperature		0 to 50 °C (32 to 122 °F)				
Storage temperature		-20 to 70 °C (-4 ° to 158 °F)				
Output	analog digital	±5 V RS 485 / 687.5 kBaud				
Supply voltage		24 VDC (±15 %), max. 500 mA				
Sensor cable	standard option	2 m (6.5 ') - integrated 5/10 m (16.5 '/33 ') - without additional calibration				
Controller	functions dimensions	auto zero / signal averaging 143 x 145 x 52 mm (5.6 " x 5.6 " x 2 ") - without mounting clips				
Electromagnetic compatibility (EMC)		EN 50081-1 and EN 50082-2				
Vibration		2 g / 20 ... 500 Hz				
Shock		15 g / 6 ms				

FSO = Full Scale Output

All specifications apply for a diffusely reflecting matt white ceramic target

sensor type: Long-Stroke Sensor for Hydraulics and Pneumatics
physical principle: eddy current principle
model range: Micro-Epsilon strokeSENSOR
typical applications: movement and stroke of hydraulic and pneumatic cylinders

Measuring range		mm inch	100 4	160 6.20	200 8	250 9.75	300 12	400 16	630 25
Models			S	S, F	S	S, F	S, F	S, F	S, F
Linearity	≤±0.3 % FSO	±mm ±inch	.30 .01	.48 .02	.60 .02	.75 .03	.90 .04	1.20 .05	1.90 .07
Resolution	0.05 % FSO	mm inch	.05 .002	.08 .003	.1 .004	.125 .005	.15 .006	.20 .008	.32 .01
Temperature range			-40 °C ... +85 °C (-40 °F ... +185 °F)						
Temperature stability			≤±0.02 % FSO / °C (±0.011 % FSO /°F)						
Frequency response	(-3 dB)		150 Hz						
Output signal			4 - 20 mA						
Load			≤500 Ω						
Supply voltage			18 - 30 VDC						
Current consumption			max. 40 mA						
Connector	Model S		4-pin Connector (Sensor cable as an option) options radial or axial output						
	Model F		5-pin Bayonet-connector with mating plug						
Pressure resistance			450 bar [1] (Sensor rod, Flange)						
Protection class			IP 67						
Electromagnetic compatibility (EMC)			EN 50 081-2 Spurious emission EN 50 082-2 Immunity to Interference						
Shock[2]	IEC 68-2-29 IEC 68-2-27		40 g, 3000 Shocks/axis 100 g radial, 300 g axial						
Vibrati(	IEC 68-2-6		5 Hz ... 44 Hz ±2,5 mm 44 Hz ... 500 Hz ±23 g						
Material			V4A-Steel 1.4571						

FSO = Full Scale Output

1) High pressure on request

2) Half sinusoid 6 ms

sensor type: Inductive Potentiometric Displacement Measurement
physical principle: distributed electromagnetic parameters
model range: Micro-Epsilon vipSENSOR
typical applications: industrial use, productions plants, OEM - sensor

Measuring range		mm inch	50 2	100 4	150 6
Linearity	Standard ±0.4 % FSO	±mm ±inch	.2 .008	.4 .016	.6 .024
	Option ±0.2 % FSO	±mm ±inch	.1 .004	.2 .008	.3 .012
Resolution	0.05 % FSO	mm inch	.025 .001	.05 .002	.075 .003
Temperature range			-40 °C to +85 °C (-40 °F to +185 °F)		
Temperature stability			±0.015 % FSO / °C (±0.006 % FSO / °F)		
Bandwidth	(-3 dB)		300 Hz		
Output signal			4 - 20 mA		
Load			≤500 Ω		
Supply voltage			18 - 30 VDC		
Current consumption			max. 40 mA		
Protection class			IP 65		
Electromagnetic compatibility (EMC)			EN 50 081-1 Spurious emission EN 50 082-2 Immunity to Interference		
Shock[1]	IEC 68-2-29 IEC 68-2-27		40 g, 3000 Shocks / axis 100 g radial, 300 g axial		
Vibration	IEC 68-2-6		5 Hz ... 44 Hz ±2.5 mm 44 Hz ... 500 Hz ±20 g		

FSO = Full Scale Output

1) *Half sinusoid 6 ms*

Special measuring ranges on request

sensor type: Draw-wire Displacement Sensors
physical principle: draw wire
model range: Micro-Epsilon wireSENSOR
typical applications: test stand, forg lifter, alevators

Model	WDS-MPM	WDS-MP/MPW	WDS-P60	WDS-Z100	WDS-Z200	WDS-P1200	WPS-MK30	WDS-.....-M
50 mm	•						•	
100 mm		•	•					
150 mm	•		•				•	
250 mm	•						•	
300 mm		•	•					
500 mm		•	•				•	
750 mm			•				•	
1000 mm		•	•					
1500 mm			•					•
2500 mm				•				•
5000 mm					•	•		•
7500 mm						•		
10.000 mm						•		
15.000 mm						•		
20.000 mm						•		•
30.000 mm						•		•
Protection class	IP 40	IP 40/66	IP 65	IP 65	IP 65	IP 65	IP 20	
Analog output, optional potentiometric, voltage or current*								
Potentiometer (P)	•	•	•	•	•	•	•	
Voltage (I)			•	•	•	•		
Current (U)			•	•	•	•		
Digital output, optional incremental or absolute*								
Incremental (E)		•	•	•	•	•	•	
Absolute (A)			•	•	•	•		
Areas of application	- suitable for fast movements - installation under limited space conditions - application in harsh environment		- large variety of different models (output signals, mounting types, measuring ranges) offer a wide field of applications - high quality and performance at a minimum cost				The ideal sensor for customized solutions also in large quantities.	For customized encoder installation.

Braun Thermoscan IRT 3000 Range

Infrared Ear Thermometer

Feature	IRT 3020	IRT 3520 Pharmacy unit	PRO 3000 Professional unit
Sensor System	Thermopile sensor with patented metal shielding		
Site Offsets	ear mode		
Patient Temp. Range	34°C .. 42.2°C		20°C .. 42.2°C
Ambient Temp. Range	10°C .. 40°C		
Display resolution	0.1 °C		
Accuracy	± 0.2°C (at 35.5°C .. 42°C patient temperature) ± 0.3°C (outside above defined range)		
Operating Lifetime	5 years / 10 000 readings		5 years / 100 000 readings
Battery Type	2 * CR 2032		
Battery lifetime at normal ambient temp.	2000 readings and 5 years shelf life		
measuring time	< 1 s		
minimum time between readings	< 2 s		
automatic power down, power down time	120 s		60 s
Back-light	no	yes	yes
Memory	1	8	1
Memory clear	no	yes	no
Controls / buttons	3	4	4
Lens Filter	Sanitary polypropylen probe covers		
Lens-filter detector	yes		yes, auto on/clear funct.
Lens-filter ejector	yes		
°F/°C selection	key sequence (disabled for Korean subtype during production by EEPROM setting)		
Error messages (display)	Err: flashing display: HI / LO: cover symbol: battery symbol:	Ambient out of range error System error Object out of range no cover attached battery low / dead	
Self-test function	yes, functional self-test		

1- channel- sensor- IC „Ee102"

EDISEN® – electronic GmbH present with the "Ee102" as a highlight the application possibilities of it's digital capacitive sensors.The "Ee102" is worldwide the first μPpwer 1- channel- sensor IC with a power consumption < 5 μA @ (3..9) VDC supply voltage.The switching output is P-MOS-open-drain with 9V / 20 mA load und protected against overcurrent.
The application specific integrated circuit „Ee102" of **EDISEN® – electronic** has a sensor channel, which detect by a patented, digital method of processing capacities and their changes at the input. It has an integrated voltage regulator with an output voltage **VDD** = 4 VDC at a supply voltage of **VHI** = (4,7 ... 9) VDC. This could be used to supply external additional electronic with a supply current < 0,5 mA. The „Ee102" could be operate at battery voltages down to **VHI** = 3 VDC. By self- adjustment max. 2 s after Power➔ On it is ready-to-operate. For the sensor IC the parameters „sampling frequency **f_X**" and „output switching mode **TYPE**" (**„Pulse" / „momentary" / „toggle mode"**) could be fixed by external wiring. The sensitivity could be varied by the value of the storage capacitor **CPC_X**. Separated sensor areas (conductive areas or foils) have to be connected by coax cable of max. 2 m length and could be sticked behind all non-conductive materials and activated in front of them.So high requirements on water- and dust proofness, design and wear and tear are very well fulfilled, because cout-outs are not necessary anymore.The method allows a compensation of static input capacities **$C_{IN\,x0}$ ≤ 200 pF** only by a resistor **R_C** vs. **GND** at the input of the „Ee102". The input should have a filter resistor **R_F** to limit the input bandwidth.

Simplified block diagram with a typical external circuit.

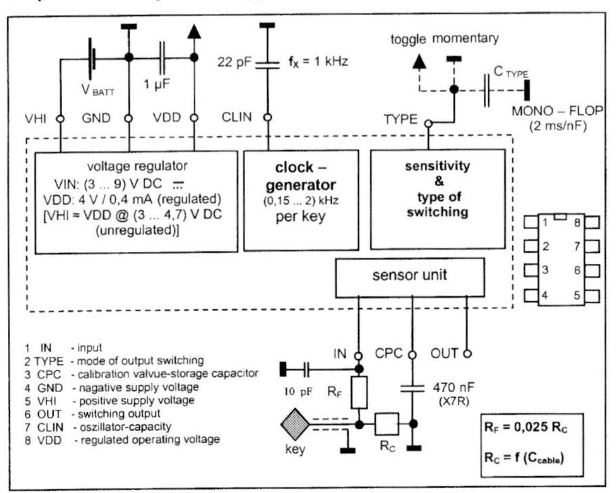

The EDISEN- method allow the compensation of static input capacities **C_{IN} ≤ 200 pF** only by a resistor **R_C** vs. **GND** at the input of the „Ee102". This resistor discharge during every sampling cycle a constant quota **C_C** of the sensor capacity **$C_{IN\text{-}X0}$**, so the input circuit of the IC has not to discharge **C_C**.

Technical features
- available in SO8 (Die at request)
- dynamic movement sensor
- digital mode of processing
- automatic calibration
- separated sensor areas up to 2 m in distance
- P- channel- open- drain- output
- μ Power (5 μA)➔ battery applications possible
- synchronisation adjustable
- 3 different switch modes (toggle, momentary, Pulse)
- internal voltage regulator

Applications
- ▶ domestic appliances
- ▶ automotive
- ▶ equipment keys
- ▶ security technology
- ▶ sanitary technology
- ▶ IT
- ▶ industrial automation

DC - characteristics

conditions: VHI = 6 VDC; T_{AMB} = 25 °C

Name	Parameter	Conditions	Min	Typ	Max	unit
VHI	positive supply voltage	VDD – regulator active	4,7			
		battery- powered	3	6	9	V
VDD	regulated IC-operating voltage	4,7 V ≤ VHI ≤ 9 V ➔ VDD – regulator active	3,4	4	4,6	V
ΔVDD	Dropout voltage	(VHI – VDD) @ inactive regulator ➔ VHI < 4,7 V		10	50	mV
I_{HI}	IC – supply current	outputs = OFF/ fx = 1 kHz		5		μA
V_{OUT}	reverse voltage	@ OUT = OFF	0	VHI	9	V
I_{OUT}	output current	working range (P-MOS)	0	10	20	mA
		over load switching off @ V_{FOUT} ≥ 0,6 V	20	30	50	mA
V_{FOUT}	output saturation voltage	@OUT = ON / I_{OUT} = + 10 mA	0,1	0,2	0,4	V
C_{VDD}	ext. VDD – blocking cap.	SMT – ceramic – chip – cap`s	0,1	1	2,2	μF
V_{CPC}	calibration voltage	linear range	0,8		VDD-1	V
I_{LCPC}	leckadge current		-1	0	1	nA
T_{AMB}	operating temperature range		- 20		+ 85	°C

AC – characteristics

conditions: VHI = 6 VDC; C_{CLIN} = 22 pF; C_{CPC}= 470 nF; T_{AMB} = 25 °C

Name	Parameter	conditions	Min	Typ	Max	unit
C_{CLIN}	oscillator condensator	useful range	0	22	100	pF
C_{CPC}	ext. storage cap.	X7R – ceramic – chip-cap`s	90	470	2500	nF
A_D	equivalent digital resolution			14		Bit
C_{MONO}	Mono Flop capacitor	useful range	0,1		500	nF
C_{IN}	INPUT capacity	without external compensation	10		60	pF
t_{TA}	loading pulse	per sample	1,4	2	3,5	μs
t_{INDIS}	discharge time input capacity	per sample	0,5	1	2	μs
t_{WAIT}	waiting time	between 2 actions		0,5	2	s
t_{MONO}	output pulse width	"Mono Flop"	1,3	2	3	ms/nF
f_X	sampling frequency	f_x= (300 μs + 33 μs/pF* C_{LIN})$^{-1}$		1		kHz
		@C_{CLIN} = 100 pF		275		Hz
		@C_{LIN} = 0		3,3		kHz
H_{FM}	sampling- frequency modulation rate	accidental 32 values	3	4	5	%
t_{SW}	reaction time	@f_X = 1 kHz	64			ms

3- channel- sensor- IC „Ee301"

With the "Ee301" **EDISEN**® **– electronic GmbH** extent the application possibilities of it's digital capacitive sensors.
The application specific integrated circuit „Ee 301" of **EDISEN**® **– electronic** has 3 identical sensor channels, which detect by a patented, digital method of processing capacities and their changes on the 3 inputs. It has an integrated voltage regulator with an output voltage **VDD** = **4** VDC at a supply voltage of **VHI** = (4,7 ... 9) VDC. This could be used to supply external additional electronic with a supply current < 0,5 mA. The „Ee301" could be operate at battery voltages down to **VHI** = 3 VDC. By self- adjustment of all channels max. 5 s after Power→ On it is ready-to-operate. For all 3 channels together the parameters „sampling frequency f_X", „weighting number **N**" and „output switching mode **TYPE**" **(„Pulse" / „momentary" / „toggle mode")** could be fixed by external wiring. The sensitivity of every channel could be varied by the value of the storage capacitor CPC_X. Separated sensor areas (conductive areas or foils) have to be connected by coax cable of max. 4m length and could be sticked behind all non-conductive materials and activated in front of them.
So high requirements on water- and dust proofness, design and wear and tear are very well fulfilled, because cout-outs are not necessary anymore.
The method allows a compensation of static input capacities $C_{IN\,X0}$ **≤ 400 pF** only by a resistor R_C vs. **GND** at the input **X** of the „Ee301". Every input **X** should have a filter resistor R_F to limit the input bandwidth.
The customer himself could built keyboards with low expense by cascade connection of several circuits. Mechanical works are not necessary anymore.
With the internal stabilized IC- operating voltage VDD e.g. opamp's could be supplied.
With these opamp's the analoge voltages stored at CPC_X can be coupled out without feedback and processed by external electronics. This analoge voltage U_X represents the value of the static capacity on the input X at very good linearity.
So with the Ee301 level and distance measurements can be designed.

Simplified block diagram with a typical external circuit.

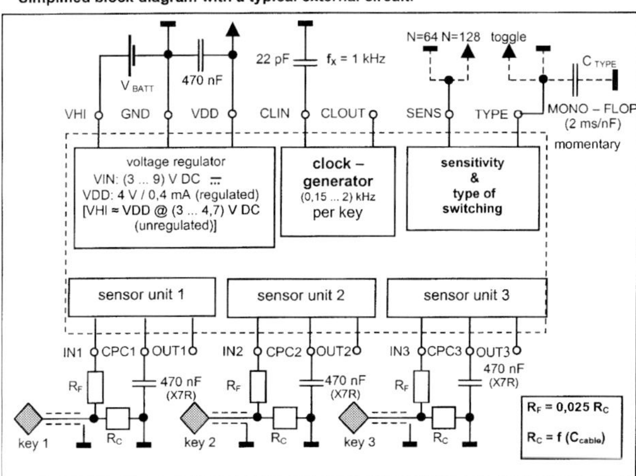

The EDISEN- method allow the compensation of static input capacities C_{IN} **≤ 400 pF** only by a resistor R_C vs. **GND** at the input **X** of the „Ee301". This resistor discharge during every sampling cycle a constant quota C_C of the sensor capacity $C_{IN\text{-}X0}$, so the input circuit of the IC- channel **X** has not to discharge C_C.

Technical features
- available in SO16 (Die or DIL16 at request)
- dynamic movement sensor
- digital mode of processing
- automatic calibration
- 3 sensor channels per IC
- separated sensor areas up to 4 m in distance
- N- channel- open- drain- outputs
- µ Power→ battery applications possible
- synchronisation adjustable
- vary sensitivities
- 3 different switch modes (toggle, momentary, Pulse)
- up to 16 IC's to connect in cascade
- internal voltage regulator
- analogue measure systems designable

Applications
- ► domestic appliances
- ► automotive
- ► installation in buildings
- ► sanitary technology
- ► IT
- ► industrial automation
- ► equipment keyboards

DC - Eigenschaften

conditions: VHI = 6 VDC; C_{CLIN} = 22 pF; T_{AMB} = 25 °C

Name	Parameter	Conditions	Min	Typ	Max	unit
	positive supply voltage	VDD – regulator active	4,7			
VHI		battery- powered	3	6	9	V
VDD	regulated IC-operating voltage	4,7 V ≤ VHI ≤ 9 V → VDD – regulator active	3,4	4	4,6	V
ΔVDD	Dropout-regulator voltage	(VHI – VDD) @ inactive regulator → VHI < 4,7 V		10	50	mV
I_{HI}	IC – supply current	outputs = OFF	100	150	200	µA
V_{OUT}	reverse voltage			VHI		
		working range (N-MOS) @„PULSE-mode"		- 10	- 20	mA
I_{OUT}	output current to GND	working range (N-MOS) @ „toggle" and „momentary"		- 3	- 5	mA
V_{FOUT}	output saturation voltage	@ I_{OUT} = - 10 mA	0,2	0,4	0,6	V
$I_{OUT\text{-}OFF}$	overload protection	@ V_{FOUT} ≥ 1,2 V	-20	-40	-50	mA
		middle level	45%	50%	55%	VDD
V_{CLOUT}	clock output voltage	SYNC – Puls	90%		100%	VDD
I_{CLOUT}	clock output current	middle level	- 12		+ 12	µA
C_{VDD}	ext. VDD – blocking cap.	SMT – ceramic – chip – cap's	0,1	0,47	2,2	µF
T_{AMB}	operating temperature range		- 20		+ 85	°C

AC – characteristic

conditions: VHI = 6 VDC; C_{CLIN} = 22 pF; T_{AMB} = 25 °C

Name	Parameter	conditions	Min	Typ	Max	unit
V_{CPC}	analog storage voltage	linear range	0,8		VDD-1	V
C_{CPC}	ext. storage cap.	only foil cap's or X7R – ceramic – chip-cap's	90	470	2500	nF
A_D	equivalent digital resolution	of the linear range of V_{CPC}		14		Bit
I_{LCPC}	CPC – leakage current		- 1	0	+ 1	nA
		as MASTER – IC @cascade connection	80	128	200	kHz
f_S	system clock frequency	without cascade connection [C_{CLIN} = (10 ... 100) pF]	20		256	kHz
f_K	channel sampling frequency	$f_K = f_S / 128$		1		kHz
	quantity of processed sequence	IC – Pin „SENS" → GND		64		
N	of samples	IC – Pin „SENS" → VDD		128		samples
	min. response time	f_K = 1 kHz / N = 64		63		
t_{SW}	[$t_{SW} = (N - 1) / f_K$]	f_K = 1 kHz / N = 128		127		ms
C_{X0}	input capacity		8		46	pF
$t_{IN\text{-}DIS}$	time of discharge of C_{X0}			1		µs
t_{PO}	Power ON – calibration time	ready for operating after power→ ON [f_K = 1 kHz]	0,25	1,2	7	s
t_H	hold time = fct. (C_{TYPE})	output signal pulse width @ „PULSE-mode"	1,2	2	3,3	ms/nF
C_{TYPE}	MONO- FLOP- capacity	capacity range of C_{TYPE} @ IC-pin TYPE	0,1		500	nF

Sensoric

Approximation switch "NAHE"

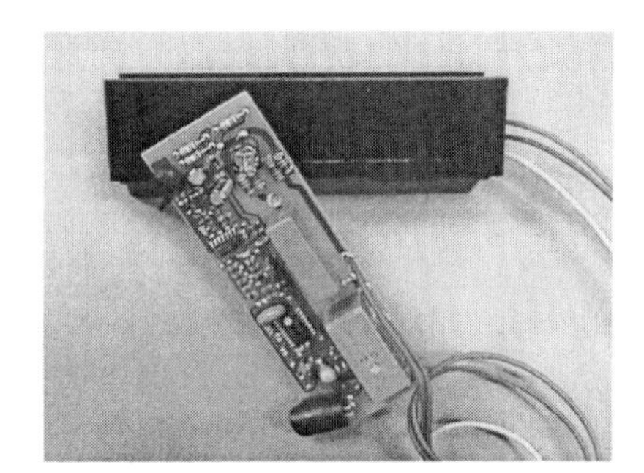

Characteristics:

- IR diffuse reflective sensor
- Switching distance 140...190 mm
- Limitation of the duty cycle
- Protection against vandalism
- Protection against external light influences
- Long durability
- Low cost product

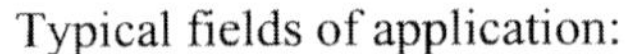

Typical fields of application:

- Hand dryer

Short description:

The electronic approximation switch "NAHE" was especially developed for devices at which it is necessary to start an electrical equipment without touching. The sensor is based on the opto-electronic principle. The beam of a transmitter is reflected diffusely on a target (e.g. hands) and registered by a receiver. The switching distance of 140...190 mm can be adjusted according to customer´s specification. The switching on load is 8 A/230 VAC or 16 A/120 VAC.

The approximation switch is supplied with 120 or 230 VAC at a line frequency of 50 until 60 cps by a 3 core line. The approximation switch offers protection against vandalism and enables a simple installation.

Technical data:

Switching distance:	140...190 mm
Supply:	120 VAC or 230 VAC 50/60 cps
Range of operating temperature	0° - 70° C
Range of storage temperature	-20° - 80° C
Switchable load:	8 A/230 VAC 16 A/120 VAC
Closed circuit current	<20 mA
Active wavelength	infra red
Output signal:	Low switched
Electrical connection:	3 core line
Dimension:	114 x 47 mm
Protective system:	IP 21
Casing material:	PES, PA

Sensoric

Position sensor „PASE“

Characteristics:

- System: magnet < > hall sensor
- contactless switch
- switching distance 5-7 mm
- resistance against humidity
- long durability
- Low cost product

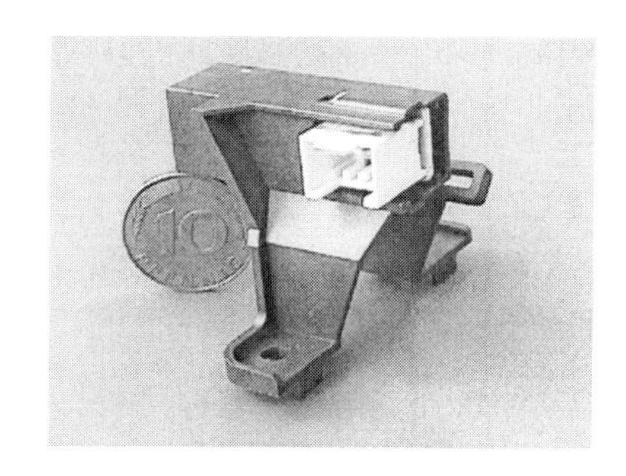

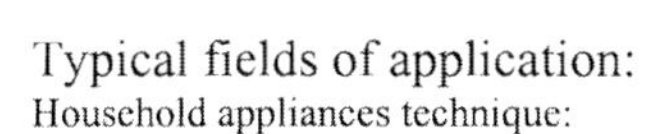

Typical fields of application:

Household appliances technique:

- e. g. recognizing of the tumble position at "top loader" washing machines.

Short description:

The position sensor PASE from emz offers a balanced cost effectiveness. The sensor is based on the principle of the hall effect. The output of the sensor, that turns to minus, results from the "open collector"- execution. The sensor is supplied with 12 VDC via a coded RAST 2.5 plug connection.
A switching distance of 5...7 mm guarantees a safe function of the sensor.
A long durability and a high resistance against humidity make the sensor especially suitable for the use in household appliances technique.
The compact design and the very simple installation permit a wide range of use. A possible field of application is, for example, the recognition of the postiton of tumbles of "top loader"– washing machines.

Technical data:

Supply voltage:	12 VDC
Range of working voltage:	5 - 24 VDC
Range of operating temperature	0° - 70° C
Range of storage temperature	-40° - 85° C
Max. current consumption:	10 mA ($R_L > 2\ k\Omega$)
Sensitivity:	12 mT
Switching distance:	5...7 mm
Output:	open collector
Hall sensor:	unipolar
Output current:	max. 10 mA
Electrical connection:	3 core RAST 2.5 plug tank
Dimensions:	56 x 43 x 36 mm

Pinning:

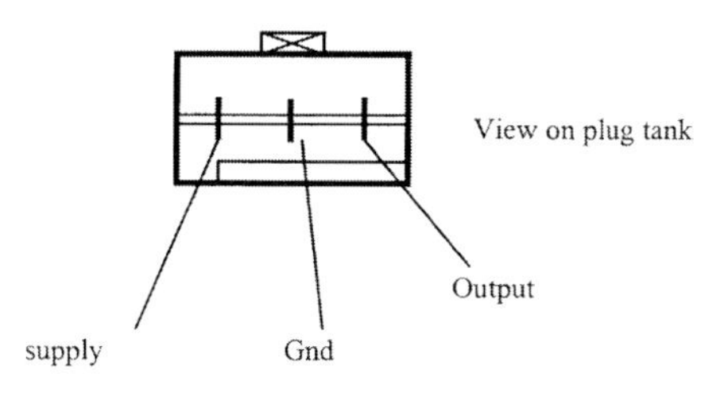

Sensoric

Spray arm detector "SADE above"

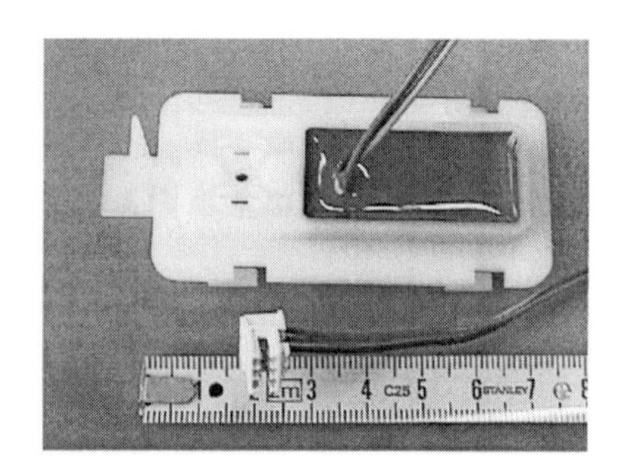

Essential characteristics:

- System: magnetoresistive sensor
- Contactless switch
- Max. switching distance:
  - horizontal: 40 mm
  - vertical: 82 mm
- Digital output signal
- Resistance against water and humidity
- Long durability
- Low cost product

Typical fields of application:

Household appliances technique:

- Washing machines (tumble)
- Dishwasher (spray arm)

Short description:

The rotation-sensor is based on the magnetoresistive principle and contains a following amplifier and comparator for preparation of the signal.
The output of the sensor system delivers a digital signal. The sensor is supplied with 5VDC via a coded RAST 2.5 plug connection.
The sensor consists of water resistant material and is therefore especially suitable for the use in household appliances like dishwashers or washing machines.

Sample of application:

A permanent magnet is fixed on a spray arm and therefor rotates with it. The recognized rotation is corresponding to the impuls, that is given by the magnet. The magnetoresitive sensor reacts on the changing of the magnetic field. This changing of the field is caused by the permanent magnet rotating past the sensor.
The sensor is useful for the touchless monitoring of the correct spray arm rotation in order to warn the user at unfavourable loading.

Technical data:

Supply voltage:	5 V
Current input:	< 5 mA
Switching distance (T=80° C):	
- horizontal	40 mm
- vertical	82 mm
Range of operating temperature:	0° - 80° C
Range of storage temperature:	-25° - 80° C
Sensitivity:	5.5 ±1.5 (mV/V)/(kA/m)
Output signal:	
- Low level	≈0 V
- High level	≈3,7 V
Rated frequency of the rotating element:	18 rpm
electrical connection:	3 core RAST 2.5 plug
Dimension:	84 x 38 x 17 mm
Length of connecting line:	24 cm

Pinning:

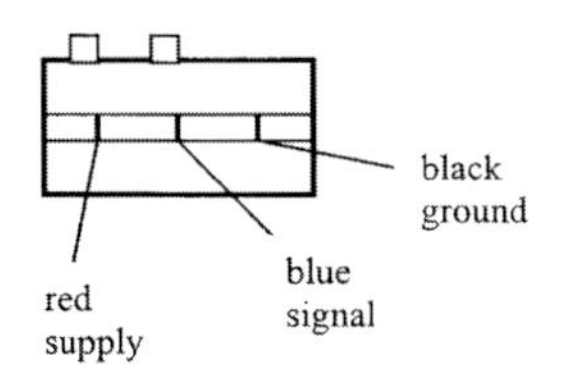

Standard Temperature Pressure Sensors
Silicon Pressure Sensors on TO-Headers

K Series (preliminary)

Description	K-series pressure sensors feature LPSi-NT Series and MPSi-NT Series pressure sensor chips mounted on TO8-headers.
Features	➢ Low pressure and temperature hysteresis ➢ Fast response ➢ High sensitivity and linearity ➢ Fatigue free monocrystaline silicon diaphragm giving high load cycle stability ➢ High long term stability ➢ Built in silicon temperature sensor ➢ Provided for further fabrication, protection cap
Application	➢ Industrial Control

Type-specific Characteristics

Type	Parameter	Min	Typ	Max	Unit
KLP035-NT-A/G-XXX	Pressure range		35		kPa
	Sensitivity	340	570	800	µV/VkPa
	Temperature coefficient of output span (TCS)	-0.19	-0.16	-0.09	%/K
	Temperature coefficient of offset voltage (TCO)	-0.08	-	+0.08	mV/K
KLP100-NT-A/G-XXX	Pressure range		100		kPa
	Sensitivity	120	200	280	µV/VkPa
	Temperature coefficient of output span (TCS)	-0.19	-0.16	-0.12	%/K
	Temperature coefficient of offset voltage (TCO)	-0.06	-	+0.06	mV/K
KMP002.5-NT-A/G-XXX	Pressure range		250		kPa
	Sensitivity	48	80	112	µV/VkPa
	Temperature coefficient of output span (TCS)	-0.19	-0.16	-0.13	%/K
	Temperature coefficient of offset voltage (TCO)	-0.05	-	+0.05	mV/K
KMP005-NT-A/G-XXX	Pressure range		500		kPa
	Sensitivity	24	40	56	µV/VkPa
	Temperature coefficient of output span (TCS)	-0.19	-0.17	-0.14	%/K
	Temperature coefficient of offset voltage (TCO)	-0.05	-	+0.05	mV/K

First Sensor Technology GmbH
Carl-Scheele-Straße 16
D-12489 Berlin, Germany
www.first-sensor.com

phone: +49 (0)30 6779 88-0
fax: +49 (0)30 6779 88-19
eMail: contact@first-sensor.com

Standard Temperature Pressure Sensors
Silicon Pressure Sensors on TO-Headers

KMP010-NT-A/G-XXX	Pressure range Sensitivity	 12	1.000 20	 28	kPa µV/VkPa
	Temperature coefficient of output span (TCS)	-0.19	-0.17	-0.14	%/K
	Temperature coefficient of offset voltage (TCO)	-0.05	-	+0.05	%/K
KMP030-NT-A/G-XXX	Pressure range Sensitivity	 4	3.000 6.6	 9.4	kPa µV/VkPa
	Temperature coefficient of output span (TCS)	-0.19	-0.17	-0.15	%/K
	Temperature coefficient of offset voltage (TCO)	-0.05	-	+0.05	%/K
KMP100-NT-A/G-XXX	Pressure range Sensitivity	 4	10.000 5	 6	kPa µV/VkPa
	Temperature coefficient of output span (TCS)	-0.19	-0.17	-0.15	%/K
	Temperature coefficient of offset voltage (TCO)	-0.05	-	+0.05	%/K

Common Characteristics[1]				
Parameter	**Min**	**Typ**	**Max**	**Units**
Bridge resistance (R_B)	2.4	3.0	3.6	kΩ
Output voltage (F.S.O.) from 35 up to3.000kPa	60	100	140	mV
Output voltage (F.S.O.) 10.000kPa	200	250	300	mV
Offset voltage $P=P_0$ (V_0)	-25	-	+25	mV
Temperature hysteresis of offset voltage (TH V_0)	-0.5	±0.1	+0.5	± % F.S.O.
Linearity error[3] $P=\ldots P_N$ (F_L)		<0,3	0.5	± % F.S.O.
Pressure hysteresis $P_1=P_0$; $P_2=P_N$; $P_3=P_0$ (P_H)	-	±0.1	-	% F.S.O.

[1] Common Characteristics at V_{in} = 5V, T_A = 25°C (unless otherwise specified)
[3] The linearity error is calculated as end point straight line linearity error.

Absolute Maximum Ratings			
Parameter	**Limit Values**[1] **Frontside**	**Rearside**	**Unit**
Pressure overload (P_{MAX})			
KLP035-NT-A/G-XXX	200	100	kPa
KLP100-NT-A/G-XXX	300	300	
KMP002.5-NT-A/G-XXX	750	750	
KMP005-NT-A/G-XXX	1.500	1.500	
KMP010-NT-A/G-XXX	3.000	3.000	
KMP030-NT-A/G-XXX	7.000	4.000	
KMP100-NT-A/G-XXX	15.000	7.000	
Operating temperature range (T_A)	-40..........+125		°C
Storage temperature range (T_{STG})	-50..........+150		°C
Supply voltage (V_{IN})	10		V

[1] Frontside coupling applies pressure onto chip face. Rearside coupling applies pressure through KOVAR® centre tube.

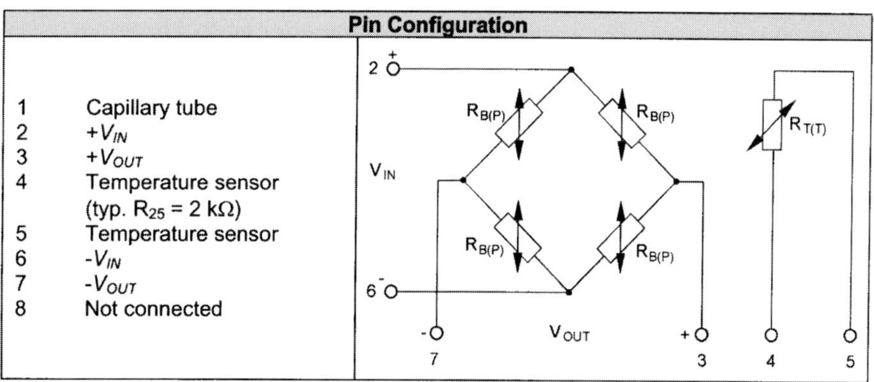

Polarity according to pressure onto frontside.

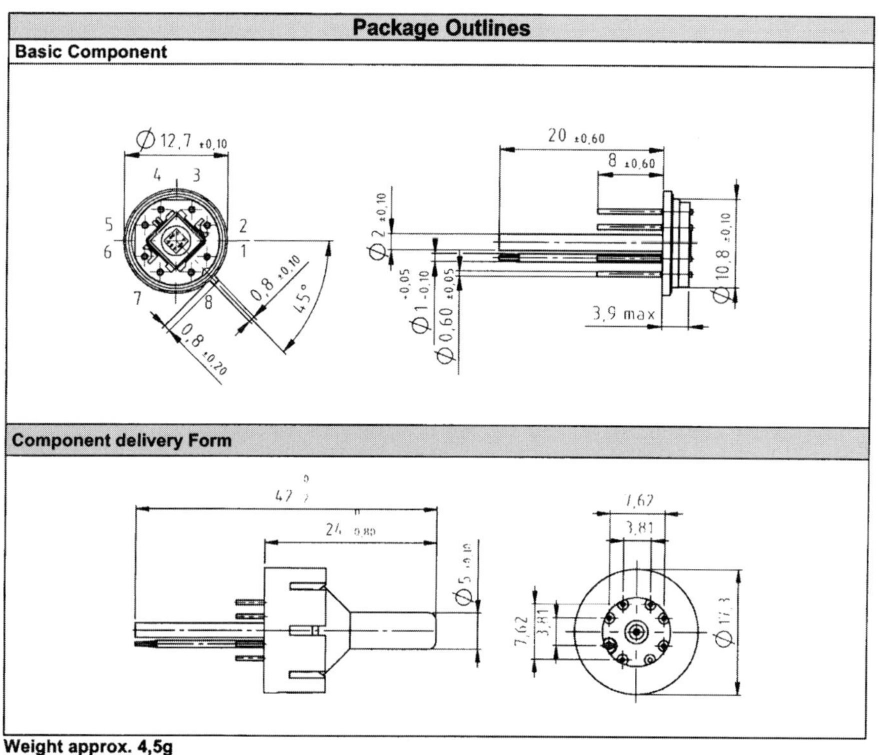

Weight approx. 4,5g

First Sensor Technology GmbH
Carl-Scheele-Straße 16
D-12489 Berlin, Germany
www.first-sensor.com

phone: +49 (0)30 6779 88-0
fax: +49 (0)30 6779 88-19
eMail: contact@first-sensor.com

Standard Temperature Pressure Sensors
Pressure Sensor Dies

MPSi-NT Series	
Description	The MPSi-NT Series is a silicon micromachined piezoresistive pressure sensor chip with a pressure-proportional voltage output signal. Using integrated ion implanted resistors in a silicon diaphragm, the stress resulting from the applied pressure is transformed into an electrical output. The devices allow absolute measurement and the application of pressure on both sides of the diaphragm for the use as gauge or differential sensor. The dies are probed and shipped on tape or in waffle packs.
Features	➢ Low Cost ➢ Compact Size ➢ Small Linearity Error ➢ High Sensitivity ➢ Additional On-Chip Resistor and Diode
Application	➢ Industrial Controls ➢ Domestic Appliances ➢ Automotive

Type-specific Characteristics

Type	Parameter	Min	Typ	Max	Unit
MPSi001-NT-A/G-001	Pressure range		100		kPa
	Sensitivity	600	1000	1400	μV/kPa
MPSi002.5-NT-A/G-001	Pressure range		250		kPa
	Sensitivity	240	400	560	μV/kPa
MPSi005-NT-A/G-001	Pressure range		500		kPa
	Sensitivity	120	200	280	μV/kPa
MPSi010-NT-A/G-001	Pressure range		10^3		kPa
	Sensitivity	60	100	140	μV/kPa
MPSi030-NT-A/G-001	Pressure range		$3 \cdot 10^3$		kPa
	Sensitivity	20	33	47	μV/kPa
MPSi100-NT-A/G-001	Pressure range		10^4		kPa
	Sensitivity	20	25	30	μV/kPa
	Span voltage (F.S.O.)	200	250	300	mV

[1] Common Characteristics at U_{supply} = 5V, T_0 = 25°C (unless otherwise specified)

Standard Temperature Pressure Sensors
Pressure Sensor Dies

Common Characteristics[1]				
Parameter	**Min**	**Typ**	**Max**	**Unit**
Bridge resistance	2400	3000	3600	Ω
Offset voltage	-10	0	+10	mV/V
Span voltage (F.S.O.)	60	100	140	mV
Temperature coefficient of offset[2]		<± 0,03		mV/K
Temperature coefficient of span[2]	-0,17	-0,19	-0,21	%/K
Temperature coefficient of bridge resistance[2]	+0,08	+0,10	+,012	%/K
Additional resistor	1200	1500	1800	Ω
Temperature coefficient of additional resistor[2]	+0,08	+0,10	0,12	%/K
Temperature coefficient of U_{Diode}[2]	-2,0	-2,1	-2,2	mV/K
Linearity error[3]		<0,3	0,5	± % F.S.O.
Hysteresis		<0,2		± % F.S.O.
Repeatability		<0,5		± % F.S.O.

[1] Common Characteristics at U_{supply} = 5V, T_0 = 25°C (unless otherwise specified)
[2] Measured from 25°C to 85°C
[3] The linearity error is calculated as end point straight line linearity error.

Maximum Ratings				
Parameter		**Limit Value** **Min**	**Max**	**Unit**
Burst pressure		3x		Rated Full Scale Pressure
Operating temperature range		-40	+125	°C
Storage temperature range		-55	+135	°C
Supply voltage		typ. 5		V

Standard Temperature Pressure Sensors
Pressure Sensor Dies

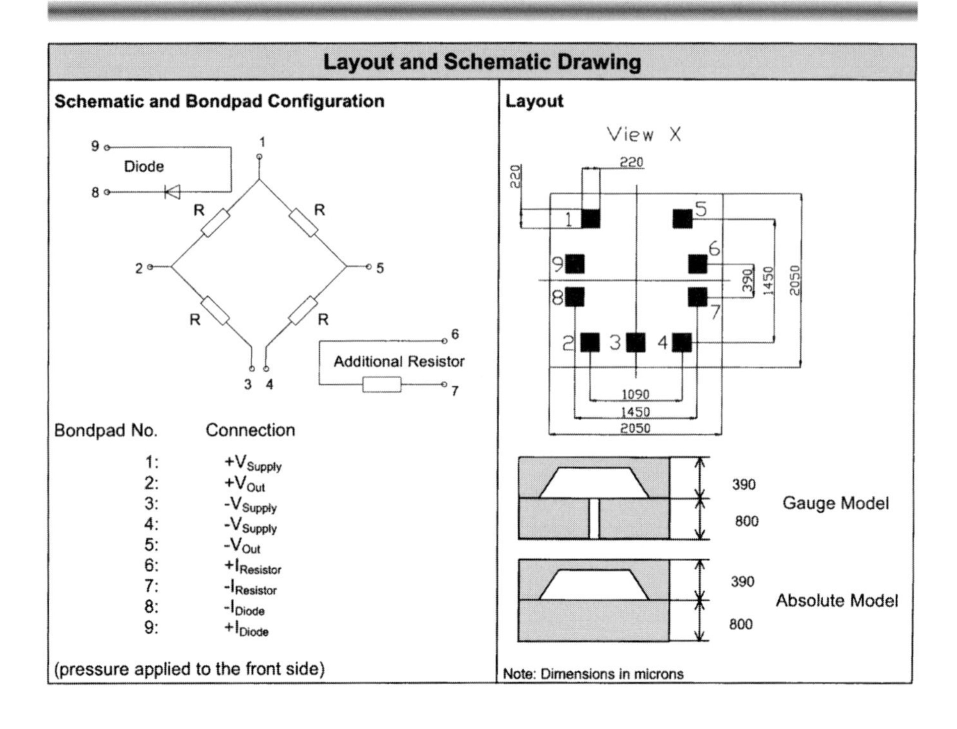

Standard Temperature Pressure Sensors
Pressure Sensor Dies

Type-specific Characteristics[1] Specification X03/X042[2]					
Type	**Parameter**	**Min**	**Typ**	**Max**	**Unit**
MPSi001-NT-A/G-X03	Pressure range		100		kPa
MPSi001-NT-A/G-X04	Sensitivity	600	1000	1400	μV/kPa
	Span voltage (F.S.O.)	60	100	140	mV
MPSi002.5-NT-A/G-X03	Pressure range		250		kPa
MPSi002.5-NT-A/G-X04	Sensitivity	240	400	560	μV/kPa
	Span voltage (F.S.O.)	60	100	140	mV
MPSi005-NT-A/G-X03	Pressure range		500		kPa
MPSi005-NT-A/G-X04	Sensitivity	120	200	280	μV/kPa
	Span voltage (F.S.O.)	60	100	140	mV
MPSi010-NT-A/G-X03	Pressure range		10^3		kPa
MPSi010-NT-A/G-X04	Sensitivity	60	100	140	μV/kPa
	Span voltage (F.S.O.)	60	100	140	mV
MPSi030-NT-A/G-X03	Pressure range		$3 \cdot 10^3$		kPa
MPSi030-NT-A/G-X04	Sensitivity	20	33	47	μV/kPa
	Span voltage (F.S.O.)	60	100	140	mV
MPSi100-NT-A/G-X03	Pressure range		10^4		kPa
MPSi100-NT-A/G-X04	Sensitivity	20	25	30	μV/kPa
	Span voltage (F.S.O.)	200	250	300	mV

[2] -X03: without passivation
-X04: with passivation

Common Characteristics[1] Specification X03/X04				
Parameter	**Min**	**Typ**	**Max**	**Units**
Bridge resistance	2400	3000	3600	Ω
Offset voltage	-10	0	+10	mV/V
Temperature coefficient of offset[2]		<± 0,05		mV/K
Temperature coefficient of span[2]	-0,17	-0,19	-0,21	%/K
Temperature coefficient of bridge resistance[2]	+0,25	+0,28	+0,31	%/K
Additional resistor	1200	1500	1800	Ω
Temperature coefficient of additional resistor[2]	+0,25	+0,28	+0,31	%/K
Temperature coefficient of U_{Diode}[2]	-2,0	-2,1	-2,2	mV/K
Linearity error[3]		<0,3	0,5	± % F.S.O.

[1] Common Characteristics at U_{supply} = 5V, T_0 = 25°C (unless otherwise specified)
[2] Measured from 25°C to 85°C
[3] The linearity error is calculated as end point straight line linearity error.

Standard Temperature Pressure Sensors
Pressure Sensor Dies

Maximum Ratings Specification X03/X04				
Parameter		**Limit Value** **Min**	**Max**	**Unit**
Burst pressure		3x		Rated Full Scale Pressure
Operating temperature range		-40	+125	°C
Storage temperature range		-55	+135	°C
Supply voltage		typ. 5		V

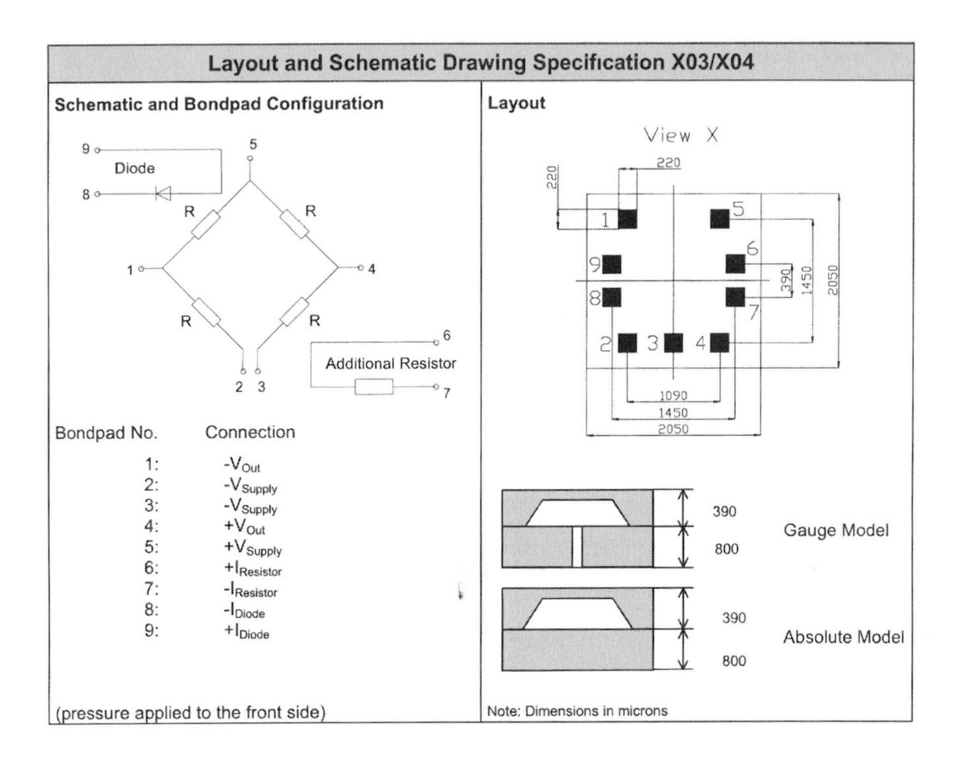

Standard Temperature Pressure Sensors
Pressure Sensors Systems

SP Series
Uncompensated Pressure Sensor in 8-lead-SOP-Package

Description	The SPXXX-NT-G-X01 is a silicon micromachined piezo-resistive pressure sensor in 8-lead-SOP-Package (Gull wing) with a pressure-proportional voltage output signal. Using integrated ion implanted resistors in a silicon diaphragm, the stress resulting from the applied pressure is transformed into an electrical output. The devices allow the application of pressure on both sides of the diaphragm for the use as gauge sensor.
Features	▪ Low Cost ▪ Compact Size ▪ Small Linearity Error ▪ High Sensitivity ▪ Additional On-Chip Diode
Application	▪ Industrial Controls ▪ Domestic Appliances

Common Characteristics[1]

Parameter	Min	Typ	Max	Unit
Pressure range (Type: **SP035-NT-G-X01**)		35		kPa
Pressure range (Type: **SP100-NT-G-X01**)		100		kPa
Pressure range (Type: **SP500-NT-G-X01**)		500		kPa
Pressure range (Type: **SP1000-NT-G-X01**)		1000		kPa
Bridge resistance	2400	3000	3600	Ω
Offset voltage	-10	0	+10	mV/V
Span voltage (F.S.O.)	60	100	140	mV
Temperature coefficient of offset[2]		<± 0,05		mV/K
Temperature coefficient of span[2]	-0,17	-0,19	-0,21	%/K
Temperature coefficient of bridge resistance[2]	+0,25	+0,28	+,031	%/K
Temperature coefficient of U_{Diode}[2]	-2,0	-2,1	-2,2	mV/K
Linearity error[3]		<0,3	0,5	± % F.S.O.

[1] Common Characteristics at U_{supply} = 5V, T_0 = 25°C (unless otherwise specified)
[2] Measured from 25 °C to 85 °C.
[3] The linearity error is calculated as end point straight line linearity error.

Maximum Ratings

Parameter	Limit Value Min	Limit Value Max	Unit
Burst pressure	3x		Rated Full Scale Pressure
Operating temperature range	-40	+125	°C
Storage temperature range	-55	+135	°C
Supply voltage	typ. 5		V

Standard Temperature Pressure Sensors
Pressure Sensors Systems

Pinout

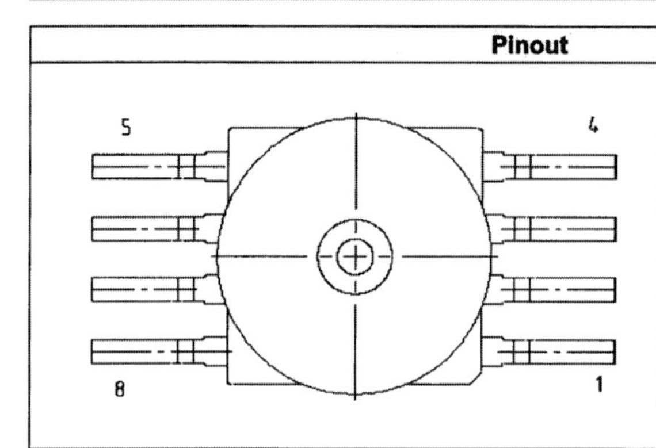

Signal	Lead-Nr.
$-V_{Out}$	7
$-V_{Supply}$	2
$-V_{Supply}$	3
$+V_{Out}$	4
$+V_{Supply}$	6
$+I_{Resistor}$	n.c.
$-I_{Resistor}$	n.c.
$-I_{Diode}$	1
$+I_{Diode}$	8

Note: Pressure applied to the front side

Schematic Drawing

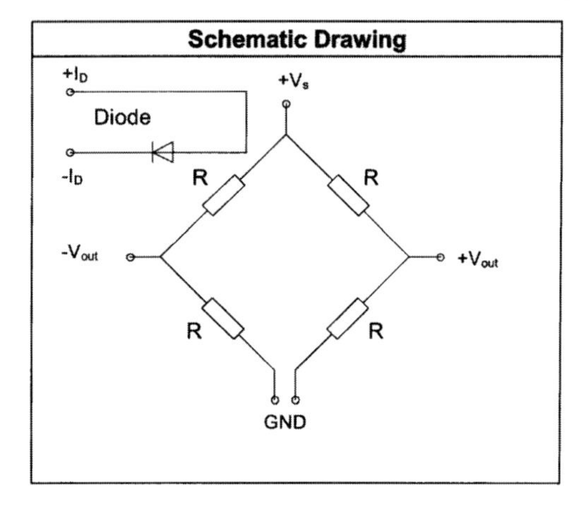

Recommended Footprint

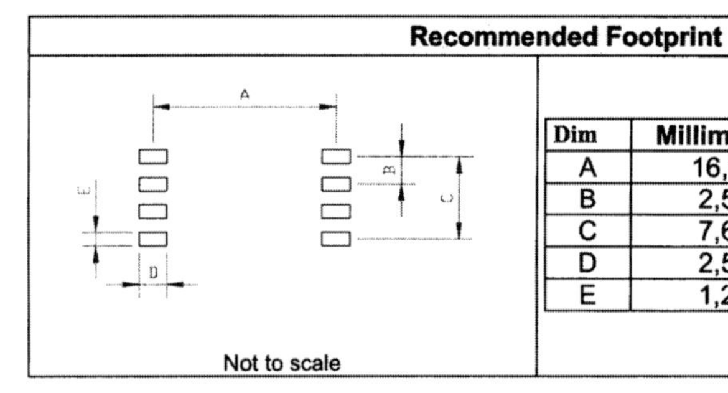

Not to scale

Dim	Millimeters	Inches
A	16,75	0,66
B	2,54	0,1
C	7,62	0,3
D	2,55	0,1
E	1,25	0,05

Recommended Soldering Profile		
	Convection / IR	Vapor Phase
Ramp up [°C/s]	**2** ... 4	**2** ... 4
Dwell time at 183°C [s]	60 ... **75** ... 150	60 ... **75** ... 85
Peak temperature [°C]	210 ... **220** ... 240	210 ... **215** ... 220
Dwell time at peak temperature [s]	10 ... **15** ... 20	**70** ... 75
Ramp down [°C/s]	**2** ... 4	**2** ... 4

Package

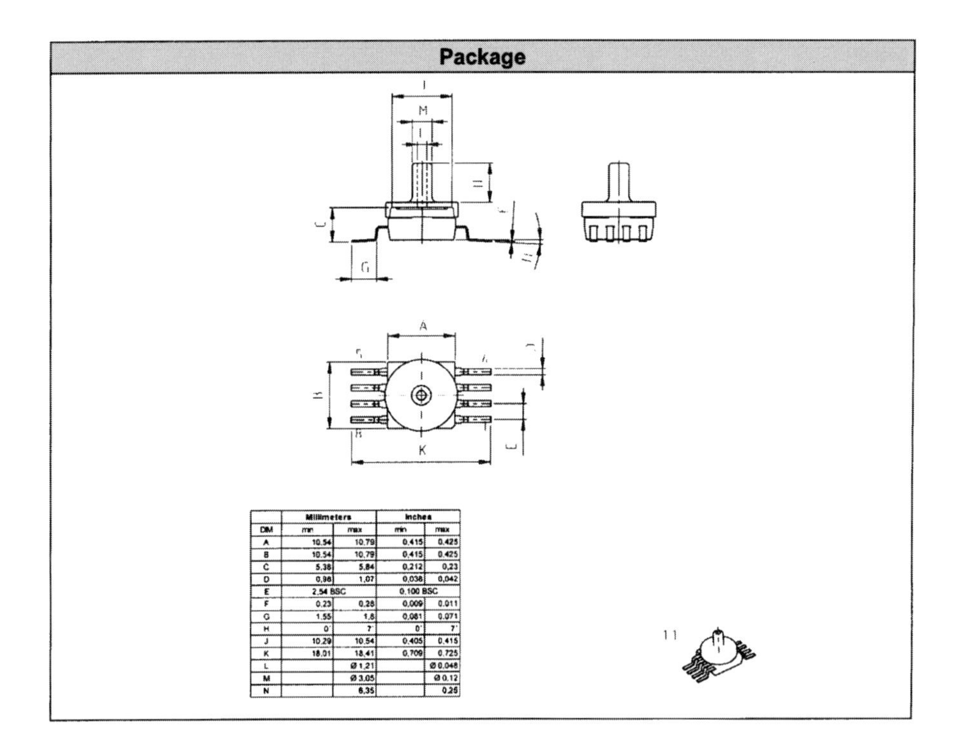

	Millimeters		Inches	
DIM	min	max	min	max
A	10,54	10,79	0,415	0,425
B	10,54	10,79	0,415	0,425
C	5,38	5,84	0,212	0,23
D	0,96	1,07	0,038	0,042
E	2,54 BSC		0,100 BSC	
F	0,23	0,28	0,009	0.011
G	1,55	1,8	0,081	0.071
H	0°	7°	0°	7°
J	10,29	10,54	0,405	0,415
K	18,01	18,41	0,709	0,725
L		Ø 1,21		Ø 0,048
M		Ø 3,05		Ø 0,12
N		6,35		0,25

First Sensor Technology GmbH
Carl-Scheele-Straße 16
D-12489 Berlin, Germany
www.first-sensor.com

phone: +49 (0)30 6779 88-0
fax: +49 (0)30 6779 88-19
eMail: contact@first-sensor.com

Steinel Solutions SGAS-Gas Sensors

SGAS-Sensor	Gas Type	Measurement Range	Measuring Principle	R_0 [kΩ] (typical) or Output Voltage	Power Consumption [mW]	Typical Sensitivity (G/G_0)* or Output Voltage at [Concentration]
210	Ethanol C_2H_5OH	0-0.5 µg/m^3	Semi-conductor	50	900	25 [0.2 µg/m^3]
220	Carbon monoxide CO	0-3000 ppm	Semi-conductor	300	900	3.5 [500 ppm]
230	Oxygen O_2	0.1-21 vol.%	Semi-conductor	10	1000	2 [1 vol.%] **
240	Ozone O_3	0-250 µg/m^3	Semi-conductor	1.5	800	6.6 [250 µg/m^3]***
250	Methane CH_4	0-1 vol.%	Semi-conductor	100	900	4 [5000 ppm]
260	Hydrogen H_2	0-1000 ppm	Semi-conductor	1000	900	100 [1000 ppm]
310	Mixed gas	unspecific	Semi-conductor	300	850	unspecific
320	Natural gas	20 % lower explosion limit	Semi-conductor	100	900	4 [5000 ppm]
330	HC (Hydrocarbons)	0-3000 ppm	Semi-conductor	300	850	3.5 [500 ppm]
500	Carbon dioxide CO_2	0-10 vol.%	Electro-chemical	0 V	1.5	375 mV [10 vol.%]
600	Carbon dioxide CO_2	0-2000 ppm	Optical	0 V	1000	10 V [2000 ppm]

* G_0=Conductivity in Air, G=Conductivity under Gas Exposure. ** Reference to Ambient Air (21 vol.% O_2).

*** G_0/G.

Index

acceleration 3, 17, 184 ff
acetate 112
acetone 63
acoustic sensors 40 f
acrolein 59
activators 83
actual control unit 196
adsorption, detergents 95
agitation washers 20 f, 83, 112
air cleaners 2, 15, 218
air quality 61, 141 ff, 154 ff
air surplus 38
Al_2O_3 46, 118, 200
alarm monitoring 233
alarm systems 159 f
alcohol ethoxylates 87
alkaline builders 83
alkyl chains 94
alkyl polyglycol ethers 98
alkyl sulfates 87
alkylbenzenesulfonate 87
alkylbetains 87
AlN 200
amine oxides 91
ammonia 59
ammoniumhydroxide 202
amperometric sensors 145, 148 f
amphoteric surfactants 87
amplifiers 76, 182
amylases 90
anionic surfactants 87
anisotopic etching 201
anologue-to-digital converter 76
appliance doors safety control 137
application-specific integrated circuit (ASIC) 76
aqua sensor system 215
array structures 56
artificial noses 3
auto-cooking functions 217
automated kitchen 224
automated washers 12, 184 ff
automatic dish washing (ADD) tablets 110
auxiliary agents, detergents 82, 89 ff
availibility 100

baking control 3, 120, 217
band gap, silicon 167
bar code readers 212
basic functions
– reed switches 125
– washers 20 f
$BaSnO_3$ semiconductor 48
beard trimming 68
beverage 68
bending
– pressure sensor 189
– reed switch 128
benzenes 59
bis-benzoxazole 92
betaines 91
biosensor 101, 106
blackbody radiator 166
bleaching agents 83, 89, 108
blood pressure equipment 2
body beauty 68
body temperature 78
boilers 150
Boltzmann fomula 78
brushless drive 23
bubble pressure tensiometer 101 ff
builders, detergents 82 ff, 88 ff, 104
building automation system (BAS) 233
building controls 230
built-in rinse aids 81
built-in sensor functions 109
bulk micromachining 204
burners 37 f, 47, 172
– control 150–158, 221
bus systems 229–241

cable insulation smoldering 63
cablevision 234
caboxymethyl cellulose (CMC) 90
caking, detergents 93
calcium-selective electrode 104
calibration, IR ear thermometers 78
camera-vision systems 177
capacitive sensors 40, 213, 252 f
– displacement 177
– pressure 189
carbon black 96
carbon dioxide 146, 154, 219
– combustion 39
carbon dioxide sensors 156
carbon monoxide 59, 141–160, 223
– combustion 39
– indoor detection 156 ff
carboxymethyl starch (CMS) 90
carpet cleaners 138
catalysts 142 ff
catalytic converters 120
catalytic oxidation 55
cationic surfactants 87
centrifugal acceleration 185
ceramic bending beam technology 189
ceramic packaging 200
CH_4 143
chemical bleaching 89, 108
chemical sensors 3, 17, 215 f
chest freezers 2
chimney cleaning 158
chlorides 87
citrates 84, 112
clarifiers, dishwashers 134
classifications 2 ff
cleaning liquids 138
cleaning robots 226
cleanliness 4, 213
climate conditioning systems 2
climate control 220 ff
coatings, detergents 112
coffeemaker 2, 69
cold contacts 74
color 4
color heavy-duty detergents 85
colored fabrics 86
combustion
– control 151
– gases 2, 222
– intelligent 37–51
– products 221
comfort 211 f, 237
compact heavy-duty detergents 83 f
computing 230
concentration, detergents 108
concentrator lenses 170
conditioning
– installation 231
– intelligent home 52–68
conductivity 100, 107
– thermal 42
conductivity sensors 213
consumer life quality 69
consumption data 11
contact resistance 125
contact sensors 177
contacts 74
converters, fluorescent 169
convex-corner underetching 204
cookers 2, 161 f, 214
coolers 2
corrosion inhibitors 92
coumarin 92
critical micelle concentration 94
cross section, IR ear thermometers 75
crystal orientation 202
crystalline wide-band gap semiconductors 168
current sensors 4
curtains 86
cutting, reed switch 128
cutting edge silicon-based micromachined sensors 198–210

damage, reed switch 130
dampers 178 ff
deep reactive ion etching (DRIE) 202
demodulators 182
dental care 68

depilation 68
detergents 226
– consumption 211 ff
– control 4
– dispensing 27, 88
– sensorics 81–116
diamond photodiodes 168
dielectric constants 40
DIN EN 60751 118
DIN IEC 68-2 179
dirt content 4, 213
discharge tubes 169
discrete electronics 182
dish washers 2, 15, 214
– detergents 81 ff, 110
disintegration systems, detergents 112
displacement sensors 208, 254 ff
– washing machines 177 ff
display technology 32
distance sensors 187
distyrylphenyl 92
domestic appliances 68
domestic white goods 230
door safety control 123, 137
dosing systems 28
draw-wire displacement sensor 260
dropping shock, reed switch 131
drum washers 20, 136, 184
dry etching 200
dryers 2, 15, 208, 213 ff
dusting 93
dyes 85 f, 93

ear thermometers, infrared 72 ff, 261
eddy current sensors 177, 181, 253, 258
effervescence 112
efficiency 233
EIB bus 231
electrical installation 231
electrical resistance 118
electricity measurement 132
electrochemical cells 43
electrochemical etching 201, 204
electrochemical pumping cells 148
electrochemical sensors 160, 164
electrolyte cells 43
electrolyte sensors 155
electronic mail 234
electronic nose 53
electronic thermometers 72, 76
emergency control 234
EN 50194 161
enablers, laundry 19–37
encapsulation, reed switch 132
end-position sensors 140
energy management system (EMS) 233
energy saving 211–228
– washers 21, 81
entrance window, UV radiation 170
enzymes 83, 90, 101
erythema 165, 171
esterquats 87
etching 200 f
ethanol 63
ethoxylated surfactants 98
ethylene tetra fluorine ethlene (EFTE) 63
ethylenediamine Pyrocatechol (EDP) 202
European market data 10
exhaust gases 2, 223
explosion limit 144
extinction 101
extrudates, detergents 82 ff

failure modes 199
fan-assisted sealed boiler 48
Faraday constant 146 f
fascia panel, washers 33
fatty acid alkanol 87
fatty acid amides 91
female depilation 68
ferromagnetics, reed switch 128
Figaro sensor 209
fillers, detergents 93
filters, UV radiation 170
fingerprint sensors 212
fire detection systems 2, 141, 173

fire gases 59
flame detection 2, 221
– burners 172
– UV radiation 166
flammable gases 143 ff
floating level indicators 123
floor care 68
flow meters 215
flowability, detergents 93
flue gas fan 47
fluorescence probe 101
fluorescent converters 169
fluorescent lamps 166
fluorescent whiteners 86, 92
foam regulators 83, 91
foaming detergents 86, 99 ff
formaldehyde 59
fragrances 92
freezers 2, 15, 216
frying processes 60, 161 f
fuel burners 47, 150, 156
function classification 4
future developments 211 ff
fuzzy algorithms, washers 32, 82, 191 ff

Ga_2O_3 sensors 45, 142
gallium arsenide 200
GaN photodiodes 168
garment care 68
gas analytical performance 57
gas boilers 37 f, 150
gas burners 172
gas cookers 2
gas hotplates 120
gas measurement, reed switch 132
gas sensor microarrays 54
gas sensors 17, 141–164, 215–225
gases 38 ff, 222
glass breakage, thermometers 73
gradient microarray 55 ff
granular swelling agents 112
Graphon 96
grease 85

hair care equipment 217
hair dryers 15, 69, 218
hand washing detergents 86
health services 227
heat conduction 40
heat exchangers 158
heating 2, 220 f, 230 f
heavy-duty detergents 82 ff, 110
Henry–Dalton law 147
high-level integrated gas sensor microarrays 54
high-speed internet access 234
high-speed laser micrometer 255
home automation 227
home entertainment 230
hot contacts 74
hotplate 120
housing
– IR ear thermometers 75
– pressure sensors 190
human skin burning 165 f
humid air 59
humidity 4, 213 ff
hydraulics 258
hydrocarbons sensitivity 57
hydrogen peroxide 89

imbalance compensation 23, 184 ff
impeller washers 83
indoor air quality 61, 141 ff, 156 ff
inductive displacement sensors 177 ff, 254
infrared carbon monoxide detector 149
infrared ear thermometers (IRET) 72 ff, 261
infrared spectral photometrics 41
inorganic gases detection 57
installation processes 231
intelligent combustion 37–51
intelligent home concept 34
intelligent kitchen 117 f
intelligent real estate 238
internet access 234 ff
intoxication symptoms 157

intrusion alarm 2, 234
ion beam-assisted deposition (IBAD) 57
ionic strength 101
ionisation signal 47
iridium catalysts 143
irons 2, 15, 218
isolation 124
isopropanol 63
isotropic etching 201

jet-dry cleaning liquid 139
junctions 167

KAMINA gas sensor 54 ff
kapton 63
kitchen
– intelligent 117 f, 224
– vapor extractor 2
knob control 137

lambda probe 38, 43, 148 ff, 153
Lambert–Beer equation 41
Laplace equation 102 f
laser micrometer 255
laser triangulation 177
laundry 2, 19–37, 225
– aids 82
– detergents 82 ff, 111
layers 200 ff
– gas sensors 56
lifetime 208, 212 f
– capacitive sensors 41
– washers 33
light control 233
light-emitting diodes 106
linear discrimination analysis (LDA) 58
lipases 90
liquid detergents 82
liquid electrolyte cells 43
liquid heavy-duty detergents 85
liquid–liquid interface, detergents 96
liquid-state electrochemical cells 155
liquid-state electrochemical gas sensors 145
load control 26, 178
LON bus 231
long-stroke sensor 258
low-pressure chemical vapor deposition (LPCVD) 203
low-temperature cofired ceramics (LTCC) 200

magnetic response 126
magnetic sensors 3, 215
magnetostrictive displacement sensors 177
male shaving 68
man-made UV radiation 167
manufacturers, laundry systems 19–37
market data 9–18
– UV sensors 174
masking 203
– gas sensors 56
mechanical shock, reed switch 132
media compatibility 199
medical mobil transport systems 226
MEMS based sensors 17, 198 ff
Mengenautomatik 191
mercury-in-glass thermometer 72
metal-oxide semiconductor gas sensors 56, 142
metal-oxide sensors 160
methane gas sensors 162
micellar aggregates 94
microarrays, gas sensors 54
micromachined sensors, silicon based 198–210
micronose 2
microprocessors 32
microsystems technology 198 ff
microwave ovens 2, 15, 216
mixers 2
motor overload 213
mounting, reed switch 131
movement detection 140
multichip modules (MCMs) 200

multigas sensors 4
multiplexer 76
multipolar tachogenerator 185

NASICON sensor 155
natural gas detection 159 ff
natural radiation 165
Nernst-type lambda probe 153
neural networks 32
neurofuzzy applications 191
Newtonian flow behavior 85
NH_3 143
nitrogen sensitivity 57
nitrotriacetic acid 88
noise reduction 23, 211 ff
nonionic surfactants 87
noses 2 ff, 52
NTC elements 119
NTC resistors 213
NTC temperature sensor 27

odor, intelligent home 52
oil, frying 61
oil burners 172
oily soils 85
olfactory sense 52
online sensorics, detergents 100 ff
operating modes/time, reed switches 125 ff
operation technology, washers 32
optical brighteners 85, 92
optical gas sensors 149
optical IR sensors 155, 160, 164
optical laser displacement sensor 256
optical sensors 215
– combustion 40 f
optical transmittance 106
optoelectronic systems 3, 215
oscillators 182
ovens 2, 15, 121, 216
oxidation, catalytic 55
oxidative bleaching 89
oxygen fraction 39
oxygen probe 44
oxygen sensor 152 f
ozone 222

packaging 198 ff
– UV sensors 170
palladium catalysts 143
passivation layers 203 f
passive compounds 201
pellistors 143
– combustion 40, 43
perhydrolyze 89
peroxides 89, 101, 108
personal care 68
personal UV exposure dosimetry 171
pH electrode 146
pH value, detergents 99 f, 109
phase behavior, surfactants 97
phosphonates 83
photo-acoustic cells 155
photodiodes 101, 106
– UV 175
physical properties, MEMS-based sensors 207
physicochemical parameters, detergents 93 ff, 100 f
piezo-resistive silicon pressure sensor 188
pigment production 171
pin junction 167
Planck's law 77
plasma-enhanced chemical vapor deposition (PECVD) 203
platinum catalysts 143
platinum-doped SnO_2 chip 57
platinum temperature sensors 117 f
platinum wires 46
p-n junction 167
pneumatics 258
pollutant emissions 37 f
polycarboxylates 84, 104
polycrystalline wide-band gap semiconductors 168
polyethylene 95

polymer solids 95
polystyrene 95
poly(vinyl alcohol) 95
poly(vinyl chloride) 95
poly(vinyl pyridine-*N*-oxide) 85
position sensors 123, 140, 208
potassium hydroxide (KOH) 202
potentiometric displacement sensors 177, 181, 259
potentiometric sensors 146 ff
powder detergents 82
power activation modes 163
power saving 214 ff
pressure sensors 17
– water level switches 3, 133, 187, 213
principal component analysis (PCA) 62
printed-circuit-boards (PCB) 131, 200
processing, MEMS based sensors 200
product-relevant physicochemical parameters 100
production data 11, 16
program sequences, washers 32
protease 90
protection layers 203 f
proximit 4
PTC resistors 213
pulsator washers 20, 112
pumping cells, electrochemical 148
pyroelectric sensors 73
pyrolyisis 161 f

quality check 79
quantum efficiency 167

radiation 165
radon 222
reaction heat 42
reed sensor 248 f
reed switches 123 ff, 216
reflectance 106
refrigerators 2, 12, 208, 214 ff
regulation processes, washers 32
reliability 199, 208
resistance, electrical 118
resistance sensor 76
resistors 213
rheological behavior 86
rhodium catalysts 143
rinsing 21, 193
roadmapping 211 f
rotation speed 4
run-up, laundry spinning 24

safety appliances 2
salts, detergents 93
sanitation 231
satellite communication 234
saturated current probe 44
SCOT method 154
secondary alkane sulfonates 87
security 2, 233, 237
self-cleaning processes 161 f
self-ignition temperature 144
semiconductor sensors 45 ff, 168
sensor criteria 71 ff
service and security center (SSC) 238 f
service robots 226
shavers 2, 68
shielding, reed switch 128
shock
– reed switch 131
– washing machines 179
SiC, photodiodes 168
sick building syndrom 141
signal evaluation 182
signal processing 32, 198 f
silicates 84
silicon-based micromachined sensors 198–210
silicon carbide 200
silicon etching 204
silicon-on-insulator (SOI) 200
silicon photodiode, UV enhanced 167
single-pole-single-throw-reed switch 125
smart buildings 229–241
smell sensors 52, 154

smoke detection 223, 234
smoldering 63
SnO_2 sensors 45, 142
SO_2 59
socioeconomics, laundry 19–37
sodium carbonate 84
sodium hydroxide (NaOH) 202
sodium *n*-dodecyl sulfate 96
sodium perborate tetrahydrate 89
sodium percarbonate 89
sodium sulfate 83, 93
sodium triphosphate 84, 88
softeners, dishwashers 134
soil removal 85 f, 101
soil repellents 83, 90
soiled water 193
soldering, reed switches 129
solid-electrolyte cells 43
solid–liquid interface, detergents 95
solid-state electrochemical sensors 147
solubility, detergents 93
solvents 63
sound velocity 41
SP chip 57
specialty detergents 82, 86 ff
spectrometers 4
spinning 22 f, 184 ff, 193
spray arms 135, 215
spring-mounted suspension 187
standard bus systems 231
standardization, home automation 230
stilbene 92
stoves, reed switch 137
suction power 217
suction speed 195
suds monitoring 29, 187
sulfobetaines 91
sulfur sensitivity 57
sunbeds 167, 171 f
sunlight 165 f
supercompact heavy-duty detergents 83 f
surface micromachining 204
surface-mounted device (SMD) 118, 200
surface tension 94 ff, 100, 215
surfactants 82, 87 ff
– detergents 94
– phase behavior 97
surveillance, sunbeds 171
suspension, washing machines 187
swelling agents, detergents 112
switches, reed 123 ff
synergy effects 207

tablets, detergents 81 ff, 110
tachogenerators 185, 213
tagging 213
Taguchi sensor 209
tanning 165, 171, 217
technical outline, laundry systems 20 ff
Teflon 63
telecommunication 230
temperature calculation, IR ear thermometers 76
temperature coefficient 118
temperature compatibility 199
temperature control 2 ff
– washing machines 25, 81, 109
temperature effects, reed switch 132
temperature gradient 55
temperature sensors 117 ff, 208, 215
temperatures, detergents 99
tetraacetyl ethylene diamine (TAED) 89
textile types 21
thermal conductivity 42
thermal coupling 75
thermometers 2
– infrared 72 ff
thermopiles 74, 218
thermoscan 261
thermostat, capillary 26
thin-film temperature sensors 117 ff
tilt sensors 3
timers 5
toasters 15, 69, 218
toluene 63
tooth-care equipment 2, 138
torque 5
toxic gases 2, 222

toxic warning level 155
triglycerine-containing soils 90
tumble dryers 208
turbidity 30, 100–108,215
two-phase tablets 85

ultrasonic displacement sensors 177
ultrasonic sensors 155, 160
ultrasonic welding, reed switch 131
universal power supply systems (UPS) 234
user comfort 211 f
UV exposure dosimetry 171
UV sensor 217, 165–180

vacuum cleaners 15, 218
valves 5
ventilation 222, 230
vibration reduction 5, 179, 211
videotext 234
volatile organic compounds (VOC) 154, 222 f

warm water boilers 2
warning levels 155
warning systems 233
wash temperature 81
washers 2, 15–21
– dampers 180
– detergency 81 ff
– production 12
– water level 213 ff
water consumption 81, 184, 211 ff
water flow detection 133
water hardness 101, 107
water intake 24
water level sensing 21 f, 133
water overflow 135
water quality 3 ff, 215
water softener/clarifier monitoring 134
water-soluble coatings, detergents 112
water sterilization 173
weight 5
weight sensors 215
welding
– pressure sensors 190
– reed switch 129
wet etching 200
wetting, detergents 88, 93 ff
Wheatstone bridge circuit 43
white goods, domestic 230
whiteners, fluorescent 86, 92
woolens 86
working principle
– bubble pressure sensor 105
– IR ear thermometers 75
laundry systems 20 ff
– platinum temperature sensor 118
world market data 10

xylene 63

yaw rate sensors 17
Young equation 95

zeolites 84, 104

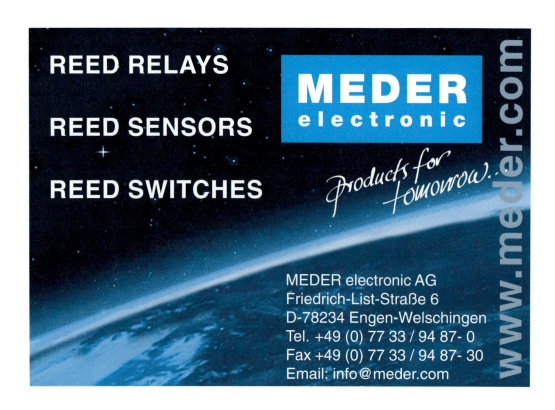

Temperature measurement with Braun *ThermoScan* : Fast, precise and uncomplicated!

Rectal temperature measurement is considered as a traditional standard of practice. However especially in the ambulant domain with regard to parents as well as children it is still extremely stressful, and entails a time delay vis-à-vis changes of body temperature which cannot always be tolerated[1]. On the other hand, axillary measurements are less reliable because environmental conditions, blood flow status and skin consistency enter into the measurements[2]. Oral measurements are also subject to variations caused among other things by thermometer placement, breathing and eating. With the Infrared Ear Thermometer by Braun, measuring exact temperature has become safer, more precise and less complicated, as Tendrup and Cotton [sic] have been able to prove in a study with children[3].

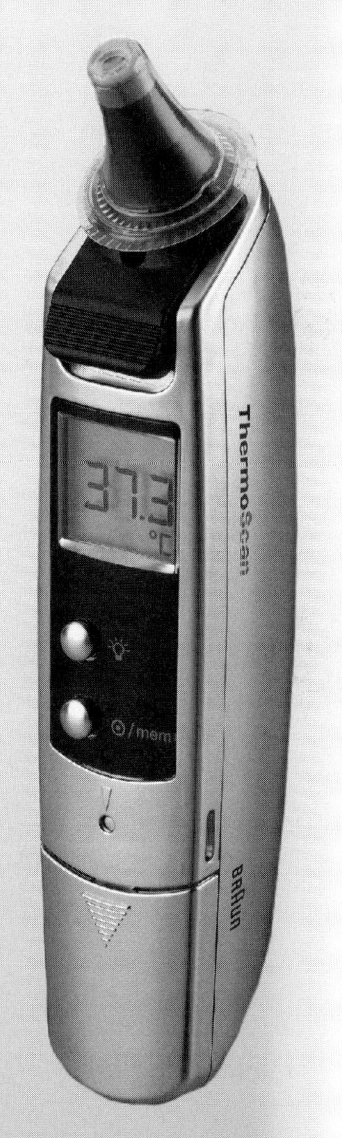

Braun *ThermoScan*

Braun ThermoScan:

- Accurate:
  Measures the heat generated by the eardrum and surrounding tissue
- Gentle:
  Infrared technology makes 1 second temperature measurement in the ear possible
- New measurement technology:
  Displays highest of 8 readings within a second
- Virtually reduces hygienic concerns, including cross-contamination, due to side location and disposable lense-filters
- Ergonomic design and built-in probe cover sensor and ejector for easy handling and use
- Meets ASTM-standards

(1) Romano et al., Critical Care Medicine, 1993
(2) Tendrup et al., American Journal of Emergency Medicine, 1989
(3) Tendrup T., Crothon D., Annals of Emergency Medicine, 1997